PYTHON FOR
BIOINFORMATICS

CHAPMAN & HALL/CRC
Mathematical and Computational Biology Series

Aims and scope:

This series aims to capture new developments and summarize what is known over the whole spectrum of mathematical and computational biology and medicine. It seeks to encourage the integration of mathematical, statistical and computational methods into biology by publishing a broad range of textbooks, reference works and handbooks. The titles included in the series are meant to appeal to students, researchers and professionals in the mathematical, statistical and computational sciences, fundamental biology and bioengineering, as well as interdisciplinary researchers involved in the field. The inclusion of concrete examples and applications, and programming techniques and examples, is highly encouraged.

Series Editors

Alison M. Etheridge
Department of Statistics
University of Oxford

Louis J. Gross
Department of Ecology and Evolutionary Biology
University of Tennessee

Suzanne Lenhart
Department of Mathematics
University of Tennessee

Philip K. Maini
Mathematical Institute
University of Oxford

Shoba Ranganathan
Research Institute of Biotechnology
Macquarie University

Hershel M. Safer
Weizmann Institute of Science
Bioinformatics & Bio Computing

Eberhard O. Voit
The Wallace H. Couter Department of Biomedical Engineering
Georgia Tech and Emory University

Proposals for the series should be submitted to one of the series editors above or directly to:
CRC Press, Taylor & Francis Group
4th, Floor, Albert House
1-4 Singer Street
London EC2A 4BQ
UK

Published Titles

Bioinformatics: A Practical Approach
Shui Qing Ye

Cancer Modelling and Simulation
Luigi Preziosi

Combinatorial Pattern Matching Algorithms in Computational Biology Using Perl and R
Gabriel Valiente

Computational Biology: A Statistical Mechanics Perspective
Ralf Blossey

Computational Neuroscience: A Comprehensive Approach
Jianfeng Feng

Data Analysis Tools for DNA Microarrays
Sorin Draghici

Differential Equations and Mathematical Biology
D.S. Jones and B.D. Sleeman

Engineering Genetic Circuits
Chris J. Myers

Exactly Solvable Models of Biological Invasion
Sergei V. Petrovskii and Bai-Lian Li

Gene Expression Studies Using Affymetrix Microarrays
Hinrich Göhlmann and Willem Talloen

Handbook of Hidden Markov Models in Bioinformatics
Martin Gollery

Introduction to Bioinformatics
Anna Tramontano

An Introduction to Systems Biology: Design Principles of Biological Circuits
Uri Alon

Kinetic Modelling in Systems Biology
Oleg Demin and Igor Goryanin

Knowledge Discovery in Proteomics
Igor Jurisica and Dennis Wigle

Meta-analysis and Combining Information in Genetics and Genomics
Rudy Guerra and Darlene R. Goldstein

Modeling and Simulation of Capsules and Biological Cells
C. Pozrikidis

Niche Modeling: Predictions from Statistical Distributions
David Stockwell

Normal Mode Analysis: Theory and Applications to Biological and Chemical Systems
Qiang Cui and Ivet Bahar

Optimal Control Applied to Biological Models
Suzanne Lenhart and John T. Workman

Pattern Discovery in Bioinformatics: Theory & Algorithms
Laxmi Parida

Python for Bioinformatics
Sebastian Bassi

Spatial Ecology
Stephen Cantrell, Chris Cosner, and Shigui Ruan

Spatiotemporal Patterns in Ecology and Epidemiology: Theory, Models, and Simulation
Horst Malchow, Sergei V. Petrovskii, and Ezio Venturino

Stochastic Modelling for Systems Biology
Darren J. Wilkinson

Structural Bioinformatics: An Algorithmic Approach
Forbes J. Burkowski

The Ten Most Wanted Solutions in Protein Bioinformatics
Anna Tramontano

Chapman & Hall/CRC Mathematical and Computational Biology Series

PYTHON FOR BIOINFORMATICS

SEBASTIAN BASSI

CRC Press
Taylor & Francis Group
Boca Raton   London   New York

CRC Press is an imprint of the
Taylor & Francis Group an **informa** business

A CHAPMAN & HALL BOOK

Chapman & Hall/CRC
Taylor & Francis Group
6000 Broken Sound Parkway NW, Suite 300
Boca Raton, FL 33487-2742

© 2010 by Taylor and Francis Group, LLC
Chapman & Hall/CRC is an imprint of Taylor & Francis Group, an Informa business

No claim to original U.S. Government works

Printed in the United States of America on acid-free paper
10 9 8 7 6 5 4 3 2 1

International Standard Book Number: 978-1-58488-929-8 (Paperback)

Library of Congress Cataloging-in-Publication Data

Bassi, Sebastian.
 Python for bioinformatics / Sebastian Bassi.
  p. cm. -- (Mathematical and computational biology series)
 Includes bibliographical references and index.
 ISBN 978-1-58488-929-8 (pbk. : alk. paper)
 1. Bioinformatics. 2. Python (Computer program language) I. Title. II. Series.

QH324.2.B387 2009
570.285--dc22   2009025700

Visit the Taylor & Francis Web site at
http://www.taylorandfrancis.com

and the CRC Press Web site at
http://www.crcpress.com

Contents

IV   Python Recipes with Commented Source Code   315

List of Tables

List of Figures

Preface

This book is a result of the experience accumulated during several years of working for an agricultural biotechnology company. As a genomic database curator, I gave support to staff scientists with a broad range of bioinformatics needs. Some of them just wanted to automate the same procedure they were already doing by hand, while others would come to me with biological problems to ask if there were bioinformatics solutions. Most cases had one thing in common: Programming knowledge was necessary for finding a solution to the problem. The main purpose of this book is to help those scientists who want to solve their biological problems by helping them to understand the basics of programming. To this end, I have attempted to avoid taking for granted any programming related concepts. The chosen language for this task is Python.

Python is an easy to learn computer language that is gaining traction among scientists. This is likely because it is easy to use, yet powerful enough to accomplish most programming goals. With Python the reader can start doing real programming very quickly. Journals such as *Computing in Science and Engineering, Briefings in Bioinformatics* and *PLOS Computational Biology* have published introductory articles about Python. Scientists are using Python for molecular visualization, genomic annotation, data manipulation and countless other applications.

In the particular case of the life sciences, the development of Python has been very important; the best exponent is the Biopython package. For this reason, Section II is devoted to Biopython. Anyhow, I don't claim that Biopython is the solution to every biology problem in the world. Sometimes a simple custom-made solution may better fit the problem at hand. There are other packages like BioNEB and CoreBio that the reader may want to try.

The book begins from the very basic, with Section I ("Programming"), teaching the reader the principles of programming. From the very beginning, I place a special emphasis on practice, since I believe that programming is something that is best learned by doing. That is why there are code fragments spread over the book. The reader is expected to experiment with them, and attempt to internalize them. There are also some spare comparisons with other languages; they are included only when doing so enlightens the current topic. I believe that most language comparisons do more harm than good when teaching a new language. They introduce information that is incomprehensible and irrelevant for most readers.

In an attempt to keep the interest of the reader, most examples are somehow related to biology. In spite of that, theses examples can be followed even if the reader doesn't have any specific knowledge in that field.

To reinforce the practical nature of this book, and also to use as reference material, Section IV is called "Python Recipes with Commented Source Code." These programs can be used as is, but are intended to be used as a basis for other projects. Readers may find that some examples are very simple; they do their job without too many bells and whistles. This is intentional. The main reason for this is to illustrate a particular aspect of the application without distracting the reader with unnecessary features, as well as to avoid discouraging the reader with complex programs. There will always be time to add features and customizations once the basics have been learned.

The title of Section III ("Advanced Topics") may seem intimidating, but in this case, advanced doesn't necessarily mean difficult. Eventually, everyone will use the chapters in this section [especially relational database management system -RDBMS- and XML]. An important part of the bioinformatics work is building and querying databases, which is why I consider knowing a RDBMS like MySQL to be a relevant part of the bioinformatics skill set. Integrating data from different sources is one of tasks most frequently performed in bioinformatics. The tool of choice for this task is XML. This standard is becoming a widely used platform for data interchange between applications. Python has several XML parsers and we explain most of them in this book.

Appendix B, "Selected Papers," introductory provide level papers on Python. Although there is some overlapping of subjects, this was done to show several points on view of the same subject.

Researchers are not the only ones for whom this book will be beneficial. It has also been structured to be used as a university textbook. Students can use it for programming classes, especially in the new bioinformatics majors.

Acknowledgments

A project such as this book couldn't be done by just one person. For this reason, there is a long list of people who deserve my thanks. In spite of the fact that the average reader doesn't care about the names, and at the risk of leaving someone out, I would like to acknowledge the following people: My wife Virginia Claudia Gonzalez (Vicky) and my son Maximo Bassi who had to contend with my virtual absence during more than a year. Vicky also assisted me in uncountable ways during manuscript preparation. My parents and professors taught me important lessons. My family (Oscar, Graciela, and Ramiro) helped me with the English copyediting, along with Hugo and Lucas Bejar. Vicky, Griselda, and Eugenio also helped by providing a development abstraction layer, which is needed for writers and developers. Thanks also to Joel Spolsky for coining this term (and providing inspirational words).

The people at the local Python community (`http://www.pyar.com.ar`): Facundo Batista, Lucio Torre, Gabriel Genellina, John Lenton, Alejandro J. Cura, Manuel Kaufmann, Gabriel Patiño, Alejandro Weil, Marcelo Fernandez, Ariel Rossanigo, Mariano Draghi, and Buanzo. I would choose Python again just for this great community. The people at Biopython: Jeffrey Chang, Brad Chapman, Peter Cock, Michiel de Hoon, Iddo Friedberg, and Andrew Dalke. Peter Cock is specially thanked for his comments on the Biopython chapter. Shashi Kumar and Pablo Di Napoli who helped me with the LaTeX 2_ε issues, Martin Albisetti who overviewed the Version Control chapter, Zachary Voase who contributed with his article "Diving into the Gene Pool with Biopython", Julius B. Lucks for his work at "OpenWetWare," Richard Gruet for the "Python Quick Reference," Luke Arno who contributed with the WSGI section, and Sunil Nair who believed in me from the first moment.

Part I

Programming

Chapter 1

Introduction

1.1 Who Should Read This Book

This book is for the life science researcher who wants to learn how to program. He may have previous exposure to computer programming, but this is not necessary to understand this book (although it surely helps).

This book is designed to be useful to several separate but related audiences, students, graduates, postdocs, and staff scientists, since all of them can benefit from knowing how to program.

Exposing students to programming at early stages in their career helps to boost their creativity and logical thinking, and both skills can be applied in research. In order to ease the learning process for students, all subjects are introduced with the minimal prerequisites. There are also questions at the end of each chapter. They can be used for self-assessing how much you've learnt. The answers are available to teachers in a separate guide.

Graduates and staff scientists having actual programming needs should find its several real world examples and abundant reference material extremely valuable.

1.1.1 What You Should Already Know

Since this book is called *Python for Bioinformatics* it has been written with the following assumptions in mind:

- The reader should know how to use a computer. No programming knowledge is assumed, but the reader is required to have minimum computer proficiency to be able to use a text editor and handle basic tasks in your operating system (OS). Since Python is multi-platform, most instructions from this book will apply to the most common operating systems (Windows, Mac OSX and Linux); when there is a command or a procedure that applies only to a specific OS, it will be clearly noted.

- The reader should be working (or at least planning to work) with bioinformatics tools. Even low scale hand made jobs, such as using the NCBI BLAST to ID a sequence, aligning proteins, primer searching, or estimating a phylogenetic tree will be useful to follow the examples. The more familiar the reader is with bioinformatics the better he will be able to apply the concepts learned in this book.

1.2   Using this Book

1.2.1   Python Versions

There are two versions of Python available for download: Python 2.6 (also called 2.x series) and Python 3 (also known as Python 3000). Python 3 is not fully compatible with the 2.x series. For that reason, at this time (mid 2009) most third party modules are not available for Python 3, in particular the Biopython package that is a "must have" if you are planning to do serious bioinformatics work. Developers are expected to test their packages in Python 3 and may port them by Python 3.1 or 3.2. The last Biopython release (1.50 at this time) works under Python 2.4, 2.5 and 2.6 in all supported platforms.

This books teaches Python fundamentals that can be applied to any Python version. When a feature is exclusive to a particular Python version, it is properly noted. All programs in this book were tested under Python 2.5 and 2.6. They are all "Python 3 aware," that is, even if they don't work because they depend on an external library that wasn't ported up to this date, they are written with Python 3 syntax in mind and are expected to work with any (or minor) modification when external libraries become available.

Regarding which version to use, if your script doesn't rely on external packages, you may try Python 3 right now. But this is an unlikely scenario. If the packages you need are not ported yet, you should use Python 2.6. The Python 2.x line will continue to be supported and improved for years to come. Python 2.6 has the "-3" command line option (Py3k warnings) that warns about incompatibilities with Python 3 and there is also a tool called "2to3" that converts Python 2.6 code to 3 compatible Python code.

1.2.2   Typographical Conventions

There are some typographical conventions I have tried to use in a uniform way throughout the book. They should aid readability and were chosen to tell apart user made names (or variables) from language keywords. This comes in handy when learning a new computer language.

Bold: Objects provided by Python and by third party modules. With this notation it should be clear that **round** is part of the language and not a user defined name. Bold is also used to highlight parts of the text. **There is no chance** to confuse one bold usage with the other.

`Mono-spaced font`: User declared variables and filenames.

Italics: In commands, it is used to denote a variable that can take different values. For example, in **len(*iterable*)**, "iterable" can take different values. Used in text, it marks a new word or concept. For example "One such fundamental data structure is a *sequence.*"

$<=$: Break line. Some lines are longer than the available space in a printed page, so this symbol is inserted to mean that what is on the next line in the page represents the same line on the computer screen.

1.2.3 Code Style

Python source code is presented as **listings**. Each line of these listings is numbered. These numbers are not intended to be typed, they are used to reference each line in the text. All code is available in the accompanying Virtual Machine.[1] Each code sample is also available on the web in a site especially crafted to show source code (**Pastebin**). You will see a URL (web address) with this form "py3.us/#" (where # is a number) next to each listing. Type this URL in your browser and you will see the same source code that is presented in the book. The source code can be downloaded by using the "download" link on its Pastebin webpage.

Code can be formatted in several ways and still be valid to the Python interpreter. This following code is syntactically correct:

```
Dna='accatcagt'
def MyFunction(X,N):
    avG=sum(X)/N
    " Calculate the average "
    return avG
```

So is this one:

```
dna = 'accatcagt'
def my_function(x,n):
    """ Calculate the average
    """
    avg = sum(x)/n
    return avg
```

The former code sample follows most accepted coding styles for Python.[2] Throughout the book you will find mostly code formatted as the second sample. Some code in the book will not follow accepted coding styles for the following reasons:

---

[1]Please refer to the instructions on the DVD on how to use the Virtual Machine.

[2]See page 553 for details on coding styles.

- There are some instances where the most didactic way to show a particular piece of code conflicts with the style guide. On those few occasions, I choose to deviate from the style guide in favor of clarity.

- Due to size limitation in a printed book, some names were shortened and other minor drifts from the coding styles have been introduced.

- To show there are more than one way to write the same code. Coding style is a guideline, so some programmers don't follow them. You should be able to read "bad" code, since sooner or later you will have to read other people's code.

1.2.4   Get the Most from This Book without Reading It All

- If you want to **learn how to program**, read the first section, from Chapter 1 to Chapter 8. The Regular Expressions (REGEX) chapter (Chapter 9) can be skipped if you don't need to deal with REGEX.

- If you know Python and just want to **know about Biopython**, read first the Section II (from page 175 to page 222). It consists in a large chapter on Biopython modules and functions. Then try to follow programs found in Section IV (from page 317 to page 391).

- If you need some **introduction to biological basics** before reading about Biopython you can start with "Diving into the Gene Pool with Biopython" from page 431 to page 447.

- To use it as **reference material**, see Section IV (Python Recipes with Commented Source Code, from page 317 to page 391), D (Python Language Reference, from 457 to 527) and F (Python Style Guide, from 553 to 576).

1.3   Why Learn to Program?

Many of the tasks that a researcher performs with his or her computer are repetitive: Collect data from a Web page, convert files from one format to another, execute or interpret 10 or hundreds of BLAST results, first design, look for restriction enzymes, etc. In many cases it is evident that these are tasks that can be performed with a computer, with less effort on our part and without the possibility of errors caused by tiredness or distractions.

An important consideration when you're evaluating whether or not to create a program is the apparent time lost in the definition and formulation of the problem, implementing it with code and then debugging it (correcting errors that surface inevitably). It is incorrect to consider problem definition and evaluation a waste of time. It is generally at this precise point in the process where we understand thoroughly the problem that we face. It is common that during the attempt to formulate a problem, we realize that many of our initial assumptions were mistaken. It also helps us to detect when it is necessary to restart the planning process. When this happens, it is better that it happens at the planning stage than when we are in the middle of the project. In these cases, the planning of the program represents time saved. Another advantage to take into account is that the time that is invested to create a program once is compensated by the speed with which the tasks are performed every time we run it.

Not only can it automate the procedures that we do manually, but it will also be able to do things that would otherwise not be possible: Personalized graphics, web applications and interaction with databases, just to name a few.

Sometimes it is not very clear if a particular task can be done by a program. Reading a book such as this one (including the examples) will help you identify which tasks are feasible to automate with a script and which ones are better done manually.

1.4   Basic Programming Concepts

Before installing Python, let's review some programming fundamentals. If you have some previous programming experience, you may want to skip this section and jump straight to page 19 (Installing Python). This section introduces basic concepts such as *instructions, data types, variables* and some other related terminology that is used throughout this book.

1.4.1   What Is a Program?

A *program* is a set of ordered instructions designed to command the computer to do something. The word "ordered" is there because is not enough to declare what to do, but the actual order of the directions should also be stated.[3]

A program is often characterized as a recipe. A typical recipe consists in a list of ingredients followed by step by step instructions on how to prepare a dish. This analogy is reflected in several programming websites and tutorials

---

[3]There are *declarative* languages that state what the program should accomplish, rather than describing how to accomplish it. Since most computer languages (Python included) are *imperative* instead of *declarative*, this book assumes that all programs are written in an *imperative* form.

with the words "recipe" and "cookbook" on it. A laboratory protocol is another useful analogy. A protocol is defined as a "predefined written procedural method in the design and implementation of experiments."

Here is a typical protocol, followed almost every day in several molecular laboratories:

Listing 1.1: Protocol for Lambda DNA digestion

```
Restriction Digestion of Lambda DNA

Materials

5.0 mcL     Lambda DNA (0.1 g/L)
2.5 mcL     10x buffer
16.5 mcL    H2O
1.0 mcL     EcoRI

Procedure

Incubate the reactions at 37°C for 1 hr.
Add 2.5 mcL loading dye and incubate for another 15 minutes.
Load 20 mcL of the digestion mixture onto a minigel
```

There are at least two components of a protocol: procedure and materials. A procedure provides specific order like incubate, add, mix, store, load and many others. The same goes for a computer program. The programmer gives specific order to the computer: print, read, write, add, multiply, assign, round, and others.

While protocol procedures correlate with program instructions, materials are the *data*. In protocols, procedures are applied to materials: Mix 2.5 μL of buffer with 5 μL of Lambda DNA and 16.5 μL of H_2O, load 20 μL onto a minigel. In a program, instructions are applied to data: print the text string "Hello", add two integer numbers, round a float number.

As a protocol can we written in different language (like English, Spanish or French), there are different languages to program a computer. In science protocols, English is the de facto language. Due to historical, commercial and practical reasons, there is no such a equivalent in computer science. There are several languages, each with its own strong points and weakness. For reasons that will make sense shortly, Python was the computer language chosen for this book.

Let's see a simple Python program:

Listing 1.2: Sample Python Program

```
seq_1 = 'Hello,'
seq_2 = ' you!'
```

```
total = seq_1 + seq_2
seq_size = len(total)
print(seq_size)
```

This small program can be read as "Name the string `Hello,` as `seq_1`. Name the string ` you!` as `seq_2`. Add the strings named `seq_1` and `seq_2` and call the result as `total`. Get the length of the string called `total` and name this value as `seq_size`. Print the value of `seq_size`." This program prints 11.

As shown, there are different types of data (often called "data types" or just "types"). Numbers (integers or float), text string, and other data types are covered in Chapter 3. In `print(seq_size)`, the instruction is `print` and `seq_size` is the name of the data. Data is often represented as *variables*. A *variable* is a name that stands for a value that may vary during program execution. With variables, a programmer can represent a generic order like "round n" instead of "round 2.9." This way he can take into account for a non fixed (hence *variable*) value. When the program is executed, "n" should take a specific value since there is no way to "round n." This can be done by assigning a value to a variable or by binding a name to a value.[4] The difference between "assign a value to a variable" and "bind a name to a value" is explained in detail in Chapter 3 (from page 65). In both cases, it is expressed as:

```
var = value
```

Note that **this is not an equality** as seen in mathematics. In an equality, terms can be interchanged, but in programming, the term of the right (`value`) takes the name of the term of the left (`var`). For example,

```
seq_1 = 'Hello,'
```

After this assignment, the variable `seq_1` can be used, like,

```
len(seq_1)
```

This is translated as "return the length of the value called `seq_1`". This command returns "6" because there are six characters in the string `Hello,`.

1.5   Why Python?

Let's have a look at some Python features worth pointing out.

---

[4]In Python the later form is used.

1.5.1   Main Features of Python

- Readability: When we talk about readability, we refer as much to the original programmer as any other person interested in understanding the code. It is not an uncommon occurrence for someone to write some code then return to it a month later and find it difficult to understand. Sometimes Python is called a "human readable language."

- Built-in features: Python comes with "Batteries included." It has a rich and versatile standard library which is immediately available, without the user having to download separate packages. With Python you can, with few lines, read an XML file, extract files from a zip archive, parse and generate email messages, handle files, read data sent from a Web browser to a Web server, open a URL as if were a file, and many more possibilities.

- Availability of third party modules: 2/3D plotting, PDF generation, bioinformatics analysis, animation, game development, interface with popular databases, and application software are only a handful of examples of modules that can be installed to extend Python functionality.

- High level built-in data structures: Dictionaries, sets, lists, and tuples help to model real world data.

- Multiparadigm: Python can be used as a "classic" procedural language or as "modern" object oriented programming (OOP) language. Most programmers start writing code in a procedural way and when they are ready, they upgrade to OOP. Python doesn't force programmers to write OOP code when they just want to write a simple script.

- Extensibility: If the built-in methods and available third party modules are not enough for your needs, you can easily extend Python, even in other programming languages. There are some applications written mostly in Python but with a processor demanding routine in C or FOR-TRAN. Python can also be extended by connecting it to specialized high level languages like R or MATLAB.

- Open source: Python has a liberal open source license that makes it freely usable and distributable, even for commercial use.

- Cross platform: A program made in Python can be run under any computer that has a Python interpreter. This way a program made under Windows Vista can run unmodified in Linux. Python interpreters are available for most computer and operating systems, and even some devices with embedded computers like the Nokia 6630 smartphone.

- Thriving community: Python is gaining momentum among the scientific community. This translates into more libraries for your projects and people you can go to for support.

Why Was Python Created in the First Place?

Here is a recounting by Guido van Rossum, Python author, about what was the motivation for "inventing" a new computer language:

"I was working in the Amoeba distributed operating system group at CWI. We needed a better way to do system administration than by writing either C programs or Bourne shell scripts, since Amoeba had its own system call interface which wasn't easily accessible from the Bourne shell. My experience with error handling in Amoeba made me acutely aware of the importance of exceptions as a programming language feature.

It occurred to me that a scripting language with a syntax like ABC but with access to the Amoeba system calls would fill the need. I realized that it would be foolish to write an Amoeba-specific language, so I decided that I needed a language that was generally extensible.

During the 1989 Christmas holidays, I had a lot of time on my hand, so I decided to give it a try. During the next year, while still mostly working on it in my own time, Python was used in the Amoeba project with increasing success, and the feedback from colleagues made me add many early improvements.

In February 1991, after just over a year of development, I decided to post to USENET. The rest is in the Misc/HISTORY file."

In January 2009, Guido opened a blog devoted to Python history. It can be found at `http://python-history.blogspot.com`.

1.5.2 Comparing Python with Other Languages

You may be wondering why you should use Python, and not more well known languages like C, Perl or JAVA. It is a good question. A programming language can be regarded as a tool, and choosing the best tool for the job makes a lot of sense.

Readability

Nonprofessional programmers tend to value the learning curve as much as the legibility of the code (both aspects are tightly related).

A simple "hello world" program in Python looks like this:

```
print("Hello world!")
```

Compare it with the equivalent code in Java:

```
public class Hello
{
```

```
    public static void main(String[] args) {
        System.out.printf("Hello world!");
    }
}
```

Let's see a code sample in C language. The following program reads a file (input.txt) and copies its contents into another file (output.txt):

```
#include <stdio.h>
int main(int argc, char **argv) {
  FILE *in, *out;
  int c;
  in = fopen("input.txt", "r");
  out = fopen("output.txt", "w");
  while ((c = fgetc(in)) != EOF) {
    fputc(c, out);
  }
  fclose(out);
  fclose(in);
}
```

The same program in Python is shorter and easier to read:

```
in = open("input.txt")
out = open("output.txt", "w")
out.writelines(in)
in.close()
out.close()
```

A one-liner could also do the job:

```
open("output.txt", "w").writelines(open("input.txt"))
```

Let's see a Perl program that calculates the average of a series of numbers:

```
sub avg(@_) {
    $sum += $_ foreach @_;
    return $sum / @_ unless @_ == 0;
    return 0;
}
print avg((1..5))."\n";
```

The equivalent program in Python

```
def avg(data):
    if len(data)==0:
        return 0
    else:
        return sum(data)/float(len(data))
print(avg([1,2,3,4,5]))
```

The purpose of this Python program could be almost fully understood by just knowing English.

Python is designed to be a highly readable language.[5] The use of English keywords, the use of spaces to limit code blocks and its internal logic (indentation), contribute to this end. Its possible to write hard to read code in Python, but it requires a deliberate effort to obfuscate the code.[6]

Speed

Another parameter to consider when choosing a programming language is code execution speed. In the early days of computer programming, computers were so slow that some differences due to language implementation were very significant. It could take a week for a program to be executed in an interpreted language, while the same code in a compiled language could be executed in a day. This performance difference between interpreted and compiled languages stays with the same proportion, but it is less relevant. This is because a program that took a week to run, now takes less than ten seconds, while the compiled one takes about one second. Although the difference seems important, it is not so relevant if we consider the development time.

This does not mean that execution speed does not need to be considered. A 10X speed difference can be crucial in some high performance computing operations. Sometimes a lot of improvements can be achieved by writing optimized code. If the code is written with speed optimization in mind, it is possible to obtain results quite similar to the ones that could be obtained in a compiled language. In the cases where the programmer is not satisfied with the speed obtained by Python, it is possible to link to an external library written in other language (like C or Fortran). This way, we can get the best of both worlds: the ease of Python programming with the speed of a compiled language.

1.5.3   How It Is Used?

Python has a wide range of applications. From cell phones to web servers, there are installed thousands of Python applications in the most diverse fields. There is Python code powering Wikipedia robots, the OLPC (One Laptop Per Child) project[7], and it is the scripting language of the OpenOffice suite.[8]

---

[5]Other languages are regarded as "write only," since once written it is very difficult to understand it.

[6]A simple **print 'Hello World'** program could be written, if you are so inclined, as **print ".join([chr((L>=65 and L<=122) and (((((L>=97) and (L-96) or (L-64))-1)+13)%26+((L>=97) and 97 or 65)) or L) for L in [ord(C) for C in 'Uryyb Jbeyq!']])** (py3.us/1).

[7]http://wiki.laptop.org/go/OLPC_Python_Environment

[8]http://wiki.services.openoffice.org/wiki/Python

Some languages are strong in one niche (like Perl and PHP for web applications, Java for desktop programs), but Python can't be typecasted easily.

With a single code-base, Python desktop applications run with a native look and feel on multiple platforms. Well known examples of this category include the **BitTorrent** p2p client/server, **Emesene** an IM client for Windows Live Messenger, media players like **Exaile** and **Tim Player** and even a CAD package, **PythonCAD**.

As a language for building web applications, Python can be found in `Zooomr.com`, a popular image sharing site as well as several other Web sites like **Google**, **Yahoo** and **Nasa.gov**. There are specialized software for building Web sites (called webframeworks) in Python like **Django**, **Pylons**, **Zope** and **TurboGears**. Tools for accessing webservices are also available in Python (Yahoo Python Developer Center,[9] Google Data API,[10] Facebook API.[11])

Python also excels in small one-use scripts. Not all programs are meant to be publicly released, some are built just to solve a user's problem. From system administration to data analysis, Python provides a wide range of tools to this end:

- Generic Operating System Services (os, io, time, curses)

- File and Directory Access (os.path, glob, tempfile, shutil)

- Data Compression and Archiving (zipfile, gzip, bz2)

- Interprocess Communication and Networking (subprocess, socket, ssl)

- Internet Data Handling (email, mimetools, rfc822)

- Internet Protocols (cgi, urllib, urlparse)

- String Services (string, re, codecs, unicodedata)

Python is gaining users in the scientific community. There is library (**SciPy**) that integrates several modules like linear algebra, signal processing, optimization, statistics, genetic algorithms, interpolation, ODE solvers, special functions, etc. Python has support for parallel programming (if you have appropriate hardware) with the pyMPI and 2D/3D scientific data plotting.

Python is known to be used in wide and diverse fields like engineering, electronic, astronomy, biology, paleomagnetism, geography, and many more.

---

[9] `http://developer.yahoo.com/python`

[10] `http://code.google.com/p/gdata-python-client`

[11] `http://wiki.developers.facebook.com/index.php/PythonPyFacebookTutorial`

1.5.4 Who Uses Python?

Python is used by several companies, from small and unknown shops up to big players in their fields like Google, Yahoo, Disney, NASA, NYSE, and many more.

Google for instance has three "official languages" for deploying in production services: JAVA, C++ and Python. They have Web sites made in Python,[12] stand-alone programs[13] and even hosting solutions.[14] As a confirmation that Google is taking Python seriously, in December 2005 they hired Guido van Rossum, the creator of Python. He is working most of the time improving Python. It may not be Google's main language, but this shows that they are a strong supporter of it.

Even Microsoft, a company not known for their support of open source programs, have developed a version of Python to run their ".Net" platform. This version is called **IronPython**.

Many well-known Linux distributions already use Python in their key tools. Red Hat's Anaconda installer, and Gentoo's Portage package manager are two examples. Ubuntu Linux (the most successful Linux distribution at this time) "... prefers the community to contribute work in Python." Python is so tightly integrated into Linux that some distributions won't run without a working copy of Python.

1.5.5 Flavors of Python

Although in this book I refer to Python as one specific programming language, Python is actually a language definition. What we use for programming is a specific implementation. Since there is an implementation that is used by most Python programmers (cPython, also known as Python), this subject is usually overlooked by some users.

The most relevant Python implementations are: cPython, PyPy,[15] Stackless,[16] Jython[17] and IronPython. This book will focus on the standard Python version (cPython), but it is worth knowing about the different versions.

- CPython: The most used Python version, so the terms CPython and Python are used interchangeably. It is made mostly in C (with some modules made in Python) and is the version that is available from the official Python Web site (`http://www.python.org`).

---

[12]See the ".py" at `http://www.google.com/support/bin/topic.py?topic=352`.

[13]`http://code.google.com/p/sitemap-generators`

[14]`http://code.google.com/appengine`

[15]`http://codespeak.net/pypy/dist/pypy/doc/home.html`

[16]`http://www.stackless.com`

[17]`http://www.jython.org/Project`

- PyPy: A Python version made in Python. It was conceived to allow programmers to experiment with the language in a flexible way (to change Python code without knowing C). It is mostly an experimental platform.

- Stackless: Is another experimental Python implementation. The aim of this implementation doesn't focus on flexibility as PyPy, instead, it provides advanced features not available in the "standard" Python version. This is done in order to overcome some design decisions taken early in Python development history. Stackless allows custom designed Python application to scale better than cPython counterparts. This implementation is being used in the EVE Online massively multi-player online game, *Civilization IV*, *Second Life*, and *Twisted*.

- Jython: A Python version written in JAVA. It works in a JVM (Java Virtual Machine). One application of Jython is to add the Jython libraries to their JAVA system to allow users to add functionality to the application. A very well known learning 3D programming environment (Alice[18]) uses Jython to let the users program their own scripts.

- IronPython: Python version adapted by Microsoft to run on ".Net" and ".Mono" platform. .Net is a technology that aims to compete with JAVA regarding "write once, runs everywhere." Another use of IronPython envisioned by Microsoft is as a script language for running in the Web browser along Silverlight (a Flash-like Microsoft technology).

1.5.6 Special Python Bundles

Apart from Python implementations, there are some special adaptations of the original cPython that are packaged for specific purposes:

- Python(x,y): It is defined as a "free scientific and engineering development software for numerical computations, data analysis and data visualization based on Python programming language, Qt graphical user interfaces (and development framework) and Eclipse integrated development environment." In other words, it is a bundle of several Python related package to ease the use and installation. The main advantage of Python(x,y) is that by installing just one program you end up with a complete development environment that includes, Eclipse, IPython, C++, Fortran, Extensive documentation, Numeric, SciPy, Mayavi, and others. It is available at http://www.pythonxy.com. Up to the moment of writing this, it was available only for Windows.[19] The main drawback of this approach is that the resulting package is about 254 Mb long (or 150 Mb without Eclipse).

---

[18]Alice is available for free at http://www.alice.org.

[19]With an "available soon..." for Linux on the downloaded page.

- Enthought Python Distribution (EPD): Another "all-in-one" Python solution. Includes over 60 additional tools and libraries, like NumPy, SciPy, IPython, 2D and 3D visualization, database adapters, and other libraries. Everything available as a single-click installer for Windows XP, Mac OS X (a universal binary for Intel 10.4 and above), and RedHat EL3 and EL4 (x86 and amd64). This bundle is suitable for scientific users, and it is made by the same people who made NumPy and SciPy. It is free for academic and nonprofit private-sector use, and for an annual fee for commercial and governmental use. It is available at `http://www.enthought.com/products/epd.php`, and since it includes so many libraries, the resulting size is about 400Mb.

- PortablePython: A Python version capable of running without the need of installation. It can be used to carry a working program environment in a pendrive or any removable storage unit. Another application of PortablePython is to distribute Python program to people that can't or don't want to install Python (like some controlled corporate and academic environment).

Chapter 2

First Steps with Python

2.1 Installing Python

2.1.1 Learn Python by Using It

This section shows how to install Python to start running your own programs. Learning by doing is the most efficient way of learning. It is even better than just passively reading a book (even this book). You will find "Python interactive mode" very rewarding in this sense, since it can answer your questions faster than a book and even faster than a search engine. The answers you get from the Python interpreter are definitive.

For these reasons I suggest installing Python before continuing to read this book.

2.1.2 Python May Be Already Installed

Python is pre-installed in Mac OS X and most Linux distributions. In Windows (XP or Vista), you have to download the Windows installer from the Python download page (`http://www.python.org/download`) and then install it. Installation is pretty straightforward and should not present any difficult if you are used to installing Windows programs.

In a few words, you should double click the installer file (with `msi` extension) and run the Python Install Wizard. Accepting the default settings and have Python installed in a few minutes without hassle.

However, there is a step-by-step guide in Appendix A (from page 393). This appendix also has instructions for users with Unix like systems (also called *nix) that want to install an extra copy of Python. Having more than one version of Python is useful for testing and for cases where the user has no administrative privileges and wants to run his own copy of Python.

2.1.3   Testing Python

Once Python is installed, you should make sure it works. On Windows, just double-click on the Python icon. Linux and Mac OS X[1] users could open a terminal and then type 'python'.

You should see a screen like this one:[2]

```
Python 2.5 (r25:51908, May  7 2007, 15:38:46)
[GCC 3.3.5 (Debian 1:3.3.5-3)] on linux2
Type "help", "copyright", "credits" or "license" for more <=
information.
>>>
```

2.1.4   First Use

There are two ways to use Python: interactive and batch mode. Both methods are complementary and they are used with different purposes. Interactive mode allows the programmer to get an immediate reply to each instruction. In batch mode, instructions are stored in one or more files and then executed. This is the standard way of running Python programs. Interactive mode is used mostly for small tests while most programs are run in batch mode. Since testing is a fundamental step when learning a new skill, interactive mode will be used thoroughly in this book.

Let's learn some Python basics using the interactive mode.

2.2   Interactive Mode

2.2.1   Baby Steps

The following code shows how to command the interpreter to print the string "Hello world!"[3]:

```
>>> print("Hello World!")
Hello World!
```

Note the three greater than characters (>>>), this is the Python prompt of the interactive mode. You don't need to type it. This means that Python is ready to execute our commands or evaluate our expressions.

---

[1]On Mac the terminal is located under the Applications/Utilities folder.

[2]This output could vary from system to system depending on Python version, base system, and options set during compilation.

[3]There is a tradition among programmers to show how a language works by printing the string "Hello world". Python programmers are not immune to this custom. See what happens when you include this statement in your programs: **import __hello__**.

2.2.2 Basic Input and Output

Output: Print

Before Python 3.0, **print** was a statement and worked like this:

```
>>> print "Hello World!"
Hello World!
```

It was very simple to use but lacked some functionality often requested by developers: Change the program output (from screen to a file for example), change the separator from space to another character, and more features not easy to implement in a statement. This was fixed in Python 3, and **print()** is now a function:[4]

```
>>> print("Hello World!")
Hello World!
```

The print function can receive several elements:

```
>>> print("Hello","World!")
Hello World!
```

Change the separator:

```
>>> print("Hello","World!",sep=",")
Hello,World!
```

Redirect the output to a file:

```
>>> print("Hello","World!",sep=",",file=filehandle)
```

Change the end of the output:

```
>>> print("Hello","World!",sep=";",end='\n\n')
Hello;World!
```

Input: raw_input and input in Python 2.x

There are two functions to accept input from the user into a program:
raw_input: Take a string of data from the user and return it:

```
>>> name = raw_input("Enter your name: ")
Enter your name: Seba
>>> name
'Seba'
```

---

[4]A *function* is a portion of code that performs a specific task. They are discussed in detail in Chapter 6.

While **input** also takes a string of data, it attempts to evaluate it as if it were a Python program:

```
>>> name = input("Enter your name: ")
Enter your name: Seba
Traceback (most recent call last):
  File "<stdin>", line 1, in <module>
  File "<string>", line 1, in <module>
NameError: name 'Seba' is not defined
```

Since `Seba` is not a defined name, it triggers an error. So this time we enter an expression that can be evaluated in Python (a string in this case):

```
>>> name = input("Enter your name: ")
Enter your name: "Seba"
>>> name
'Seba'
```

Input: input in Python 3

There is no **raw_input** in Python 3, it was renamed to **input**:

```
>>> name = input("Enter your name: ")
Enter your name: Seba
>>> name
'Seba'
```

To evaluate an expression in Python 3, use the **eval()** function:

```
>>> input("Operation: ")
Operation: 2+2
'2+2'
>>> eval(input("Operation: "))
Operation: 2+2
4
```

The old **input** was deprecated since it was considered insecure.

2.2.3   More on the Interactive Mode

Interactive mode can be used as a calculator:

```
>>> 1+1
2
```

When '+' is used on strings, it returns a concatenation:

```
>>> '1'+'1'
'11'
>>> "A string of " + 'characters'
'A string of characters'
```

Note that single (') and double (") quotes can be used in an indistinct way, as long as they are used with consistency. That is, if a string definition is started with one type of quotes, it must be finished with the same kind of quote.[5]

Different data types can't be added:

```
>>> 'The answer is ' + 42
Traceback (most recent call last):
  File "<stdin>", line 1, in ?
TypeError: cannot concatenate 'str' and 'int' objects
```

Only elements of the same type can be added. To convert this into a sum of strings, the number must be converted into a string, this is done with the **str()** function:

```
>>> 'The answer is ' + str(42)
'The answer is 42'
```

The same final result can be archived with "String Formatting Operations"[6]:

```
>>> 'The answer is %s'%42
'The answer is 42'
```

You can assign **names** to any Python element, and then refer to them later:

```
>>> n = 42
>>> 'The answer is %s'%n
'The answer is 42'
```

Names should contain only letters, numbers, and underscores (_), but they can't start with numbers. In other programming languages names are refered as variables. There is a more detailed description on rules and naming conventions on page 64 and in Appendix F.

---

[5]In Chapter 3 there is a detailed description of strings.

[6]See page 474 for reference on how to use String Formatting. In Python 2.6 and 3, there is also a new String Formatting Operation described in PEP-3101 (http://www.python.org/dev/peps/pep-3101).

TABLE 2.1:
Arithmetic-Style Operators

Symbol	Description
+	Addition
-	Subtraction
*	Multiplication
/	Division
**	Exponentiation
%	Modulus (remainder)

2.2.4 Mathematical Operations

Any standard mathematical operation can be done in the Python shell:

```
>>> 12*2
24
>>> 30/3
10
>>> 2**8/2+100
228
```

Double star (**) stands for "elevated to the power of" and the inverted slash (/) is the division operation. So this expression means: $2^8 : 2 + 100$. In Table 2.1 there is a list of Arithmetic-Style operators supported by Python.

Note that the operator precedence is the same as used in math. An easy way to remember precedence order is with the acronym **PEMDAS**:

P **P**arentheses have the highest precedence and are used to set the order of expression evaluation. This is why 2 * (3-2) yields 2 and (3-1) ** (4-1) yields 8. Parentheses can also be used to make expressions easier to read.

E **E**xponentiation is the second in order, so 2**2+1 is 5 and not 8.

MD **M**ultiplication and **D**ivision share the same precedence. 2*2-1 yields 3 instead of 2.

AS **A**ddition and **S**ubtraction also share the same (latest) order of precedence.

Last but not least, operators with the same precedence are evaluated from left to right. So 60/6*10 yields 100 and not 1. In Table D.4 (page 464) there is a list with operators precedence order.

Something to take into account for mathematical operations in Python versions prior to 3.0 is how to handle integer values.

Division in Python 2.x

This may not be expected:

```
>>> 10/3
3
```

Division returns the floor, that is, the integer part of the result. To get the floating point result, at least one operand must be float:

```
>>> 10.0/3
3.3333333333333335
>>> 10/3.
3.3333333333333335
```

Division in Python 3

In Python 3, any division is a floating point division:

```
>>> 10/3
3.3333333333333335
>>> 10/2
5.0
```

To get the previous behavior, use //:

```
>>> 10//3
3
>>> 10//2
5
```

2.2.5   Exit from Python Shell

You can exit from any version of Python with CRTL-D (that is pressing Control and D simultaneously). Since Python 2.5, there is also the **exit()** function:[7]

```
$ python2.5
Python 2.5 (r25:51908, May  7 2007, 15:38:46)
[GCC 3.3.5 (Debian 1:3.3.5-3)] on linux2
Type "help", "copyright", "credits" or "license" for more <=
information.
>>> exit()
$
```

---

[7]For exiting from a previous version of Python, use **sys.exit()**.

2.3   Batch Mode

Although the interactive interpreter is very useful, most nontrivial programs are stored in files. The code used in an interactive session can be accessed only when the session is active. Each time that an interactive session is closed, all typed code is gone. In order to have code persistence, programs are stored in text files. When a program is executed from such a text file, rather than line by line in an interactive interpreter, it is called **batch mode**.

These are regular text files usually with the ".py" extension. These files can be generated with any standard text editor (as Windows Notepad).[8]

An optional feature of python scripts under a Unix-like system is a first line with the path to the python interpreter. If the python interpreter is located at /usr/bin/python (a typical location in Linux), the first line will be:

```
#!/usr/bin/python
```

This is called **shebang** and it is a Unix convention that allows the operating system to know what is the interpreter for the program and this interpreter can be executed without the user having to explicitly invoke the python interpreter.[9] Invoking a Python program without this line causes the operating system to try to execute the program as a shell script.

Let's suppose that you have this very simple program:

Listing 2.1: A "Hello World!" program (hello.py)

```
print("Hello World!")
```

This program will work from the command line only if it is called as an argument of the Python interpreter:

```
$ python hello.py
Hello World!
```

But if you want to run it as a standalone program, you will see something like this:

```
$ ./hello.py
./hello.py: line 1: syntax error near unexpected token <=
''Hello world!''
./hello.py: line 1: 'print('Hello world!')'
```

---

[8]Any text editor can be used for Python programming, but it is highly advisable to use a programmer editor. At the end of this chapter there is a section devoted to choosing an editor.

[9]To specify interpreter path can also be used to select a particular Python version when there is more than one version installed.

This error message is sent by the shell when trying to execute the program as a system script (without invoking Python). It can be avoided by editing the first line of the program:

Listing 2.2: Hello World! with python path

```
#!/usr/bin/python
print("Hello World!")
```

This version works as it were an executable binary file:[10]

```
$ ./hello2.py
Hello World!
```

If you want to invoke the first available Python interpreter instead of a spefic interpreter, use `#!/usr/bin/env python`. Use the path to a specific Python interpreter only when you want to run your program with a particular version (like `/mnt/hda2/py252/bin/python2.5`).

In Windows, this line is ignored since the interpreter is launched according to the file extension (`.py`).

Python can also be executed from within the programming editor, provided that the editor has this functionality. In Python IDLE (and most other editors) you can launch a program with the *F5* key.

Other line that is usually found in Python programs is the "encoding comment". This line defines the character econding for the rest of the document and it takes this form:

```
# -*- coding: ENCODING -*-
```

where `ENCODING` may be for example `ascii`, `latin1`, `8859-1`, `UTF-8` and others. So a encoding line for a source code with Spanish characters will have this line:

```
# -*- coding: latin1 -*-
```

Without encoding comment, Python's parser will assume ASCII (that is the default encoding). If your source code includes a non-ASCII character, you should assign an encoding.

---

[10]In Linux and Mac OSX you have to make sure that the file has executable permission, this is done with *chmod a+x hello.py*.

2.3.1 Comments

If you tried IDLE or any other editor with syntaxis coloring capabilities for this program, you may have noted that the first line (`#!/usr/bin/python`) has a particular color. That is due to the use of the "#" symbol. This character has special significance for Python. It is used to identify lines that aren't executed by the interpreter. As a result, the lines that begin with that symbol are called "comments." Comments don't add functionality to the program but help the programmer or other possible readers of the code. Let's look at the previous program with a comment:

Listing 2.3: Hello World! with comments

```
#!/usr/bin/env python
# The next line prints the string "Hello World!"
print("Hello World!")
```

The comment in this particular code is rather pointless since there is no doubt about what is the funcion of "print". In other programs there is code that is not so easy to understand and where a comment can improve code readability. It is customary to put the comments before the code you are refferring to.[11] Comments are made mostly thinking in helping someone else to understand our code, but they can be useful even for the same programmer who see the code some time after writing and doesn't remember the purpose of a routine.

Comments can also be used to disable part of the code (this is called "comment-out" in programming jargon). This is usually done for debugging purposes. When trying alternative codes to accomplish a task, it is better to have an inactive part of the code until you are sure which code you will use. It is easier to uncomment an inactive code than retype something that was deleted. This is such a common task that all Python editors have tools to comment-out or to uncomment entire block of texts.[12]

Tip: Extensions in Python.
Python files have the **.py** extension, but you could also find other extensions that are related to Python:

- **py**: Standard Python files.

- **pyc**: "Compiled" Python files. When you import a Python module for the first time, it gets compiled into byte code, so the next time it

---

[11] There is a guide of code style to standardize the way code is written. An adapted version of such a guide can be found in Appendix F.

[12] In Python default editor (IDLE) this tool is under the *Format* menu.

starts faster. Compile can be forced from Python with the **compile_dir** function in the **compileall** module. Note that **.pyc** files load faster but do not run faster.

- **pyo**: "Optimized" code. It is generated by running the Python interpreter with the *-o flag*. Don't be fooled with the name, most code will run at same speed even with the *-o flag* enabled.

- **pyw**: It is a standard Python file with an extension that makes Windows to execute them with pythonw.exe instead of python.exe. Pythonw.exe doesn't launch the DOS console so it is prefered for graphical programs under Windows.

2.3.2   Indentation

One of the first things that stands out to programmers about Python is its code indentation system. Non-programmers must be wondering at this point what is indentation of the source code. Here is some C code that is not indented:

```
if (attr == -1){while (x<5){
printf("Waiting...\n");wait(1);
x = x+1;}printf("Everything is OK\n");}
else {printf("There is an error\n");}
```

Indented version of the same program portion (or "code snippet"):

```
if (attr == -1) {
    while (x<5) {
        printf("Waiting...\n");
        wait(1);
        x = x+1;}
    printf("Everything is OK\n");}
else {
    printf("There is an error\n");}
```

Even not knowing C we can say that the second program is "clearer" than the first. In a programming language like C or Java code blocks that are executed as an entity, are separated with braces. This way the interpreter knows that for example `printf("Everything is OK\n");` is within the **if** structure but not within **while**. The logical relations between the elements are clearer in an indented program than one without indentation. Inspect the following code snippet in Python, where there are no braces but there are code blocks that are defined by indentation.

```
if attr==-1:
    while x<5:
        print("Waiting...")
        wait(1)
        x = x + 1
    print("Everything is OK")
else
    print("There is an error")
```

It is not important if you don't understand this program. The purpose of this example is to show one of the most striking aspect of language. This is considered an advantage because when the structure of the code is clearly enough, there will be less chance to introduce coding mistakes. Some say that it is annoying having to maintain the code this way, but this is not the case. Most text editors deal with code indentation in an automatic way so there is no burden to the programmer. Another criticism to the mandatory indentation is the deep nesting of the code; some statements are placed at the far right. There are programming tools to avoid writing code with too many levels of indentation (such as writing modular code). Using these tools appropriately is a desired skill to have and it is independent of the programming language that you use.[13] Forcing the programmer to use indentation is a feature that goes along with one the Python's design philosophy: Readability counts.[14] As Oliver Fromme wrote in "Python: Myths about Indentation":[15] *Python forces you to use indentation that you would have used anyway, unless you wanted to obfuscate the structure of the program.*

2.4   Choosing an Editor

In principle any text editor can be used to program in Python. Nothing prevents you to program in Notepad (if you are masochistic enough) or any lightweight text editor, although the specifics of the syntax of Python make it convenient to use an editor that takes into account their characteristics.

Taking this into account, text editors can be classified in terms of their relationship with Python.

- **Regular editors**: Those without any kind support for Python, i.e., Notepad and Mousepad.

---

[13]Linus Torvalds, the creator of the Linux kernel has said "If you need more than 3 levels of indentation, you're screwed anyway, and should fix your program."

[14]Please see http://www.python.org/dev/peps/pep-0020 for more information on the guiding principles for Python's design.

[15]http://www.secnetix.de/~olli/Python/block_indentation.hawk

- **"Python aware" editors**: General use or programmer oriented editors that recognize Python syntax. It may have native support or added by a plug-in or any other extension mechanism. The following editors falls in this category: Kate, vim, Eclipse, and many others. These "all purpose" editors have the advantage that what you learn can be used to edit different types of documents.

- **Pure Python**: Editors created to program with Python. IDLE, the official python editor is a clear representative of this group, although there are much more like Dr Python, Eric, SPE, and others. The advantage of these types of editors is that they have exclusive features like Code completion, context sensitive help, and integrated debugger.

In this chapter we deal with editors from the last two groups since they are the only ones that are worthwhile for serious programming. Particularly in this chapter I am going to review certain editors (chosen arbitrarily) to evaluate their pros and cons.

Following is a short review of some popular editors.

2.4.1 Kate

Kate is one of the most popular multi-use text editors for Linux. It is installed in KDE based linux distributions, although it can be installed as a separate program. Kate is a very versatile editor, it has support for editing files for tens of programming languages and other file types (such as MySQL, LaTeX, CSS, configuration files, etc.). Its most prominent features are:

- Multiple document interface: It supports having multiple open documents, each one overlapping, within the same window. Each one of these windows can in turn be divided to look simultaneously at different parts of the same document. All of this is configurable, to the point of being able to override this functionality, forcing each document to have its own interface (SDI, Single Document Interface).

- Save several files in named sessions: It is rare for a program to consist only of one text file. Usually we have to manage a series of files (XML, HTML, CSV, etc.) and this functionality allows us to group various files under a single project.

- Block text selection: This allows us to select portions of text in arbitrary forms, like a column of data.

- Change the end of line text sequence (EOF): This is useful for working with several operating systems.

- Integrated Terminal Emulation: It incorporates a window where we can execute operating system commands or start an interactive Python session. It can be used for trying things out.

- Extensible with plugins : You can add new plugins created with javascript.[16]

Concerning its Python support, it has "syntax coloring," automatic tabs and code folding capabilities. Those of us who are already Kate users and want to begin programming with Python will feel at home.

Availability

Kate is a free software released under the GPL. Since Kate is part of the KDE project, it runs on any system where you can run KDE. Most Linux systems run KDE natively. There are ports for Mac OS X (`http://mac.kde.org`) and Windows (`http://windows.kde.org`).

2.4.2   Eric

Eric is an Integrated Development Environment for Python and Ruby. It is writen in Python and it is one of the most complete products of those reviewed in this chapter. However this "completeness" may be a bit intimidating at first (see Figure 2.1). Most of the features are not visible just by looking at the icons on the toolbar. It requires some dedication to learn how to use it effectively. Investing time in this IDE is worth it, as you will benefit from an increase in productivity that surpasses the time you invested to learn how to use it.

Most prominent features are:

- Advanced project management facilities

- Sourcecode autocompletion

- Error highlighting (a red line tells you where the problem is)

- Integrated Python debugger

- Integrated class browser

- Integrated version control interface for VCS and Subversion repositories

If you are looking for a complete editor and you don't mind having several buttons on the screen,[17] there is no doubt that Eric is a great option.

---

[16]`http://www.kate-editor.org/article/scripting_katepart_with_javascript`

[17]The number of buttons and windows can be reduced substantially from the configuration section of the program. For this it has an extensive configuration menu in *Settings, Preferences*.

FIGURE 2.1: Eric Python Editor.

Availability

Eric is a free software released under GPL. Eric is available from its web site (`http://www.die-offenbachs.de/eric`) and runs in most common platforms.

2.4.3 Eclipse

Eclipse is not just a text editor. It is certainly the "500-pound gorilla" of programming environments. According to its own definition it is an "extensible development platform for building, deploying and managing software across the entire software lifecycle." Eclipse is known as a JAVA Integrated Development Environment (IDE),[18] but there are plugins[19] that extend it to support other languages like C, Perl, Colobo, and of course, Python.[20]

The most prominent features of Eclipse are:

- Code completion

- Code folding

- Syntax highlighting

---

[18]`http://www.eclipse.org/jdt`
[19]`http://www.eclipseplugincentral.com`
[20]`http://pydev.sourceforge.net`

FIGURE 2.2:   PyDeb: Easy Eclipse for Python.

- Refactoring

- Integrated debugger

- Support for plugins

- Very active community

- Supported by industry leaders such as Borland, Ericsson, Red Hat, SuSE, HP, IBM, Intel, SAP, and others.

The definition of 500 pound gorilla fits perfectly to this program. If you don't have at least 1 Gb of RAM, don't try to run Eclipse (unless you have a lot of patience). Another inconvenience is the complexity of the software. In this area, it is similar to Eric but with a better interface (see Figure 2.2). In Eclipse, you don't open a file and begin coding. First you have to create a project, select the type of project, give it a name and then add files. Once you've learned this routine and have a computer powerful enough, you will enjoy programming with Eclipse, especially if you have some programming experience.

Availability

Eclipse is a free software released under GPL and it is available for most platforms. You can download the "Classic" version from the official website[21] or download PyDev,[22] a version of Eclipse preconfigured for Python and Jython development.

2.4.4 IDLE

IDLE is the "official" Python editor. On certain systems IDLE is included with Python. As a result it is the most used Python editor. It is built with Tcl/tk, for which its interface does not necessarily reflect the native look and feel of the underlying platform. However do not be fooled by its appearance. IDLE has all the features desired for a programmer's editor:

- Multi-window text editor with multiple undo

- Python syntax colorizing with smart indent and call tips

- Basic debugger

- Integrated shell

IDLE can be recommended as a first editor, especially if you've never programmed with Python.

Availability

IDLE is available under a GPL compatible license. In Windows, IDLE is part of Python, so once Python in installed, IDLE is available. In a Debian based Linux distribution, IDLE can be installed with `apt-get install idle-pythonx` (where x is the Python version number, like 2.6). If you want to manually install IDLE, install first `TK-devel` and `TC-devel`.

2.4.5 Final Words about Editors

There is no editor that is unquestionably better than all others in all areas. IDLE for example is good as a first Python editor, although it may not be the prettiest or most stable. Eric and Eclipse are the most complete, although not all machines can run them efficiently. Despite not being specialized Python editor, they very versatile and it not to be left behind in terms of features.

There are a lot more editors which, for space reasons have not been reviewed but not because they are worse or less complete than those shown here. For example SPE, DrPython, KomodoIDE, WingIDE, Emacs, NetBeans, among the multi-platform editors, PythonWin (Windows only), and TextMate (Mac OS X).

---

[21]http://www.eclipse.org/downloads

[22]http://pydev.sourceforge.net

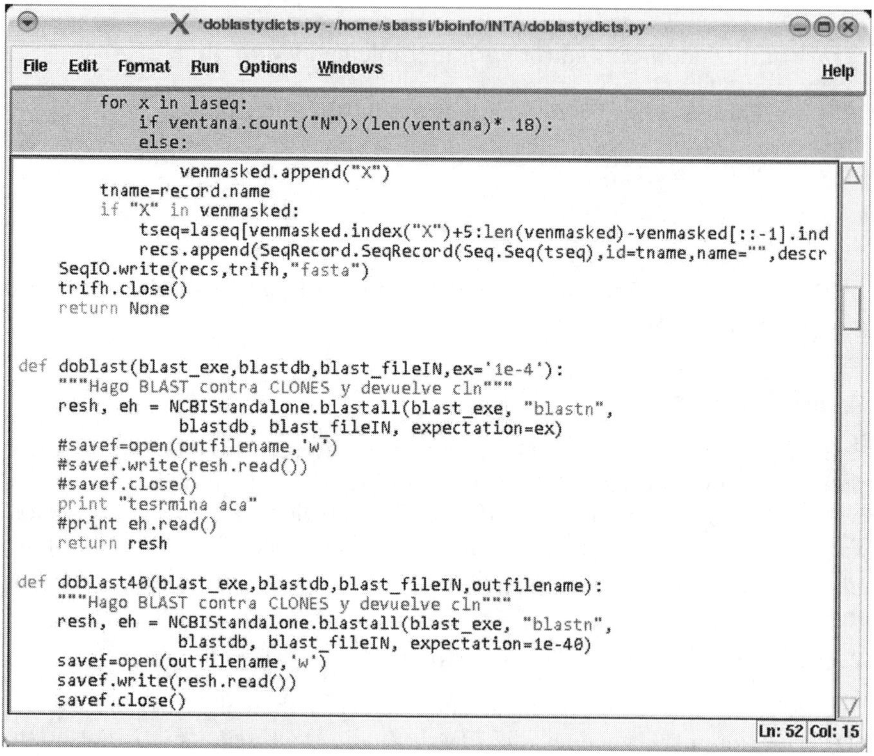

FIGURE 2.3: IDLE.

Although some illustrations of the book are based on IDLE, I am not going to recommend a particular editor as I consider this a personal choice. My recommendation is that you try all that you can and choose the one that best fits your needs.

2.5   Additional Resources

- Python implementations:
  http://www.python.org/dev/implementations

- IPython: an interactive computing environment.
  http://ipython.scipy.org/moin

- bpython: a fancy interface to the Python interpreter for Unix-like operating systems:
  http://www.bpython-interpreter.org/home

- Wikipedia article: "Comparison of text editors."
  http://en.wikipedia.org/wiki/Comparison_of_text_editors

- "Some notes on text editors for Stata users." This article is oriented towards Stata programers (Stata is a statistical software package), but it is full of fundamental concepts about text editors so it is worth mentioning here.
  http://fmwww.bc.edu/repec/bocode/t/textEditors.html

- "Python Mode for Emacs."
  http://www.emacswiki.org/emacs/PythonMode

- "Introduction to Python EA in NetBeans IDE"
  http://www.netbeans.org/kb/docs/python/temperature-converter.html

- "Useful VIM Settings for working with Python."
  http://www.vex.net/~x/python_and_vim.html

- vimrc file for following the coding standards specified in PEP 7 & 8.
  http://svn.python.org/projects/python/trunk/Misc/Vim/vimrc

2.6   Self-Evaluation

1. Define: Program, instruction, and variable.

2. What is the difference between Python and cPython?

3. Name some Python implementations.

4. What is the advantage of having both single and double quotes?

5. What is the difference between input and raw_input in Python 2.x?

6. How do you replace Python 2.x input functionality in Python 3?

7. How do you make a float division in Python 2.x?

8. What is indentation? Why is it mandatory in Python?

9. What is a comment in a source code?

10. Is there a valid reason to comment out working source code?

11. What is a "shebang"?

12. What is an "encoding comment." and when should you use it?

Chapter 3

Basic Programming: Data Types

As said in the previous chapter, some data structures are shared between different computer languages, but some of them are language specific. That is why data types somehow define a computer language. Python has its own characteristic data types.

One such fundamental data structure is a *sequence*. Inside sequence, we have those data types having a sequential order. A well-known example is the **string** which is nothing other than an ordered sequence of characters. Other sequences are **lists** and **tuples**.[1] Although fundamental differences exist between these types of sequences, they share common properties. Sequence elements have an order, can be indexed, can be sliced, and can be iterated. Don't worry if you don't understand some of these terms. Just keep on reading. We'll see all these points during this chapter.

Apart from *sequences*, there are also *unordered* data types: *dictionaries* and *sets*. A *dictionary*[2] stores relationships between a key and a value, while a *set* is just an unordered collection of values. The next pages are focused on ordered (string, list, and tuple) and unordered types (dictionary and set).

3.1 Strings

A string is a type of sequence of symbols delimited by a single quote ('), double quotes ("), single triple quotes ('''), or double triple quotes (""""). Therefore, the following strings are equivalent:

```
"This is a string in Python"
'This is a string in Python'
'''This is a string in Python'''
"""This is a string in Python"""
```

---

[1] There are more sequence types not covered in this book. For more information on other sequence types see *Additional Resources* at the end of the chapter.

[2] Also classified as a mapping data type.

It may seem a little bit redundant to have so many ways to delimit a string. The advantage of having both single (') and double (") quote delimiters is that we can insert a single quote in a string delimited for double quote and vice versa:

```
"A single quote (') inside a double quote"
'Here we have "double quotes" inside single quotes'
```

The important thing to remember is that if we begin a string with a type of quote, we must finish it with the same type of quote. The following string is not valid:

```
>>> "Mixing quotes leads to the dark side'
  File "<stdin>", line 1
    "Mixing quotes leads to the dark side'
                                         ^
SyntaxError: EOL while scanning single-quoted string
```

Note: EOL stands for end-of-line.

Regarding strings enclosed by triple quotes, we can use them to indicate multi-line strings (also known as block string):

```
"""Hi! I'm a
multiline
        string"""
```

The character '\n' represents an end-of-line (EOL) character. Therefore, the code above could be written in one line as:

```
"Hi! I'm a\nmultiline\n        string"
```

You can use triple quotation marks to build and format a string just like you'd expect to see it displayed. There are other uses for triple quoted strings such as documentation. It is covered in section (6.1).

3.1.1 Not All Strings Are Created Equal

Python 2.x

There are two types of strings: **byte strings** and **Unicode strings**. Byte-strings contain bytes while Unicode is intended for text. In practice, English characters are stored as **byte strings**, while "international" characters are stored as **Unicode strings**:

```
'I am a byte string in Python 2.5!!!'
u'I am UNICODE: π'
```

Note that Unicode strings have support for characters such as π, ℵ, ñ, ρ, χ, and any other imaginable character.[3]

byte and **unicode** are interconvertible as long as you specify an ecoding scheme:

```
>>> s = u'Sebastián'
>>> s.encode('utf-8')
'Sebasti\xc3\xa1n'
>>> b=s.encode('utf-8')
>>> x=unicode(b,'utf-8')
>>> print x
Sebastián
>>> x==s
True
```

Python 3

In Python 3, strings are Unicode by default:

```
>>> 'Python 3, strings are Unicode:  ñ'
'Python 3, strings are Unicode:  ñ'
```

But there is a **byte** object that is defined as "an immutable array of bytes" and it is equivalent to the default string (byte-string) of Python 2.x:

```
>>> b'Bytes in Python 3'
b'Bytes in Python 3'
```

Another new feature in Python 3 regarding strings are **bytearray**, that are "a mutable array of bytes." We will see the difference between mutable and immutable later on this chapter.

3.1.2 String Manipulation

Strings are immutable (with the exception of *bytearray* in Python 3). Once a string is created, it can't be modified. If you need to change a string, what you can do is to make a derived string. This is done using the string as a parameter in a function and then get the returned value. In the follow-

---

[3]For more information on available Unicode characters please see http://www.unicode.org/charts.

ing example there is a string that holds the amino-acid sequence of a signal peptide,[4] and it is called, with originality, *signal_peptide*:

```
>>> signal_peptide="MASKATLLLAFTLLFATCIA"
```

To get a lower case version of the string, use the method **lower()**:

```
>>> signal_peptide.lower()
'maskatlllaftllfatcia'
```

In spite of having obtained the string lower case, the original string has not been modified:

```
>>> signal_peptide
'MASKATLLLAFTLLFATCIA'
```

If we want this new lower case string to have the same name as the previous one, all we need to do is rename it:

```
>>> signal_peptide=signal_peptide.lower()
>>> signal_peptide
'maskatlllaftllfatcia'
```

The net effect is like we had modified the string. Bearing this in mind, it's time to see some methods associated with strings.

3.1.3   Methods Associated with Strings

replace(old,new[,count]): Allows us to replace a portion of a string (*old*) with another (*new*). If the optional argument *count* is used, only the first *count* occurrences of *old* will be replaced. Usage,

```
>>> DNAseq="TTGCTAG"
>>> mRNAseq=DNAseq.replace("T","U")
>>> mRNAseq
'UUGCUAG'
```

count(sub[, start[, end]]): Counts how many times the substring *sub* appears, between *start* and *end* position (if available). Let's see how it can be used to calculate the CG content[5] of a sequence:

---

[4]A signal peptide is a short amino acid chain (about 20 amino acids) that is recognized by certain organelles to direct the transport of the nascent protein to a specific subcellular location or to the secretory pathway.

[5]CG content is the amount of cytosine and guanine in a DNA sequence. CG content is related to the DNA melting temperature and other physical properties.

```
>>> c=DNAseq.count("C")
>>> g=DNAseq.count("G")
>>> float(c+g)/len(DNAseq)*100
48.387096774193552
```

Note that the float function is used to force a floating point result, this is not required in Python 3.

find(sub[,start[,end]]): Returns the position of the substring *sub*, between *start* and *end* position (if available). If the substring is not found in the string, this method returns the value −1:

```
>>> mRNAseq.find("AUG")
17
```

index(sub[,start[,end]]): Works like *find()*. The difference is that *index* will raise a *ValueError* exception when the substring is not found. This method is recommended over *find()* because the value -1 could be interpreted as a valid value, while a *ValueError* returned by *index()* can't be taken as a valid value.

split([sep [,maxsplit]]): Separates the "words" of a string and returns them in a list. If a separator (*sep*) is not specified, the default separator will be the white space:

```
>>> "This string has words separated by spaces".split()
['This', 'string', 'has', 'words', 'separated', 'by', 'spaces']
```

When white space is not the data separator, we have to specify a custom separator:

```
>>> "Alex Doe,5555-2333,nobody@example.com".split()
['Alex', 'Doe,5555-2333,nobody@example.com']
```

In this case the separator is a comma (","), so we have to state it explicitly:

```
>>> "Alex Doe,5555-2333,nobody@example.com".split(",")
['Alex Doe', '5555-2333', 'nobody@example.com']
```

Bioinformatic Application: Parsing BLAST Files. One of the most used bioinformatics programs is NCBI-BLAST (this program is reviewed from page 191).

BLAST standalone executable can generate output as a "tab separated file" (by using -m 8 argument). This output file can be parsed by using split('\t').

The inverse function of **split()** is **join()**:

join(seq): Joins the sequence using a string as a "glue character":

```
';'.join(['Alex Doe', '5555-2333', 'nobody@example.com'])
'Alex Doe;5555-2333;nobody@example.com'
```

To join a sequence without any glue character, use empty quotes (""):

```
>>> ''.join(['A','C','A','T'])
'ACAT'
```

For a complete description of string methods, see Table D.4.8 in page 472 (Appendix D).

3.2   Lists

3.2.1   List Is the Workhorse Datatype in Python

Lists are one of the most versatile object types in Python. A list is an ordered collection of objects. It is represented by elements separated by commas and enclosed between square brackets.

We already have seen a list as a result of applying the **split()** function:

```
>>> "Alex Doe,5555-2333,nobody@example.com".split(",")
['Alex Doe', '5555-2333', 'nobody@example.com']
```

This is a three element list, 'Alex Doe', '5555-2333' and 'nobody@example.com', all of them strings.

The next code shows how to define and name a list:

```
>>> first_list=[1,2,3,4,5]
```

This is a list with five elements. In this case, all the elements are of the same type (integer). A list can hold different kinds of elements:

```
>>> other_list=[1,"two",3,4,"last"]
```

A list can even contain another list:

```
>>> nested_list=[1,"two",first_list,4,"last"]
>>> nested_list
[1, 'two', [1, 2, 3, 4, 5], 4, 'last']
```

An empty list is defined as empty brackets:

```
>>> empty_list=[]
>>> empty_list
[]
```

An empty list doesn't have any use of its own, but sometimes we may want to define an empty list to add elements at a later time.

3.2.2 List Initialization

If you know in advance that a list is going to have five elements, we can initialize it with a default value:

```
>>> codons = [None] * 5
>>> codons
[None, None, None, None, None]
```

This type of list initialization can be useful when working with big lists and the number of elements is known beforehand. Defining a list with a fixed size is more efficient than creating an empty list expanding it as needed. Fixed sized lists don't have the overhead of lists that change positions in memory.

3.2.3 List Comprehension

There is another way to define a list. A list can be created from another list. As in mathematics where you can define a set by enumerating all its elements (enumeration) or by describing properties enjoyed exclusively by its members (comprehension), in Python a list can be created by both methods.

A set defined by enumeration,

$$A = \{0, 1, 2, 3, 4, 5\}$$

A list defined by enumeration in Python,

```
>>> A = [0,1,2,3,4,5]
```

A set defined by comprehension,

$$B = \{3 * x / x \in A\}$$

This is equivalent to,

$$B = \{0, 3, 6, 9, 12, 15\}$$

A list defined by comprehension in Python,

```
>>> [3*x for x in A]
[0, 3, 6, 9, 12, 15]
```

Any Python function or method can be used to define a list by comprehension:

```
>>> animals = [' king kong', ' godzilla ', 'gamera ']
>>> [y.strip() for y in animals]
['king kong', 'godzilla', 'gamera']
```

The resulting list can be narrowed by using a conditional statement:

```
>>> animals = [' king kong', ' godzilla ', 'gamera ']
>>> [y.strip() for y in animals if 'i' in y]
['king kong', 'godzilla']
```

3.2.4   Accessing List Elements

As one of the other sequence data types, you access list elements by an index starting at zero.

```
>>> first_list=[1,2,3,4,5]
>>> first_list[0]
1
>>> first_list[1]
2
```

You can also access lists from the right by using negative numbers:

```
>>> first_list=[1,2,3,4,5]
>>> first_list[-1]
5
>>> first_list[-4]
2
```

Another way of obtaining lists is by turning a non list object into a list by using the built-in function **list()**:

```
>>> aseq="atggctaggc"
>>> list(aseq)
['a', 't', 'g', 'g', 'c', 't', 'a', 'g', 'g', 'c']
```

The function **str()** converts the input parameter into a string. Can we use it to revert the effect of **list()**?

```
>>> str(['a', 't', 'g', 'g', 'c', 't', 'a', 'g', 'g', 'c'])
"['a', 't', 'g', 'g', 'c', 't', 'a', 'g', 'g', 'c']"
```

Clearly the result is not for what we were expecting. The **str()** function turned the list into a string, but not into the original string. Instead, the result was a literal representation of the list. To have the original string back, we have to join the elements of the list. This can be done with the **join()** function:

```
>>> "".join(['a', 't', 'g', 'g', 'c', 't', 'a', 'g', 'g', 'c'])
'atggctaggc'
```

3.2.5   Copying a List

Copying a list can be tricky:

```
>>> a=[1,2,3]
>>> b=a
>>> b.pop()
3
>>> a
[1, 2]
```

As seen, "=" doesn't copy the values, but works as it copy a reference to the original object[6]. There are two ways to make an independent copy of a list:

Using the copy module:

```
>>> import copy
>>> a=[1,2,3]
>>> b=copy.copy(a)
>>> b.pop()
3
>>> a
[1, 2, 3]
```

Without the copy module:

```
>>> a=[1,2,3]
>>> b=a[:]
>>> b.pop()
3
>>> a
[1, 2, 3]
```

3.2.6   Modifying Lists

Unlike strings, lists can be modified[7] by either adding, removing, or changing their elements:

Adding

There are three ways to add elements into a list: **append**, **insert**, and **extend**.

append(*element*): Adds an element at the end of the list.

---

[6]For a detailed review of what is going on under the hood, please see page 65.
[7]Lists are called "mutables" in Python jargon.

```
>>> first_list.append(99)
>>> first_list
[1, 2, 3, 4, 5, 99]
```

insert(*position,element*): Inserts the element *element* at the position *position*.

```
>>> first_list.insert(2,50)
>>> first_list
[1, 2, 50, 3, 4, 5, 99]
```

extend(*list*): Extends a list by adding a list to the end of the original list.

```
>>> first_list.extend([6,7,8])
>>> first_list
[1, 2, 50, 3, 4, 5, 99, 6, 7, 8]
```

This is the same as using the + symbol:

```
>>> [1,2,3]+[4,5]
[1, 2, 3, 4, 5]
```

Removing

There is also three ways to remove elements from a list.

pop(*[index]*): Removes the element in the index position and returns it to the point where it was called. Without parameters, it returns the last element.

```
>>> first_list
[1, 2, 50, 3, 4, 5, 99, 6, 7, 8]
>>> first_list.pop()
8
>>> first_list.pop(2)
50
>>> first_list
[1, 2, 3, 4, 5, 99, 6, 7]
```

remove(*element*): Removes the element specified in the parameter. In the case where there is more than one copy of the same object in the list, it removes the first one, counting from the left. Unlike **pop()**, this function does not return anything.

```
>>> first_list.remove(99)
>>> first_list
[1, 2, 3, 4, 5, 6, 7]
```

TABLE 3.1: Common List Operations

Properties	Description
s.append(x)	Adds the x element to list s
s.count(x)	Counts how many times x is in s
s.index(x)	Returns where is x in list s
s.remove(x)	Removes the element x from list s
s.reverse()	Reverse list s
s.sort()	Sort list s

Trying to remove a nonexistent element raises an error:[8]

```
>>> first_list
[1, 2, 3, 4, 5, 6, 7]
>>> first_list.remove(10)
Traceback (most recent call last):
  File "<stdin>", line 1, in ?
ValueError: list.remove(x): x not in list
```

Another way of removing an element of a list is using the command **del**, for what:

```
del first_list[0]
```

Has a similar effect to:

```
first_list.pop(0)
```

With the difference that **pop()** returns the extracted element where it was called, while **del** just deletes it.[9]

Table 3.1 summarizes other properties of lists.

3.3   Tuples

3.3.1   Tuples Are Immutable Lists

Recall that a list is a collection of ordered objects. The main characteristic of the tuple is that once created, it cannot be modified. That is why they are

---

[8]In Chapter 7 there is a more detailed description of exceptions.

[9]The object is not deleted. What actually happens is that the reference between the object and its name is lost. For the programmer, this action has the same effect as if it were deleted (it is not possible to gain access to the object). At some time in the future, the "python garbage collector" will eliminate it in a transparent and automatic way.

referred to as "immutable lists." Python objects are sometimes divided into mutable and immutable. As the name implies, immutable objects cannot be modified after they are created. You can easily tell a tuple from a list because the tuple's elements are enclosed between parentheses instead of square brackets:

```
>>> point=(23,56,11)
```

`point` is a tuple with three elements (23, 56 and 11).

There is a particular case that you should use a trailing comma, when defining a tuple with one element:

```
lone_element_tuple = (5,)
```

This is done to sort the ambiguity of having (5) that means 5 (five) since round brackets around an expression are ignored. With the trailing comma and parentheses it is clear that it is a tuple and not an expression.

You are not allowed to add or to remove elements from a tuple:

```
>>> point.append(3)
Traceback (most recent call last):
  File "<stdin>", line 1, in ?
AttributeError: 'tuple' object has no attribute 'append'
>>> point.pop()
Traceback (most recent call last):
  File "<stdin>", line 1, in ?
AttributeError: 'tuple' object has no attribute 'pop'
```

In a certain way, a tuple is like a limited list (limited in the sense that we cannot modify it). So what are tuples good for? Why not just use a list instead?

There is a conceptual difference between the data types stored as a tuple and the data stored as a list. Lists should hold a variable quantity of objects of the same data type. A list containing the file names of all the files in a directory can be stored in a list. They are all of the same type (string), and the number of elements in the list changes according to each directory. The element ordering inside this list is not relevant.

On the other hand, a typical example of a tuple is a coordinate system. In a three-dimensional coordinate system, each point is referred to by a three element tuple (x, y, z). The number of elements for each tuple does not change (since there are always three coordinates), and each position is important since each point corresponds to a specific axis.

We can say the same thing regarding the elements that are returned from a function or a dictionary key.[10] Another advantage of the tuple is it can

---

[10]This will become clear after seeing functions and dictionaries.

be used to make safer code – the information we don't want to change stays "write-protected" in an immutable tuple.

Under some conditions, the processing speed of tuples is faster than that of lists. While this true in most cases, this shouldn't be a major consideration when choosing between a list or a tuple.

3.4 Common Properties of the Sequences

Since sequences share common properties, we've seen them together. You can apply these properties indifferently to lists, tuples, and strings.

Indexing

Indexing was discussed when covering list, but for the sake of completeness, it is also here. Since the elements in the sequences are ordered, we can gain access to any element through an index that begins at zero:

```
>>> point=(23,56,11)
>>> point[0]
23
>>> point[1]
56
>>> sequence="MRVLLVALALLALAASATS"
>>> sequence[0]
'M'
>>> sequence[5]
'V'
>>> parameters=['UniGene','dna','Mm.248907',5]
>>> parameters[2]
'Mm.248907'
```

We can also gain access to the elements of a sequence from the right by using negative numbers:

```
>>> point[-1]
11
>>> point[-2]
56
>>> my_sequence[-2]
'T'
>>> my_sequence[-4]
'S'
```

```
>>> my_sequence[-1]
5
```

To access an element that is inside a sequence, which is itself inside another sequence, you need to use another index:

```
>>> seqdata=("MRVLLVALALLA",12,"5FE9EEE8EE2DC2C7")
>>> seqdata[0][5]
'V'
```

The first index (0) indicates we're accessing the first element of `seqdata`. The second index (5) refers to the 6th element ('V') of the first element ('MRVLLVALALLA')

Slicing

You can select a portion of a sequence using *slice notation*. Slicing consists of using two indexes separated by a colon (:). These indexes represent a position in the existing space **between** the elements. The string "Python" can be represented as,

```
+---+---+---+---+---+---+
| P | y | t | h | o | n |
+---+---+---+---+---+---+
0   1   2   3   4   5   6
```

```
>>> my_sequence="Python"
>>> my_sequence[0:2]
'Py'
```

When omitting the first sub index, the index value defaults to the first position (0):

```
>>> my_sequence[:2]
'Py'
```

On the other hand, when the second sub index is omitted, the index value defaults to the last position (-1):

```
>>> my_sequence="Python"
>>> my_sequence[4:6]
'on'
>>> my_sequence[4:]
'on'
```

There is a third, optional index to skip positions (step argument):

```
>>> my_sequence[1:5]
'ytho'
>>> my_sequence[1:5:2]
'yh'
```

A step with a negative number is used to count backwards. So -1 (in the third position) can be used to invert a sequence:

```
>>> my_sequence[::-1]
'nohtyP'
```

Note that slicing always returns another sequence.

Membership Test

You can verify whether an element belongs to a sequence, using the **in** keyword:

```
>>> point=(23,56,11)
>>> 11 in point
True
>>> my_sequence="MRVLLVALALLALAASATS"
>>> "X" in my_sequence
False
```

This check can be used to avoid trying to remove an element that does not belong to the sequence:

```
if "X" in my_sequence:
    my_sequence.remove("X")
else:
    pass # "X" is not an element of my_sequence, so do nothing.
```

Concatenation

You can concatenate two or more sequences of the same class using the "+" sign:

```
>>> point=(23,56,11)
>>> point2=(2,6,7)
>>> point+point2
(23,56,11,2,6,7)
>>> DNAseq="ATGCTAGACGTCCTCAGATAGCCG"
>>> TATAbox="TATAAA"
>>> TATAbox+DNASeq
'TATAAAATGCTAGACGTCCTCAGATAGCCG'
```

Sequences of different types can't be concatenated:

```
>>> point+TATAbox

Traceback (most recent call last):
  File "<pyshell#48>", line 1, in <module>
    point+TATAbox
TypeError: can only concatenate tuple (not "str") to tuple
```

len, max, and min

len() returns the amount of elements of a sequence:

```
>>> point=(23,56,11)
>>> len(point)
3
>>> my_sequence="MRVLLVALALLALAASATS"
>>> len(my_sequence)
19
```

The use of **max()** and **min()** is a no-brainer:

```
>>> point
(23, 56, 11)
>>> max(point)
56
>>> min(point)
11
```

max() and **min()** applied to strings returns a character according to the maximum or minimum value of its ASCII code:

```
>>> MySequence="MRVLLVALALLALAASATS"
>>> max(MySequence)
'V'
>>> min(MySequence)
'A'
```

Turn a Sequence into a List:

To convert a sequence (like a tuple or a string) into a list, use **list()**:

```
>>> TATAbox="TATAAA"
>>> list(TATAbox)
['T', 'A', 'T', 'A', 'A', 'A']
```

Using list provides us with methods to indirectly modify a string. Since lists, unlike strings, are mutable, we can convert a string to a list, modify this list and then convert it back into a string (with **str()**).[11] This process is not efficient, so I suggest that whenever possible, use string properties to obtain another string.

3.5 Dictionaries

3.5.1 Mapping: Calling Each Value by a Name

Dictionaries are a special data type not present in all programming languages. The main characteristic of a dictionary is that it stores arbitrary indexed unordered data types.

This example shows us why this data type is called a dictionary:

```
>>> IUPAC = {'A':'Ala','C':'Cys','E':'Glu'}
>>> print("C stands for the amino acid "+IUPAC['C'])
C stands for the amino acid Cys
```

IUPAC is the name of a dictionary with three elements. It was defined by enclosing is *key:value* pairs between curly brackets ({}).

This dictionary works as a translation table that allows us to translate between the one-letter amino acid code to a three-letter code. Every element consists of a pair *key:value*. The key is the index used to retrieve the value:

```
>>> IUPAC['E']
'Glu'
```

Not every object can be used as a dictionary key, only immutable objects like strings, tuples and numbers can be used as keys. If the tuple contains any mutable object, it cannot be used as a key.

A dictionary can also be created from a sequence with **dict**:

```
>>> rgb=[('red','ff0000'),('green','00ff00'),('blue','0000ff')]
>>> colors_d = dict(rgb)
>>> colors_d['green']
'00ff00'
```

dict also accepts name=value pairs in the keyword argument list:

```
>>> rgb = dict(red='ff0000',green='00ff00',blue='0000ff')
>>> rgb
{'blue': '0000ff', 'green': '00ff00', 'red': 'ff0000'}
```

---

[11]By using the method **join()** as it was described on page 46.

Another way to initialize a dictionary is to create an empty dictionary and add elements as needed:

```
>>> rgb = {}
>>> rgb['red'] = 'ff0000'
>>> rgb['green'] = '00ff00'
>>> rgb
{'green': '00ff00', 'red': 'ff0000'}
```

len(), returns the number of elements in the dictionary:

```
>>> len(IUPAC)
3
```

To add values to a dictionary,

```
>>> IUPAC['S']='Ser'
>>> len(IUPAC)
4
```

Dictionaries are labeled as unordered because they don't keep track of the order of it elements. In this interactive session, the keys are shown in the same order as entered:

```
>>> IUPAC = {'A':'Ala','C':'Cys','E':'Glu'}
>>> for aa in IUPAC:
...     print aa
...
A
C
E
```

But after entering a new element, this order seems to disappear:

```
>>> IUPAC['X']='Xaa'
>>> for aa in IUPAC:
...     print aa
...
A
X
C
E
```

Don't rely on a diccionary to keep track of element order[12].

---

[12]Since there were a lot of demand for ordered dictionaries, it was implemented in Python 3.1 with the `collections.OrderedDict` method. See PEP-372 for more information (`http://www.python.org/dev/peps/pep-0372/`).

3.5.2 Operating with Dictionaries

As lists, dictionaries have their own methods.

Dictionaries Are Made of Keys and Values

In Python 2.x, you can list the keys and values of dictionaries with the methods **keys()** and **values()** respectively:[13]

```
>>> IUPAC.keys()
['A', 'C', 'E', 'S']
>>> IUPAC.values()
['Ala', 'Cys', 'Glu', 'Ser']
```

The order of the keys correspond to the order of the values. Note that value order is not guaranteed to be kept next time you retrieve data from the dictionary.

These methods can be used to check the presence of a key in a dictionary. **keys()** returns a list . You could use **keys()** as a possible check for the presence of an element in a list:

```
>>> 'Z' in IUPAC.keys() # Method not recommended!
False
```

There is no need to generate a list for checking for the presence of an element. You can check directly on the dictionary. This is more effective, because it does not generate a temporary list of keys built and discarded just for this check.

```
>>> 'Z' in IUPAC
False
```

The **has_key()** method is also available, but its use is not recommended.[14] I show it here anyway since a lot of existing codes still use it:

```
>>> IUPAC.has_key('Z') #use 'Z' in IUPAC instead.
False
```

Another way of gaining access to the elements of a dictionary is by using **items()**:

```
>>> IUPAC.items()
[('A', 'Ala'), ('C', 'Cys'), ('E', 'Glu'), ('S', 'Ser')]
```

items() returns a list with a tuple for every key/value pair.

---

[13]This functionality is replaced by **Dictionary views** in Python 3. This is explained on page 58.

[14]The rationale behind this is that if keyword **in** works for lists, it should also work for dictionaries. There's no reason to remember a separate function (**has_key()**), which does the same thing as **in**.

Safe Query of Dictionary Values

A way to query a value from a dictionary, without the risk of invoking an exception, is to use **get(k,x)**. *K* represents the key of the element to extract, while *x* is the element that will be returned in case *k* is not found as a key of the dictionary.

```
>>> IUPAC.get('A','No translation')
'Ala'
>>> IUPAC.get('Z','No translation')
'No translation'
```

If you omit *x*, and there is no *k* present in the dictionary, the method returns **None**.

```
>>> IUPAC.get('Z')
None
```

Erasing Elements

To erase elements from a dictionary, use the **del** instruction:

```
>>> del IUPAC['A']
>>> IUPAC
[('C', 'Cys'), ('E', 'Glu'), ('F', 'Phe')]
```

Table 3.2 summarizes the properties of dictionaries.

3.5.3   New in Python 3: Dictionary Views

In Python 3, *keys()*, *values()* and *items()* methods don't return a list as in previous versions. They return a special kind of object called *dict view*:

```
>>> IUPAC.values() # Under Python 3.0
<dict_values object at 0xb7d145a0>
```

This object can be iterated as a list (it's an iterable object), so the following code is still good:

```
>>> for x in IUPAC.values():
...        print(x)
...
Ala
Cys
Glu
Ser
```

TABLE 3.2: Methods Associated with Dictionaries

Properties	Description
len(a)	Number of elements of a
a[k]	The element from a that has a k key
a[k] = v	Set a[k] to v
del a[k]	Remove a[k] from a
a.clear()	Remove all items from a
a.copy()	A copy of a
k in a	True if a has a key k, else False
k not in a	Equivalent to not k in a
a.has_key(k)	Equivalent to k in a, use that form in new code
a.items()	A copy of a's list of (key, value) pairs
a.keys()	A copy of a's list of keys
a.update([b])	Updates (and overwrites) key/value pairs from b
a.fromkeys(seq[, value])	Creates a new dictionary with keys from *seq* and values set to *value*
a.values()	A copy of a's list of values
a.get(k[, x])	a[k] if k in a, else x
a.setdefault(k[, x])	a[k] if k in a, else x (also setting it)
a.pop(k[, x])	a[k] if k in a, else x (and remove k)
a.popitem()	Remove and return an arbitrary (key, value) pair

This is more memory efficient than Python 2.x,[15] since Python doesn't store the whole list in memory. If you need to get the list anyway, use *list()*:

```
>>> list(IUPAC.values())
['Ala', 'Cys', 'Glu', 'Ser']
```

Another difference worth pointing out is that the *dict view* are kept in sync with the current content of the dictionary. In Python 2.x, the result of *keys()*, *values()*, and *items()* were not modified by subsequent changes in the dictionary:

```
>>> d={1:'a',2:'b',3:'c'}
>>> k=d.keys()
>>> k
[1, 2, 3]
>>> d[6]='p'
>>> k
[1, 2, 3]
```

Dict view objects from Python 3.0 keeps track of dictionary contents:

---

[15]To get an iterator in older versions of Python you had to use **iterkeys()**, **itervalues()** and **iteritems()** instead of **keys()**, **values()** and **items()**.

```
>>> d={1:'a',2:'b',3:'c'}
>>> k=d.keys()
>>> list(k)
[1, 2, 3]
>>> d[6]='p'
>>> list(k)
[1, 2, 3, 6]
```

3.6   Sets

3.6.1   Unordered Collection of Objects

This type of data is also not commonly found in other programming languages. Even in Python, sets are not very popular as they were implemented as a native data type only recently.[16] A **set** is a structure frequently found in mathematics. It is similar to a list, with two outstanding differences: *its elements do not preserve an implied order and every element is unique.*

The most common uses of sets are membership testing, duplicate removal, and the application of mathematical operations: intersections, unions, differences, and symmetrical differences.

Creating a Set

Sets are created with the instruction **set()**:

```
>>> first_set = set(['CP0140.1','EF3613.1','EF3616.1'])
```

It is also possible to create an empty set and then add the elements as needed:

```
>>> first_set = set()
>>> first_set.add('CP0140.1')
>>> first_set.add('EF3613.1')
>>> first_set.add('EF3616.1')
>>> first_set
set(['CP0140.1','EF3613.1','EF3616.1'])
```

---

[16]In Python 2.3, sets were not available unless imported as **from sets import Set**. Note that old versions uses **Set** with uppercase 'S'.

New in Python 3: New Syntax for Sets

In Python 3 you don't need to create an iterable and then apply a *set function*. A set can be declared by using {}:

```
>>> first_set = {'CP0140.1','EF3613.1','EF3616.1'}
>>> first_set
{'EF3616.1', 'EF3613.1', 'CP0140.1'}
```

Python 3 also supports defining a set by comprehension, as in list comprehension (see page 45):

```
>>> {2*x for x in [1,1,2,2,3,3]}
{2, 4, 6}
```

Since a **set** does not accept repeated elements, there is no effect when you try to add an element that is already in the set:

```
>>> first_set.add('CP0140.1')
>>> first_set
set(['CP0140.1','EF3613.1','EF3616.1'])
```

This property can be used to remove duplicated elements from a list:

```
>>> uniqueIds = set([2,2,3,4,5,3])
>>> uniqueIds
set([2, 3, 4, 5])
```

3.6.2   Set Operations

Intersection

To get the common elements in two sets (as shown in Figure 3.1), use the operator **intersection()**:

```
>>> first_set = set(['CP0140.1','EF3613.1','EF3616.1'])
>>> other_set = set(['CP0140.2','EF3613.1','EF3616.2'])
>>> common = first_set.intersection(other_set)
>>> common
set(['EF3613.1'])
```

Instead of **intersection()**, it is possible to use the shortcut **&**:

```
>>> common = first_set & other_set
>>> common
set(['EF3613.1'])
```

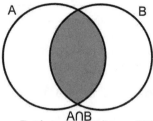

 A∩B
Python notation: A&B

FIGURE 3.1:   Intersection.

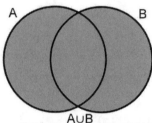

A∪B
Python notation: A|B

FIGURE 3.2:   Union.

Union

The union of two (or more) sets is the operator **union** (as seen in Figure 3.2) and its abbreviated form is |:

```
>>> first_set.union(other_set)
set(['EF3616.2', 'EF3613.1', 'EF3616.1', 'CP0140.1', 'CP0140.2'])
>>> first_set | other_set
set(['EF3616.2', 'EF3613.1', 'EF3616.1', 'CP0140.1', 'CP0140.2'])
```

Difference

A **difference** is the resulting set of elements that belongs to one set but not to the other (See Figure 3.3). Its shorthand is −:

```
>>> first_set.difference(other_set)
set(['CP0140.1', 'EF3616.1'])
>>> first_set - other_set
set(['CP0140.1', 'EF3616.1'])
```

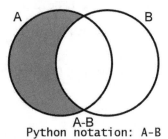

A-B
Python notation: A-B

FIGURE 3.3: Difference.

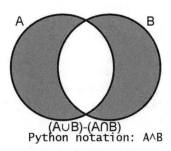

(A∪B)-(A∩B)
Python notation: A^B

FIGURE 3.4: Symmetric difference.

Symmetric Difference

A symmetric difference refers to those elements that are not a part of the intersection (see Figure 3.4); its operator is **symmetric_difference** and it is shorted as ^:

```
>>> first_set.symmetric_difference(other_set)
set(['EF3616.2', 'CP0140.1', 'CP0140.2', 'EF3616.1'])
>>> first_set ^ other_set
set(['EF3616.2', 'CP0140.1', 'CP0140.2', 'EF3616.1'])
```

3.6.3   Shared Operations with Other Data Types

Maximum, Minimum, and Length

Sets share some properties with the sequences, as **max**, **min**, **len**, **in**, etc. As we can expect, these properties work in the same way.

Converting a Set into a List

As with strings, sets can be turned into lists with the function **list()**:

```
>>> first_set = set(['CP0140.1','EF3613.1','EF3616.1'])
>>> list(first_set)
['EF3616.1', 'EF3613.1', 'CP0140.1']
```

Table D.16 in page 478 summarizes all properties of sets.

3.6.4   Immutable Set: Frozenset

Frozenset is the immutable version of set. It contents cannot be changed, so methods like add() and remove() are not available:

```
>>> fs = frozenset(['a','b'])
>>> fs
frozenset({'a', 'b'})
>>> fs.remove('a')
Traceback (most recent call last):
  File "<stdin>", line 1, in <module>
AttributeError: 'frozenset' object has no attribute 'remove'
>>> fs2 = frozenset(['c','d'])
>>> fs.add(fs2)
Traceback (most recent call last):
  File "<stdin>", line 1, in <module>
AttributeError: 'frozenset' object has no attribute 'add'
```

Since frozensets are immutable, they can be used as a dictionary key.

3.7   Naming Objects

Names rules are straightforward: Valid names should contain letters, numbers and undercores (_), but they can't start with numbers. Another restriction is the use of a "language reserved word" like:

and	del	from	not	try
as	elif	global	None	while
assert	else	if	or	with
break	except	import	pass	yield
class	exec	in	print	
continue	finally	is	raise	
def	for	lambda	return	

Here is a sample of invalid names, with an explanation in a comment:

```
>>> 23crm = "1"      # Start with a number
>>> 23 = "1"         # Start with a number
>>> Var? = "value"   # Has an invalid character (?).
>>> $five = 5        # Has an invalid character ($)
>>> for = 123        # Has a reserved word
>>> if = "data"      # Has a reserved word
```

We've seen several name assignments up to this point:

```
>>> MySequence="MRVLLVALALLALAASATS"
>>> first_list=[1,2,3,4,5]
>>> d={1:'a',2:'b',3:'c'}
>>> k=d.keys()
>>> point=(23,56,11)
>>> first_set = set(['CP0140.1','EF3613.1','EF3616.1'])
>>> fs = frozenset(['a','b'])
```

Those are valid names. Appart from valid names, there are naming conventions that must be followed to improve code readability. These conventions are part the Python Style Guide and this book has chapter devoted to it (F, from page 553 to 576). According to this guide, names should be lowercase, with words separated by underscores as necessary to improve readability.

3.8 Assigning a Value to a Variable versus Binding a Name to an Object

The following statements can be thought of as a variable assignment:

```
>>> a=3
>>> b=[1,2,a]
```

Translated into English, they mean: "Let the variable a have a value of 3" and "Let the variable b have a list with three elements: 1, 2 and a (that has the value 3)".

Printing b seems to confirm both statements:

```
>>> print b
[1, 2, 3]
```

So by changing the value of a, the value of b should also change:

```
>>> a=5
>>> print b
[1, 2, 3]
```

What happened here? If you know another programming language, you may think that "Python is storing the value instead of the reference to the value." That is not exactly the case so I urge you to keep on reading.

The following statements seem to work in a different way:

```
>>> c=[1,2,3]
>>> d=[5,6,c]
```

Translated into English, they mean: "Let the variable c have a list with three elements: 1, 2 and 3" and "Let the variable d has a list with three elements: 5, 6 and c (that has three elements: 1, 2, and 3)".

This can be confirmed by printing both variables:

```
>>> print c
[1, 2, 3]
>>> print d
[5,6,[1, 2, 3]]
```

Let's change the value of c to see what happens with d:

```
>>> c.pop()
3
>>> print c
[1, 2]
>>> print d
[5,6,[1, 2]]
```

In this case, changing one variable, does change the other variable. It seems like an inconsistent behavior. If we think all these variable assignment as a binding names with objects, what seems inconsistent, starts to make sense. Try following next explanation using Figure 3.5.

```
>>> a=3
>>> b=[1,2,a]
```

Translated into English, they mean: "Let the object 3 be called a" and "Let the list with three elements (1, 2 and a) be called b".

Printing b seems to confirm both statements:

```
>>> print b
[1, 2, 3]
```

Then we create a new object (5) and name it a. So the previous reference (a=3) is destroyed (this is represented by a cross in the arrow from a to 3). The name a is not bound to 3 anymore, now a is bound to 5. What about b?

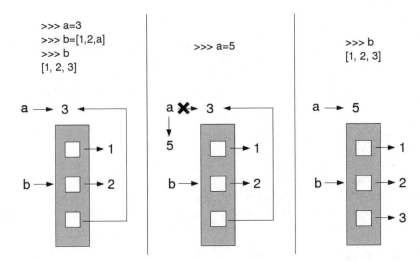

FIGURE 3.5: Case 1.

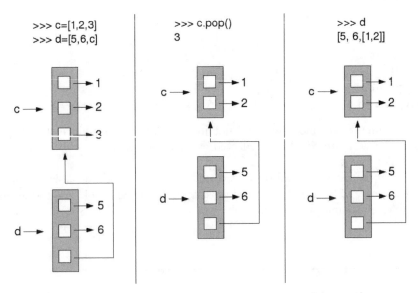

FIGURE 3.6: Case 2.

```
>>> a=5
>>> print b
[1, 2, 3]
```

Since the third position in the list called b was not altered, b remains unmodified. We only changed the binding between a and 3.

The next sample case can also be explained by taking into account that there is no variable assignments in Python, but names that bind objects. In this case you should follow Figure 3.6.

```
>>> c=[1,2,3]
>>> d=[5,6,c]
```

Translated into English, they mean, "Let the list with three elements: 1, 2 and 3 be called c" and "Let the list with three elements: 5, 6 and c (which is the name of a list of three elements: 1, 2 and 3) be called d." This can be confirmed by requesting the contents or both names:

```
>>> print c
[1, 2, 3]
>>> print d
[5,6,[1, 2, 3]]
```

In the next step, modify the list called c by removing the last element and see what happens with d:

```
>>> c.pop()
3
>>> print c
[1, 2]
>>> print d
[5,6,[1, 2]]
```

This time, c was modified (and not just a relationship). Since the actual value of c was altered, this is reflected every time it is called. See Figure 3.6 in case of doubt.

Even if names are bound to objects and there is no variable assignment in Python, force of habit is strong and most texts (even this book) use the terms variables and names interchangeably.

3.9   Additional Resources

- "Learn to Program Using Python: Variables and Identifiers."
  http://www.developer.com/lang/other/article.php/626321

- "Python 101—Introduction to Python"
  http://ascii-world.wikidot.com/python-101

- "Beginning Python for Bioinformatics"
  http://www.onlamp.com/lpt/a/2727

- String Methods
  http://docs.python.org/lib/string-methods.html

- Unicode HOWTO
  http://www.amk.ca/python/howto/unicode

- "Adding a Built-in Set Object Type"
  http://www.python.org/dev/peps/pep-0218/

- "Set Types—set, frozenset"
  http://docs.python.org/lib/types-set.html

- "Revamping dict.keys(), .values(), and .items()"
  http://www.python.org/dev/peps/pep-3106/

3.10   Self-Evaluation

1. Which are the principal data types in Python?

2. What is the difference between a list and a tuple? When would you use each one?

3. What is a set and when would you use it?

4. How do you test if an element is inside a list?

5. What is a dictionary?

6. What data type can be used as key in a dictionary?

7. What is a "dictionary view"?

8. Can you iterate over an unordered sequence?

9. Sort the data types below according to the following criteria:

   - Mutable–inmutable
   - Sorted–unsorted
   - Sequence–mapping

   Data types to sort: Lists, tuples, dictionaries, sets, strings.

10. What is the difference between a set and a frozenset?

11. How do you convert any iterable data type into a list?

12. How do you create a dictionary from a list?

13. How do you create a list from a dictionary?

Chapter 4

Programming: Flow Control

In order to be able to do something useful, programs must have some mechanism to manage how and when instructions are executed. In the same way that traffic lights control vehicular flow in a street, flow control structures direct that code portion is executed at a given time.

For greater simplicity, Python has only three flow control structures. There is one conditional and two iteration structures. A conditional structure (**if**) determines, after an expression evaluation, whether a block of code is executed or not. Iterative structures allow multiple execution of the same code portion. How many times is the code associated to a iterative structure executed? It depends the kind of cycle. A **for** cycle executes a code block many times as elements on are available in a specified iterable element, while the code under a **while** cycle is executed until a given condition turns false.[1]

4.1 If-Else

The most classic control structure is the conditional one. It acts upon the result of an evaluation. If you know any other computer language, chances are that you are familiar with **if-else**.

If evaluates an expression. If the expression is true, the block of code just after the **if** clause is executed. Otherwise, the block under **else** is executed.

A basic schema of an **if-else** condition,

```
if EXPRESSION:
    Block1
else:
    Block2
```

EXPRESSION must be an expression that returns **True** or **False**. This is the case of all comparison operators: $x < y$ (less than), $x > y$ (greater than), $x == y$ (equal to), $x! = y$ (not equal to),[2] $x <= y$ (less than or equal

---

[1]This is equivalent to saying that the condition is executed **while** the condition is true.
[2]There is an outdated equivalent operator: <>. You may find it in old Python code.

to), $x >= y$ (greater than or equal to).

Let's see an example:

Listing 4.1: Basic if-else sample

```
1 a = 8
2 if a>5:
3     print("a is greater than 5")
4 else:
5     print("a is smaller than 5")
```

Program output,

```
a is greater than 5
```

Another example,

Listing 4.2: if-else in action

```
1 trans = {"A":"Ala","N":"Asn","D":"Asp","C":"Cys"}
2 aa = raw_input("Enter one letter: ")
3 if aa in trans:
4     print("The three letter code for "+aa+" is: "+trans[aa])
5 else:
6     print("Sorry, I don't have it in my dictionary")
```

Program output,

```
Enter one letter: A
The three letter code for A is: Ala
```

To evaluate more than one condition, use **elif**:

```
if EXPRESSION1:
    Block1
elif EXPRESSION2:
    Block2
elif EXPRESSION3:
    Block3
else:
    Block4
```

You can use as many **elif** as conditions you want to evaluate. Take into account that once a condition is evaluated as true, the remaining conditions are not checked.

The following program evaluates more than one condition using **elif**:

Listing 4.3: Using elif

```
1 dna = raw_input("Enter your DNA sequence: ")
2 seqsize = len(dna)
3 if seqsize < 10:
4     print("The primer must have at least ten nucleotides")
5 elif seqsize < 25:
6     print("This size is OK")
7 else:
8     print("The primer is too long")
```

This program ask for a DNA sequence entered with the keyboard at run-time. This sequence is called *dna*. In line 2 its size is calculated and this result is binded to the name *seqsize*. In line 3 there is an evaluation. If *seqsize* is lower than ten, the message "The primer must have at least ten nucleotides" is printed. The program flows goes to the end of this **if** statement, without evaluating any other condition in this **if** statement. But if it is not true (for example if the sequence length was 15), it would execute the next condition and its associated block in case that this condition is evaluated as true. If the sequence length were of a value greater than 10, the program would skip line 4 (the block of code associated with the first condition) and would evaluate the expression in line 5. If this condition is met, it will print "This size is OK." If there is no expression that evaluates as true, the **else** block is executed.

Tip: What Is True?

Remember that the statement after the *if condition* is executed only when the expression is evaluated as **True**. So the question "What is True?" (and "What is False?") is relevant.

What is True?:

- Nonempty data structures (lists, dictionaries, tuples, strings, sets). Empty data structures count as *True*,

- 0 and *None* count as *False* (while other values count as *True*,

- Keyword *True* is *True* and *False* is *False*.

If you have a doubt if an expression is *True* or *False*, use **bool()**:

```
>>> a=1
>>> b='1'
>>> bool(a==b)
False
```

Conditionals can be nested:

Listing 4.4: Nested if

```
1 dna = raw_input("Enter your DNA sequence: ")
2 seqsize = len(dna)
3 if seqsize < 10:
4     print("Your primer must have at least ten nucleotides")
5     if seqsize==0:
6         print("You must enter something!")
7 elif seqsize < 25:
8     print("This size is OK")
9 else:
10     print("Your primer is too long")
```

In line 5 there is a condition inside another.

Note the double equal sign ("==") instead of the single equal. Double equal is used to compare values, while the equal sign is used to assign values:

```
>>> answer=42
>>> answer
42
>>> answer==3
False
>>> answer==42
True
```

The nested if introduced in code 4.4, can be avoided:

Listing 4.5: Nested if

```
1 dna = raw_input("Enter your DNA sequence: ")
2 seqsize = len(dna)
3 if seqsize==0:
4     print("You must enter something!")
5 elif 0<seqsize<10:
6     print("Your primer must have at least ten nucleotides")
7 elif seqsize < 25:
8     print("This size is OK")
9 else:
10     print("Your primer is too long")
```

See how the expression is evaluated in line 5. This leads us to think about inserting multiple statements in one **if**, like in code 4.6:

Listing 4.6: Multiple part condition

```
1 x = 'N/A'
2 if x!='N/A' and 5<float(x)<20:
```

```
3     print('OK')
4 else:
5     print('Not OK')
```

This expression is evaluated from left to right. If one part of the expression is false, the following parts are not evaluated. Since x='N/A', the program will print 'Not OK' (because the first condition is false). Look what happens when the same expression is written in reverse order.

This program,

Listing 4.7: Multiple part condition, inverted

```
x='N/A'
if 5<float(x)<20 and x!='N/A':
    print('OK')
else:
    print('Not OK')
```

returns:

```
Traceback (most recent call last):
  File "<pyshell#2>", line 1, in <module>
    if 5<float(x)<20 and x!='N/A':
ValueError: invalid literal for float(): N/A
```

The **ValueError** is produced because the string 'N/A' can't be converted to float. In code 4.7, x is also evaluated as 'N/A', but there is no **ValueError** because this part of the expression is skipped before evaluation.

4.1.1 Pass Statement

Sometimes there is no need of an alternative choice in an **if statement**, in this case you just can avoid using **else**:

```
if EXPRESSION:
    Block
Rest of the program...
```

To make the same code more readable, Python provides the **pass** statement. This statement is like a placeholder, it has any other pupose than put something when a statement is required syntactically. The following code produces the same output as the former code:

```
if CONDITION:
    BLOCK
else:
    pass
Rest of the program...
```

Advanced Tip: Conditional Expressions.

Sometimes comes in handy the availability of a special syntax to write an **if condition** in one line. Since Python 2.5, the following structure is available:

```
expression1 if condition else expression2
```

This line will take the value of *expression1*, if *condition* is true; otherwise, it will take the value of *expression2*.

This syntax allows us to write:

```
>>> print("Average = %s"%(t/n if n!=0 else "N/A"))
```

instead of,

```
if n!=0:
    print("Average = %s"%(t/n))
else:
    print("Average = N/A")
```

4.2   For Loop

This control structure allows code to be repeatedly executed while keeping a variable with the value of an iterable object.[3] The generic form of a **for loop** is,

```
for VAR in ITERABLE:
    BLOCK
```

Note the colon at the end of *ITERABLE*. It is mandatory. As the indentation of the block of code that is part of the **for loop**. This structure results on the repetition of *BLOCK* as many times as elements are in the iterable object. On each iteration, *VAR* takes the value of the current element in *ITERABLE*. In the following code **for** walk through a list (**bases**) with four elements. On each iteration, x takes the value of one of the elements in the list.

---

[3]The most common iterable objects are: lists, tuples, strings and dictionaries. Files and custom made objects can also be iterable.

```
>>> bases = ["C","T","G","A"]
>>> for x in bases:
...      print(x)
...
C
T
G
A
```

In other languages, the **for** loop is used to allow a block of code to run a number of times while changing a counter variable. This behavior can be reproduced in Python by iterating over a list of numbers:

```
>>> for x in [0,1,2,3]:
...      print(str(x)+"*"+str(x)+" = "+str(x*x))
...
0*0 = 0
1*1 = 1
2*2 = 4
3*3 = 9
```

A shortcut to generate the list is by using the built-in function **range(n)**. This function returns a list with many elements as the first parameter entered in the function,[4]

```
>>> for x in range(4):
...      print str(x)+"*"+str(x)+" = "+str(x*x)
...
0*0 = 0
1*1 = 1
2*2 = 4
3*3 = 9
```

The following code calculates the molecular weight of a protein based on its individual amino acids.[5] Since the amino acid is stored in a string, the program will walk through each letter by using can **for**,

Listing 4.8: Using **for** to calcule the weight of a protein (py3.us/55)

```
1 protseq = raw_input("Enter your protein sequence: ")
```

---

[4] All built-in functions are described in section D.6 (page 494).

[5] Amino acids are the building blocks of the proteins. Each amino acid (represented by a single letter) has an individual weight. Since each amino acid bond release a water molecule (with a weight of 18 iu), the weight of all the water molecules released is subtracted from the total.

```
 2 protweight = {"A":89,"V":117,"L":131,"I":131,"P":115,"F":165,
 3                "W":204,"M":149,"G":75,"S":105,"C":121,"T":119,
 4                "Y":181,"N":132,"Q":146,"D":133,"E":147,
 5                "K":146,"R":174,"H":155}
 6 totalW = 0
 7 for aa in protseq:
 8     totalW = totalW + protweight.get(aa.upper(),0)
 9 totalW = totalW-(18*(len(protseq)-1))
10 print("The net weight is: "+str(totalW))
```

Code explanation: On the first line the user is requested to enter a protein sequence (like for example MKTFVLHIFIFALVAF). The string returned by **raw_input** is named protseq. From line 2 to 5, a dictionary (protweight) with the amino acid weights is initialized. A **for loop** is used in line 7 to iterate over each element in protseq. In each iteration, aa takes a value from an element from protseq. This value is used to seach in the protweight dictionary. After the cycle, totalW will end up with the sum of the weight of all amino acids. In line 9 there is a correction due to the fact that each bond involves the loss of a water molecule (with molecular weight of 18). The last line prints out the net weight.

4.3   While Loop

A loop very similar to **for** since it also executes a code portion in a repeated way. In this case there is no iteration over an object, so this loop doesn't end when the iteration object is traversed, but when a given condition is not true.

Model of **while loop**,

```
while EXPRESSION:
    BLOCK
```

It is very important to take into account that there should be an instruction inside the block to make the while condition false. Otherwise, we could enter into an infinite loop.

```
>>> a=10
>>> while a<40:
...     print(a)
...     a = a+10
...
10
20
30
```

A way to exit from a **while** loop is using **break**. In this case the loop is broken without evaluating the loop condition. **break** is often used in conjunction with a condition that is always true:

```
>>> a=10
>>> while True:
...     if a<40:
...             print a
...     else:
...             break
...     a += 10
...
10
20
30
```

This is done to ensure the block inside the loop is executed at least once. In other languages there is a separate loop type for these cases (**do while**), but it is not present in Python.[6]

4.4   Break: Breaking the Loop

Break is used to escape from a loop structure. We've just seen a usage example with **while** but it also can be used under **for**.

It is not easy at first to realize where using a break statement actually makes sense.

Take, for example, code 4.9:

Listing 4.9: Searching a value in a list of tuples

```
1  cc = [('red',1), ('green',2), ('blue',3), ('black',4)]
2  name = 'blue'
3  for colorpair in cc:
4      if name==colorpair[0]:
5          code = colorpair[1]
6  print code
```

In this code there is a **for loop** to iterate over cc list. For each element, that is, for each tuple, it checks for the first element. When it matches our query (**name**), the program stores the associated code into **code**.

---

[6]There is a proposal to add this structure into Python filled under PEP-315 (http://www. python.org/dev/peps/pep-0315) but it has no implementation date, so don't rely on it.

So the output of this program is just "3."

The problem with this program is that the whole sequence is walked over, even if we don't need to. In this case, the condition in line 4 is evaluated once per each element in cc when it is clear that once the match is positive there is no need to keep on testing. You can save some time and processing power by breaking the loop just after the positive match:

Listing 4.10: Searching a value in a list of tuples

```
1  cc = [('red',1), ('green',2), ('blue',3), ('black',4)]
2  name = 'blue'
3  for colorpair in cc:
4      if name==colorpair[0]:
5          code = colorpair[1]
6          break
7  print code
```

This code is identical to 4.9 with the exception of the break statement in line 6. The output is the same as before, but this time you don't waste CPU cycles iterating over a sequence without a reason. The time saved in this example is negligible, but if the program has to do it several times over a big list or file (you can also iterate over a file), **break** can speed it up in a significant way.

The use of **break** can be avoided, but the resulting code is not so legible as in program 4.10:

Listing 4.11: Searching a value in a list of tuples

```
1  cc = [('red',1), ('green',2), ('blue',3), ('black',4)]
2  name = 'blue'
3  i = 0
4  while name!=cc[i][0]:
5      i += 1
6  code = cc[i][1]
7  print code
```

In a case like this, with a list that can easily fit in memory, it is a better idea to create a dictionary and query it:

Listing 4.12: Searching a value in a list of tuples using a dictionary

```
1  cc = [('red',1), ('green',2), ('blue',3), ('black',4)]
2  name = 'blue'
3  cc_d = dict(cc)
4  print cc_d[name]
```

4.5 Wrapping It Up

Combining **if, for, while** and the data type seen up to this point. Here I present some small programs made with the tools we've just learned:

4.5.1 Estimate the Net Charge of a Protein

At a fixed pH, it is possible to calculate the net charge of a protein summing the charge of its individual amino acids. This is an approximation since it doesn't take into account if the amino acids are exposed or buried in the protein structure. This program also fails to take into account the fact that cysteine add charge only when it is not part of a disulfide bridge. Since it is an approximate value the obtained value should be regarded as an estimation. Here is the first version of `protnetcharge.py`:

Listing 4.13: Net charge of a protein (py3.us/3)

```
 1 protseq = raw_input("Enter protein sequence: ")
 2 charge = -0.002
 3 AACharge = {"C":-.045,"D":-.999,"E":-.998,"H":.091,
 4            "K":1,"R":1,"Y":-.001}
 5 for aa in protseq:
 6     if aa in AACharge:
 7         charge += AACharge[aa]
 8     else:
 9         pass
10 print(charge)
```

The problem with this program is that it recognizes amino acids only in uppercase. If the user enters an amino acid in lowercase, it is ignored. A way to fix this is by extending the AACharge dictionary with the lowercase letters as keys. A better option is to convert all amino acid into uppercase using `upper()`.

The **if** statement in line 6 can be avoided with **get()**:

Listing 4.14: Net charge of a protein using **get** (py3.us/4)

```
1 protseq = raw_input("Enter protein sequence: ").upper()
2 charge = -0.002
3 AACharge = {"C":-.045,"D":-.999,"E":-.998,"H":.091,
4            "K":1,"R":1,"Y":-.001}
5 for aa in protseq:
6     charge += AACharge.get(aa,0)
7 print charge
```

4.5.2 Search for a Low Degeneration Zone

To find PCR primers, it is better to use a DNA region with less degeneration (or more conservation). This is made in order to have a better chance to find the target sequence. The aim of this program is to search for this region. Since a PCR primer has about 16 nucleotides, to give room for the primer design, the search space should be at least 45 nucleotides long. We should find a 15 amino acid region in the input sequence. 15 amino acids provides a search region of 45 nucleotides (3 nucleotides per amino acid).

Each amino acid is encoded by a determined amount of codons. For example valine (V) can be encoded by four different codons (GTT, GTA, GTC, GTG), while tryptophan (W) is encoded only by one codon (TGG). It is clear that a region rich in valines will have more variability than a region with lots of tryptophan.

A program that finds a low degeneration region,

First Version

Listing 4.15: Search for a low degeneration zone (py3.us/5)

```
 1 protseq = raw_input("Protein sequence: ").upper()
 2 protdeg = {"A":4,"C":2,"D":2,"E":2,"F":2,"G":4,"H":2,
 3            "I":3,"K":2,"L":6,"M":1,"N":2,"P":4,"Q":2,
 4            "R":6,"S":6,"T":4,"V":4,"W":1,"Y":2}
 5 segsvalues = []
 6 for aa in range(len(protseq)):
 7     segment = protseq[aa:aa+15]
 8     degen = 0
 9     if len(segment)==15:
10         for x in segment:
11             degen += protdeg.get(x,3.05)
12         segsvalues.append(degen)
13     else:
14         pass
15 min_value = min(segsvalues)
16 minpos = segsvalues.index(min_value)
17 print protseq[minpos:minpos+15]
```

Code explanation: lessdeg.py takes a string (protseq) entered by the user. The program uses a dictionary (protdeg) to store the amount of codons that corresponds to each amino acid. From line 6 to 8, we generate sliding windows of length 15. For each 15 amino acid segments, the amount of codons is evaluated, then we select the segment with less degeneration (line 15). Note that in line 9 there is a check of the size of segment, since when the sequence slide away of protseq, the subchain has less than 15 amino acids.

Version with While

Listing 4.16: Searching for a low degeneration zone, version with while (py3.us/6)

```
1 ProtSeq = raw_input("Protein sequence: ").upper()
2 ProtDeg = {"A":4,"C":2,"D":2,"E":2,"F":2,"G":4,"H":2,
3          "I":3,"K":2,"L":6,"M":1,"N":2,"P":4,"Q":2,
4          "R":6,"S":6,"T":4,"V":4,"W":1,"Y":2}
5 SegsValues=[]; SegsSeqs=[]; segment=ProtSeq[:15]; a=0
6 while len(segment)==15:
7     degen = 0
8     for x in segment:
9         degen += ProtDeg.get(x,3.05)
10    SegsValues.append(degen)
11    SegsSeqs.append(segment)
12    a += 1; segment = ProtSeq[a:a+15]
13 print SegsSeqs[SegsValues.index(min(SegsValues))]
```

Code explanation: This version don't use **for** for walk over protseq; instead, it uses **while**. Code will be executed as long as the sliding windows is inside protseq. Another difference is lines 15, 16, and 17 of listing 4.15 are consolidated in line 13.

Version without List of Subchains

Listing 4.17: Searching for a low degeneration zone without subchains (py3.us/7)

```
1 ProtSeq = raw_input("Protein sequence: ").upper()
2 ProtDeg = {"A":4,"C":2,"D":2,"E":2,"F":2,"G":4,"H":2,
3          "I":3,"K":2,"L":6,"M":1,"N":2,"P":4,"Q".2,
4          "R":6,"S":6,"T":4,"V":4,"W":1,"Y":2}
5 SegsValues=[]; i=0
6 while len(ProtSeq[i:i+15])==15:
7     degen = 0
8     for x in ProtSeq[i:i+15]:
9         degen += ProtDeg.get(x,3.05)
10    SegsValues.append(degen); i += 1
11 print ProtSeq[SegsValues.index(min(SegsValues)):
12            SegsValues.index(min(SegsValues))+15]
```

Code explanation: In this version, there is no list with all subchains SegsSeqs and the variable that stored each evaluated segment segment. Since there is no subchain list, the answer is evaluated using subindexes in ProtSeq.

Code 4.17 uses a list (`SegsValues`) to store the degeneration values of all possible subchains. This is better than previous code since it doesn't store every possible subchain, but still is storing a list with a size proportional to the input sequence. This can be avoided:

Listing 4.18: Searching for a low degeneration zone without subchains (py3.us/8)

```
 1 ProtSeq = raw_input("Protein sequence: ").upper()
 2 ProtDeg = {"A":4,"C":2,"D":2,"E":2,"F":2,"G":4,"H":2,
 3            "I":3,"K":2,"L":6,"M":1,"N":2,"P":4,"Q":2,
 4            "R":6,"S":6,"T":4,"V":4,"W":1,"Y":2}
 5 degen_tmp = max(ProtDeg.values())*15
 6 for n in range(len(ProtSeq)-15):
 7     degen = 0
 8     for x in ProtSeq[n:n+15]:
 9         degen += ProtDeg.get(x,3.05)
10     if degen <= degen_tmp:
11         degen_tmp = degen
12         seq = ProtSeq[n:n+15]
13 print seq
```

Code explanation: In this case every degeneration value is compared with the last one (line 10), and if the current value is lower, it is stored. Note that the first time a degeneration value is evaluated, there is no value to compare it with. This problem is sorted in line 5 where a maximun theoretical value is provided.

The next program is the Python 3 version of code 4.18:

Listing 4.19: Searching for a low degeneration zone without subchains (py3.us/9)

```
 1 ProtSeq = input("Protein sequence: ").upper()
 2 ProtDeg = {"A":4,"C":2,"D":2,"E":2,"F":2,"G":4,"H":2,
 3            "I":3,"K":2,"L":6,"M":1,"N":2,"P":4,"Q":2,
 4            "R":6,"S":6,"T":4,"V":4,"W":1,"Y":2}
 5 degen_tmp = max(ProtDeg.values())*15
 6 for n in range(len(ProtSeq)-15):
 7     degen = 0
 8     for x in ProtSeq[n:n+15]:
 9         degen += ProtDeg.get(x,3.05)
10     if degen <= degen_tmp:
11         degen_tmp = degen
12         seq = ProtSeq[n:n+15]
13 print(seq)
```

Code explanation: There are only two differences, with respect to 4.18, line 1 and 13. Note that **input** fit the role **raw_input** in Python 2.x, while **print** is a function.

4.6   Additional Resources

- "Python Tutorial: More Control Flow Tools."
  http://www.python.org/doc/2.5.2/tut/node6.html

- "Python Programming: Flow control."
  http://en.wikibooks.org/wiki/Python_Programming/Flow_control

- "Python in a Nutshell, Second Edition." By Alex Martelli. Chapter 4.
  Excerpt at http://www.devshed.com/c/a/Python/The-Python-Language.

- "Beginning Python: Controlling the Flow." By Al Lukaszewski, About.com.
  http://python.about.com/od/tutorial1/ss/begpyctrl.htm

4.7 Self-Evaluation

1. What is a control structure?

2. How many control structures has Python? Name them.

3. When would you use **for** and when would you use **while**?

4. Some languages have a **do while** control structure. How can you get a similiar function in Python?

5. Explain when you would use **pass** and when you would use *break*.

6. In line 6 of listing 4.17, the condition under the **while** can be changed from $len(ProtSeq[i : i + 15]) == 15$ to $i < (len(ProtSeq) - 7)$. Why?

7. Make a program that outputs all possible IP addresses, that is, from 0.0.0.0 to 255.255.255.255.

8. Make a program to solve a linear equation with two variables. The equation must have this form:

   $a_1.x + a_2.y = a_3$
   $b_1.x + b_2.y = b_3$

   The program must ask for a_1, a_2, a_3, b_1, b_2, and b_3 and return the value of x and y.

9. Make a program to check if a given number is a palindrome (that is, it remains the same when its digits are reversed, like 404).

10. Make a program to convert temperature Fahrenheit to Celsius and write the result with only one decimal value. Use this formula to make the conversion: $T_c = (5/9) * (T_f - 32)$

11. Make a program that converts everything you type into Leetspeak, using the following equivalence: 0 for O, 1 for I (or L), 2 for Z (or R), 3 for E, 4 for A, 5 for S, 6 for G (or B), 7 for T (or L), 8 for B and 9 for P (or G and Q). So "Hello world!" is rendered as "H3770 w02ld!"

12. Given two words, the program must determine if they rhyme or not. For this question "rhyme" means that the last three letters are the same, like wiz**ard** and liz**ard**.

13. Given a protein sequence in the one letter code, calculate the percentage of methionine (M) and cysteine (C). For example from MFKFASAVILC-LVAASSTQA the result must be 10% (1 M and 1 C over 20 amino acids).

14. Make a program like 4.18 but without using a predefined maximun value.

Chapter 5

Dealing with Files

Saving the program output to a file is important for archival purposes. It is also useful to ensure reproducibility of your program. Having a copy of your program output always comes in handy. Reading a previously saved file is a must-have feature on almost every computer program.

This chapter shows how to read and write any text file. For the purposes of this book, "reading a text file" is the process of entering the data from a file into a program. The process of understanding the meaning of the data units in the program is called **parsing**. A simple example will clarify this:

A file can have four data units in a line like this:

```
1,Joe,Doe,1976
```

When this line is **read**, Python sees it as one string. So there is a need to take an extra step for the program to recognize each of the four data on it. This step is the **parsing**. The parsing step depend on the format of the data, so there is no universal method for text parsing. This chapter covers how to parse data separated by a special character such as a comma or the tab character.[1]

5.1 Reading Files

Reading a file is a three step process in Python:

1. Open the file: There is a built-in function called **open**, that creates a *filehandle*. This *filehandle* is used to refer to the file during all the file lifetime. The **open** function takes two parameters: Name of the file and opening mode. The file name is a string with the file name, in most cases including system path. When the system path is included, this **absolute path** is used by the program. In case you enter just the file name (without any path), a relative path is assumed.[2] The second

---

[1]These kinds of file are often called CSV files and they are covered on page 92.

[2]Use **os.getcwd()** in case you need to know the current path.

parameter has the following valid parameters: "r" to **r**ead, "w" to **w**rite and "a" to **a**ppend data at the end of a file. The default value is "r". If you want to open a file for both read and write, use "r+".

Using **open**,

Create a file handle to read a file:

```
>>> fh=open('/home/sb/Readme.txt')
>>> fh
<open file '/home/sb/Readme.txt', mode 'r' at 0xb7d41380>
```

As you see, **fh is not the file, but a reference to it**. Since reading mode is the default option, it was omitted.

2. Read the file: Once the file is opened, we can read it contents. There are several ways to read a file, here are the most used:

   read(n) : Reads *n* bytes from the file. Without parameters, it reads the whole file.[3]

   readline() : Returns a string with only one line from the file, including '\n' as an end of line marker. When it reaches the end of the file, it returns an empty string.

   readlines() : Returns a list where each element is a string with a line from the file.

3. Close the file. Once we are done with the file, we close it by using: `filehandle.close()`. If we don't close it, Python will do it after program execution. However it is considered a good programming practice to close it in an explicit way.

5.1.1   Example of File Handling

Let's suppose we have a file called **seqA.fas** that contains:[4]

```
>000626|HUMAN Small inducible cytokine A22.
MARLQTALLVVLVLLAVALQATEAGPYGANMEDSVCCRDYVRYRLPLRVVKHFYWTSDS<=
CPRPGVVLLTFRDKEICADPR
VPWVKMILNKLSQ
```

---

[3]Due to the amount of memory it could take, it is not advisable to read the whole file in this way, unless you are sure of the file size. To process big files, there are better strategies like reading one line at a time.

[4]It is a FASTA file with one entry, the first line have a > followed by sequence name and description. The following lines has the sequence (DNA or amino acids). For more information on FASTA files, please see http://www.ncbi.nlm.nih.gov/BLAST/fasta.shtml.

From this file we need the name and the sequence. A first approach is to read the file with **read()**:

```
>>> fh=open('/home/sb/bioinfo/seqA.fas')
>>> fh.read()
'>000626|HUMAN Small inducible cytokineA22.\nMARLQTALLVVLVL<=
LAVALQATEAGPYGANMEDSVCCRDYVRYRLPLRVVKHFYWTSDSCPRPGVVLLTFRDK<=
EICADPR\nVPWVKMILNKLSQ\n'
```

In this case my goal is to have two variables, one with the sequence name and the other with the sequence itself. In code 5.1 we can see a way to do it using **read()**:

Listing 5.1: FastaRead.py: First try to read a FASTA file

```
1 fh = open('/home/sb/bioinfo/seqA.fas')
2 myfile = fh.read() #myfile is a string
3 name = myfile.split('\n')[0][1:]
4 sequence = ''.join(myfile.split('\n')[1:])
5 print("The name is: %s"%name)
6 print("The sequence is: %s"%sequence)
7 fh.close()
```

The first line opens the file in read mode and creates a file handle that we call fh. On line two, the whole file is read with **read()** and the resulting string is stored in system memory with the name `myfile`. The next step is to separate the names from the sequences. Since the name is after the ">" symbol and before the '\n', this information can be used to get the data we want (line 3). The sequence is obtained by joining the elements resulting of spliting `myfile` string, but without the first element.

The problem with this code is that it uses the **read()** function to read the file at once. This is a potential problem if there is not enough memory available to accommodate the file's contents. This is why it is better to use **readline()** (unless you know that you can handle the size of the file):

Listing 5.2: Read a FASTA file using readline()

```
1 fh = open('/home/sb/seqA.fas')
2 FirstLine = fh.readline()
3 name = FirstLine[1:-1]
4 sequence = ""
5 while True:
6     line = fh.readline()
7     if line=="":
8         break
9     sequence += line.replace('\n','')
```

```
10 print("The name is: %s"%name)
11 print("The sequence is: %s"%sequence)
12 fh.close()
```

Code explanation: The first line is identical to the first line of the previous code listing (code 5.1). In the second line we use **readline()** function to read the first line of the FASTA file. From this line we take the substring between the ">" and the first '\n' (line 3). In this case we don't need to use the index function to search for the '\n' character because we know it is at the end of the line, returned by **readline()**. From line 5 to 9, there is a loop to execute the **readline()** function, several times to finish reading the file. The exit condition is `line==""` that is returned at the end of the file.

Although this version is more efficient than code 5.1, it could be rewriten to make it easier to read:

Listing 5.3: FastaRead.py: Reads FASTA file, sequentially

```
1 fh = open('/home/sb/seqA.fas')
2 name = fh.readline()[1:-1]
3 sequence = ""
4 for line in fh:
5     sequence += line.replace('\n','')
6 print "The name is: %s"%name
7 print "The sequence is: %s"%sequence
8 fh.close()
```

Code explanation: The FistLine variable that was present in listing 5.2 is omitted and the result of `fh.readline()[1:-1]` is called `name`. The formula **for *x* in *filehandle*** (line 4) is the clearest and most efficient way to iterate through all the lines of a file. At this point we may add to our protein net charge calculation program (code listing 4.14) the ability to use as input data, a FASTA format sequence, instead of entering it manually.

Listing 5.4: Calculate the net charge, reading the input from a file

```
1 fh = open('/home/sb/prot.fas')
2 fh.readline()
3 sequence = ""
4 for line in fh:
5     sequence += line[:-1].upper()
6 charge = -0.002
7 AACharge = {"C":-.045,"D":-.999,"E":-.998,"H":.091,
8             "K":1,"R":1,"Y":-.001}
9 for aa in sequence:
10    charge += AACharge.get(aa,0)
11 print charge
12 fh.close()
```

Code explanation: The code is essentially the same as that in listing 4.14, with the difference that the first 5 lines are similar to those of listing 5.3 and are used to fill the sequence variable with the string that is read from the FASTA file. The only difference is on line 2, where the first line of the file is read as input, but not stored in any variable.

5.2 Writing Files

Writing a file is very similar to reading it. The steps 1 and 3 are similar. The change is at the second state. Let's have a look at the entire process anyway:

1. Open the file. It is similar to opening a file for reading, only that it is necessary to take into consideration the use of the open mode that corresponds to the operation that we are going to do. To create a new file, use "w" as the open mode. To append data to the end of the file, use "a."

   Creating a file handle for a new file:

   ```
   >>> fh=open('/home/sb/newfile.txt','w')
   >>> fh
   <open file '/home/sb/newfile.txt', mode 'w' at 0xb7d413c8>
   ```

   Creating a new file handle to append information to a file:

   ```
   >>> fh=open('/home/sb/error.log','a')
   >>> fh
   <open file '/home/sb/error.log', mode 'a' at 0xb7d41380>
   ```

2. Write data to the file. The method to write data to a file is called **write**. It accepts as a parameter a string, which will be written to the file represented by the filehandle on which the function will be applied. Schematically: *filehandle.write(string)*. Take into consideration that **write** does not add line feeds, for which they must be added as needed.

3. Close the file, the same way as done previously: **filehandle.close()**.

5.2.1 File Reading and Writing Examples

The code that follows will save to a file the numbers from 1 to 5, each one on a separate line. Between each number the respective line feeds are indicated.

Listing 5.5: Newfile.py: Write to a file.

```
1 fh=open('/home/sb/numbers.txt','w')
2 fh.write("1\n2\n3\n4\n5")
3 fh.close()
```

Program in listing 5.4 can be modified to write the result to a file, instead of displaying it on the screen:

Listing 5.6: Net charge calculation, saving results into a file (py3.us/10)

```
1 fh = open('prot.fas')
2 fh.readline()
3 sequence = ""
4 for line in fh:
5     sequence += line[:-1].upper()
6 fh.close()
7 charge = -0.002
8 AACharge={"C":-.045,"D":-.999,"E":-.998,"H":.091,
9           "K":1,"R":1,"Y":-.001}
10 for aa in sequence:
11     charge += AACharge.get(aa,0)
12 fhout = open('out.txt','w')
13 fhout.write(str(charge))
14 fhout.close()
```

Code explanation: The code is similar to listing 5.4, with the addition of the functionality on the three final lines to write the result to the file.

5.3 A Special Kind of File: CSV

While doing data processing work, it's very common to encounter a file type called CSV. CSV stands for "Comma Separated Values". These are files where the data are separated by commas, although sometimes other separators are used (such as colons, tabs, etc.). Another feature of this text file format in particular is that each line represents a separate record. All spreadsheets can be read and write this file format which helps to explain their popularity. Take, for example, the following file:

```
MarkerID,LenAmpForSeq,MotifAmpForSeq
TK0001,119,AG(12)
TK0002,255,TC(16)
TK0003,121,AG(5)
```

```
TK0004,220,AG(9)
TK0005,238,TC(17)
```

The line contains a description of each field. Like the information it stores, the descriptions are also separated by commas. The following lines contain the data, following the same order of the description. To get the average of the value in the second column, we can do something like this:

Listing 5.7: Reading data from a CSV file (py3.us/11)

```
1 tlen = 0; n = 0
2 fh = open('B1.csv')
3 fh.readline()
4 for line in fh:
5     data = line.split(",")
6     tlen += int(data[1])
7     n += 1
8 print(tlen/float(n))
9 fh.close()
```

Code explanation: Is a program that walks through a file, like code 5.6, but this time the method **split()** is used to split components of each line. In line 6 the sum of the second field is stored (this field has the length of the sequence).

These files are so extended that Python has a module to deal with them: **csv**.

Listing 5.8: Reading data from a CSV file, using **csv** module (py3.us/12)

```
1 import csv
2 tlen=0;n=0
3 lines = csv.reader(open('/home/sb/B1.csv'))
4 lines.next()
5 for line in lines:
6     tlen += int(line[1])
7     n += 1
8 print tlen/float(n)
```

Code explanation: This program is very similar to the previous one with the difference being the use of the **csv** module allows us access to the contents of each line without having to use the **split** method. The csv module has other advantages that we will see further in the text but for now let's analyze the program. Line 4 is equivalent to line 3 of the previous listing, it is used to skip the first line, where the header/descriptions are located. On line 5, the object returned by the csv module is traversed, instead of traversing the file directly as was done in the previous listing.

One way of using the csv module is to convert the object returned by the reader method to a list. Doing this, we generate something similar to a matrix from a csv file, with one line of code:

```
>>> data = list(csv.reader(open('B1.csv')))
>>> data[0][2]
'MotifAmpForSeq'
>>> data[1][1]
'119'
>>> data[1][2]
'AG(12)'
>>> data[3][0]
'TK0003'
```

This way we have a two-dimensional array of the type *name[row, column]*. Taking this into consideration we can rewrite the program from listing 5.8:

Listing 5.9: Reading a CSV file using the **csv** module

```
1 import csv
2 tlen = 0
3 data = list(csv.reader(open('B1.csv')))
4 for x in range(1,len(data)):
5     tlen += int(data[x][1])
6 print float(tlen)/(len(data)-1)
```

Code explanation: This code is similar to listing 5.8, with a few differences. On line 4 the data is stored as a list in `data`, for which the for loop traverses only the part of the data that interests us.

5.3.1   More Functions from the CSV Module

The field delimiter is changed with the **delimiter** attribute. By default it is ",", but any string can be used to delimit the fields:

```
rows = csv.reader(open("passwd"), delimiter=':')
```

For some files it is better to specify what is the CSV "dialect" that we are interested in. This is important because not all csv files are the same. There may be subtle differences that may spoil our data processing. In some cases the data is enclosed between quotations, in others the quotations are reserved for text data only. These are just a few of the possible variations. For the csv files generated by Excel, we have the Excel "dialect":

```
rows = csv.reader(open("data.csv"), dialect='excel')
```

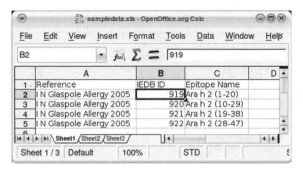

FIGURE 5.1:   Excel formatted spreadsheet called `sampledata.xls`.

Additionally there is a diaclect for Excel csv files that use a "tab" instead of the comma to separate data. If we aren't sure of the dialect that our code will have to handle, the **csv** module has a class that tries to guess it: **Sniffer()**:

```
dialect = csv.Sniffer().sniff(open('data.csv').read())
rows = csv.reader(open("data.csv"), dialect=dialect)
```

The **csv** module provides some more functions, but I stop here because this is enough to deal with most type of CSV files. For other uses, I recommend the module documentation were you will see more information.

Tip: Reading and Writing Excel Files.

The csv module allows to read files, provided that the file is converted first to csv. This step can be avoided with the **xlrd** module. This module has to be downloaded from `http://www.lexicon.net/sjmachin/xlrd.htm`.

Listing 5.10 retrieves data from an Excel file called `sampledata.xls` (see Figure 5.1). We want to make a dictionary (`iedb`) out of column B (keys) and Column C (values), so this program walks over both columns and fills the dictionary:

Listing 5.10: Reading an XLS file with xlrd (py3.us/13)

```
1 import xlrd
2 iedb = {} # Empty dictionary
3 book = xlrd.open_workbook('sampledata.xls')
4 sh = book.sheet_by_index(0)
5 for i in range(1,sh.nrows): #skips fist line.
6     iedb[sh.cell_value(rowx=i, colx=1)] = \
7         sh.cell_value(rowx=i, colx=2)
```

To write Excel files, there are two modules: **pyExcelerator** and **xlwt**. Code 5.11 writes `list1` and `list2` in column A and B using **xlwt**.

Listing 5.11: Write an XLS file with xlwt (py3.us/14)

```
1 import xlwt
2 list1 = [1,2,3,4,5]
3 list2 = [234,267,281,301,331]
4 wb = xlwt.Workbook()
5 ws = wb.add_sheet('First sheet')
6 ws.write(0,0,'Column A')
7 ws.write(0,1,'Column B')
8 i = 1
9 for x,y in zip(list1,list2): #Walk two list at the same time.
10     ws.write(i,0,x) # Row, Column, Data.
11     ws.write(i,1,y)
12     i += 1
13 wb.save('mynewfile.xls')
```

For sample usage of **pyExcelerator**, see code 18.2 on page 336.

5.4 Pickle: Storing the Contents of Variables

All variables created during the lifetime of a program are temporarily stored in memory and they disappear when you turn off the computer. Python provides a module to store and retrieve the contents of these variables: Pickle[5] (and its C version, cPickle).

Suppose that a program generates a dictionary (SPdict) and we want it to be available from another program (or from the same program in another run). For this, we first need to save its contents into a file (spdict.data):

```
>> import cPickle
>> fh = open("spdict.data", 'w')
>> cPickle.dump(SPdict, fh)
>> fh.close()
```

With this command the file handle (fh) was created and then the variable SPdict was saved to the file referenced by this handle. The function that saves the variable is called **dump** and accepts three parameters: The object to save, filehandle of the file where it will be saved and an optional protocol.

---

[5]Pickle has other features than those described in this book; in order to have a more extensive view of what **pickle** has to offer, see http://docs.python.org/lib/module-pickle.html.

The protocol is a code that represents the way in which the information will be encoded. If no protocol is specified, it is assumed to be 0 which is the original ASCII protocol. This has the advantage that it can be read without any special program, although it is slower than the binary protocols (1 and 2). The higher the protocol number, the faster it is. Although it is necessary to take into consideration that the reader for these types of objects (**load**) can read the same protocol with which it was saved or an earlier version.

Retrieving a Stored Object

The **load** method requires the filehandle of the object we want to pick up:

```
>> fh = open("spdict.data")
>> SPdict = cPickle.load(fh)
>> fh.close()
```

New in Python 3.0: New Behavior for Pickle

pickle and *cPickle* modules are consolidated as pickle. From Python 3.0, there is no cPickle module since the interpreter uses automatically an optimized C implementation of *Pickle* when available. Otherwise the pure Python implementation is used.

There is a new protocol (version 3) that has explicit support for bytes and cannot be unpickled by Python 2.x pickle modules.

5.5   File Handling: os Module

There are more actions with files apart from reading and writing. Python **os** module allows us to do with the files the same operations that are available from the operating system. Copy, move, delete, list, change directory, set file properties, and so on.

Let's see some important functions on file handling:

getcwd(): Return a string representing the current working directory.

```
>>> import os
>>> os.getcwd()
'/home/sb'
```

chdir(path): Change the current working directory to path.

```
>>> os.getcwd()
```

```
'/home/sb'
>>> os.chdir('..')
>>> os.getcwd()
'/home'
```

listdir(dir): Return a list containing the names of the entries in the directory. To know if a name returned from **listdir** is a file or a directory, use other either **os.path.isdir()** or **os.path.isfile()**.

```
>>> os.listdir('/home/sb/bioinfo/seqs')
['ms122.ab1','readme.txt','ms115.ab1','ms123.ab1']
```

path.isfile(string) and **path.isdir(string)**: Check if the string passed as argument is a file or a directory. Returns True or False.

```
>>> os.path.isfile('/home/sb')
False
>>> os.path.isdir('/home/sb')
True
```

remove(file): Remove a file. The file should exist and you should have write permission on it.

```
>>> os.remove('/home/sb/bioinfo/seqs/ms115.ab1')
```

rename(*source, destination*): Rename the file or directory *source* to *destination*.

```
>>> os.rename('/home/sb/seqs/readme.txt','/home/sb/Readme')
```

mkdir(*path*): Create a directory named *path*.

```
>>> os.mkdir('/home/sb/processed-seqs')
```

path.join(*directory1,directory2,...*): Join two or more pathname components, inserting the operating system path separator as needed. In Windows it will add "\", while in Linux and OSX it will insert "/". **path.join** will not try the check if the created path is valid.

```
>>> os.path.join(os.getcwd(), "images")
'/home/images'
```

path.exists(*path*): Checks if given *path* exists.

```
>>> os.path.exists(os.path.join(os.getcwd(), "images"))
False
```

path.split(*path*): Returns a tuple splitting the file or directory name at the end and the rest of the path.

```
>>> os.path.split('/home/sb/seqs/ms2333.ab1')
('/home/sb/seqs', 'ms2333.ab1')
```

path.splitext(*path*): Splits out the extension of a file. It returns a tuple with the dotted extension and the original parameter up to the dot.

```
>>> os.path.splitext('/home/sb/seqs/ms2333.ab1')
('/home/sb/seqs/ms2333', '.ab1')
```

Tip: The shutil Module.

Some functions related to file handle are in another module: **shutil**.

The most important functions are *copy*, *copy2* and *copytree*.

copy(source,destination): Copy the file *source* to *destination*.

copy2(source,destination): Copies also the last access time and last modification (like the Unix command `cp -p`).

copytree(source,destination): Recursively copy an entire directory tree from *source* directory to a destination directory that must not already exist.

For more information on **shutil**, see the documentation on `http://docs.python.org/lib/module-shutil.html` (or with `help(shutil)` on the Python shell).

5.5.1 Consolidate Multiple DNA or Protein Sequences into One FASTA File

The following program asumes that we have a directory with several sequences in FASTA format we want to consolidate them in a single FASTA file. This file can be used for example as an input file for a BLAST run.

Listing 5.12: Consolidating several files in one (py3.us/15)

```
 1 import os
 2 mypath = '/home/sb/bioinfo/test/'
 3 pathout = 'out.fas'
 4 fout = open(pathout,"w")
 5 for x in os.listdir(mypath):
 6     fh = open(os.path.join(mypath,x))
 7     data = fh.read()
 8     fh.close()
 9     fout.write(data)
10 fout.close()
```

Code explanation: The program retrieves the list of files in the directory mypath (defined in line 2) and walks over each file, reading its contents (line

7). In line 9 the content of the file is written to the output filehandle. The last line closes the output filehandle and the data is actually writen to the disk.

5.5.2   Estimating Net Charge of Several Proteins

This program calculates the net charge from a group of FASTA files. It scans a directory, checks for the file extension (the program assumes that FASTA files has .fas as extension) and calculates the net charge. The result of each calculation is written to a file, as one line for FASTA file.

Listing 5.13: Get net charge value from several files (py3.us/16)

```
 1 import os
 2 mypath='/home/sb/bioinfo/test/'
 3 AACharge = {"C":-.045,"D":-.999,"E":-.998,
 4             "H":.091,"K":1,"R":1,"Y":-.001}
 5 for x in os.listdir(mypath):
 6     if os.path.splitext(x)[1]=='.fas':
 7         fh = open(os.path.join(mypath,x),'U')
 8         name = fh.readline()[1:-1]
 9         seq = ""
10         for line in fh:
11             seq = seq + line[:-1].upper()
12         fh.close()
13         charge = -0.002
14         for aa in seq:
15             charge += AACharge.get(aa,0)
16         fh = open(os.path.join(mypath,'netvalues.txt'),'a')
17         fh.write("%s,%s\n"%(name,charge))
18         fh.close()
```

Code explanation: On line 5 a for loop begins that iterates through all the files returned by **os.listdir**. It checks that the file extension is ".fas" (line 6) as we want to perform the calculation of the net charge only on these file types. In this case, the **os.path.splitext()** function is used, which separates a filename in two parts: filename and extension. An alternative could have been to use **x.endswith("fas")**. On line 8 we obtain the name that we will use to identify the sequence. Take note that the first character is removed (which is the ">" symbol in FASTA files) and the last character (which is the end of line character). From 9 to 15, the program is similar to listing 5.4. From 16 to 18 the output file is written, using the mode "append" in order to add a line to the final file for each processed FASTA file.

5.6 With: An Alternative Way to Open Files

From version 2.6, Python provides a new tool that can be used to open files securely:[6] **with**.

with is a statement that takes this form:

```
with EXPRESSION as VARIABLE:
    BLOCK
```

When this statement is executed, *EXPRESSION* is evaluated. A special method (*__enter__*) of the object returned by *EXPRESSION* is called. Whatever is returned by *__enter__* is called as *VARIABLE*. The code in *BLOCK* will be executed. When is it finished (by extanaution of the block or by an error), a method called *__exit__* is called. This method can handle exceptions, so if there is an error inside *BLOCK*, the programmer is assured that a specific code will be executed. This code could be used to close an open resource.

In the case of the file object, it has both *__enter__* and *__exit__* methods. *__enter__* returns the file object itself, and *__exit__* closes the file. If an error occurs that causes the program to terminate before closing an open file, the data will be lost and the file will not be accessible from other applications. With **with** we can be sure that whatever happens with the code in *BLOCK* the file will be closed.

Let's look at code listing 5.13 using **with**:

Listing 5.14: Similar to code 5.13 with the use of *with* (py3.us/17)

```
1 from __future__ import with_statement
2 import os, glob
3 mypath = '/home/sb/bioinfo/test/'
4 AACharge = {"C":-.045,"D":-.999,"E":-.998,
5             "H":.091,"K":1,"R":1,"Y":-.001}
6 for x in glob.glob(mypath+'*.fas'):
7     with open(x,'U') as fh:
8         name = fh.readline()[1:-1]
9         seq = ""
10        for line in fh:
11            seq += line[:-1].upper()
12    charge = -0.002
13    for aa in seq:
14        charge += AACharge.get(aa,0)
15    with open(os.path.join(mypath,'netvalue.txt'),'a') as fh:
```

---

[6]In Python 2.5 **with** is available by importing it from "the future": from __future__ import with.

```
16        fh.write("%s,%s\n"%(name,charge))
```

Code explanation: This program is similar to 5.13. The most important changes are in line 8 and 16. In line 9 is where the file is opened using **with**. Note that **with** is imported in line 1, this is required in Python 2.5 since **with** is available directly only from Python 2.6. Another difference is the use of **glob.glob**. It finds all the pathnames matching a specified pattern.

This topic was included in this chapter because **with** can be used with file objects. Of course, the use of **with** is not limited to opening files, but for any objects that support the previously mentioned methods. We will not elaborate further on its use because to be able to take full advantage, it is necessary to have knowledge of object creation and error handling. For more information about **with** check the links in **Additional Resources**.

5.7 Additional Resources

- File and Directory Access
  http://docs.python.org/library/filesys.html

- Generate temporary files and directories
  http://docs.python.org/library/tempfile.html

- CSV File API
  http://www.python.org/dev/peps/pep-0305/

- Working with Excel Files in Python
  http://www.python-excel.org

- The "with" Statement
  http://www.python.org/dev/peps/pep-0343/

5.8 Self-Evaluation

1. What is the difference between "w" and "a" modes if both allow to write files?

2. Why all files that are not longer in use must be closed?

3. Program 5.9 estimates the average in line 6. Instead of dividing over the total number of rows, it does on the total less one. Why?

4. Make a program that asks a name, and then write it to a file called MyName.txt.

5. Is it possible to parse csv files without csv module? If so, how it is done?

6. Why it is not recommended to read a file using **read()**?

7. What is the most efficient way to walk through a file line by line?

8. What is the difference between Pickle and cPickle in Python 2.x? Why there is no cPickle in Python 3?

9. Make a program to detect in a text which lines have two consecutive identical words. To detect typos like "the the."

10. Make a program that reads all the numbers from the second column of an Excel file and prints the average of these values.

Chapter 6

Code Modularizing

With what we have seen so far we have an interesting portfolio of resources for Python programming. We can read files, do some data processing and store its results. Although programs made so far are very short, it is easy to imagine that they could grow up to a size that it may be difficult to manage.

There are several resources that can used to modularize source code in a way that we may end up with a small program that calls pre-made code blocks (also called *routines* in other computer languages). This approach favors code re-usability and readability. Both features also help maintenance, since you have to debug only one code implementation, regardless of how many times this code is used. As an additional advantage, it helps to improve performance, since any optimization on a modularized code benefits all the code that calls it.

For some authors, code modularizing is "The Greatest Invention in Computer Science,"[1] I don't know if this is the "greatest invention" or not, but sure it is a fundamental concept that you can't live without if you plan to do any serious programming.

Python provides several ways to modularize the source code: Functions, modules, packages and classes. This chapter covers all of them, with the exception of objects, that have their own chapter.

6.1 Functions

6.1.1 Standard Way to Modularize Python Code

Functions are the most used way to modularize code. A function takes values (called arguments or parameters), executes some operation based on those values and it returns a value. We have already seen several Python built-in function.[2] For example **range()**, first mentioned on page 77, it takes

---

[1]Read the Steve McConnell column at http://www.stevemcconnell.com/ieeesoftware/bp16.htm.

[2]A list of all available functions in Python are available at: http://docs.python.org/lib/built-in-funcs.html.

a number as parameter and returns a list:

```
>>> range(5)
[0, 1, 2, 3, 4]
```

Let's see how to make our own functions. The general syntax of a function is:

```
def FunctionName(argument1, argument2, ...):
    """ Optional Function description (Docstring) """
    ... FUNCTION CODE ...
    return DATA
```

The code in listing 4.14 can be rewritten as a function:

Listing 6.1: Function to calculate the net charge of a protein

```
1 def protcharge(AAseq):
2     """ Returns the net charge of a protein sequence """
3     protseq = AAseq.upper()
4     charge = -0.002
5     AACharge = {"C":-.045,"D":-.999,"E":-.998,"H":.091,
6                 "K":1,"R":1,"Y":-.001}
7     for aa in protseq:
8         charge += AACharge.get(aa,0)
9     return charge
```

To "use" the function, it must be called with the parameter:

```
>>> protcharge("QTALLVVLVLLAVALQATEAGPYGA")
-1.001
>>> print protcharge("EEARGPLRGKGDQKSAVSQKPRSRGILH")
4.094
```

If we forget to pass the parameter, or if we pass an incorrect number of parameters, we get an error:

```
>>> protcharge()

Traceback (most recent call last):
  File "<pyshell#3>", line 1, in <module>
    protcharge()
TypeError: protcharge() takes exactly 1 argument (0 given)
```

In this example, the function returns a number (of float type). If we want that it returns more than one value, we can make it return a list or a tuple.[3]

---

[3]It makes more sense to return a tuple instead of a list since for a given function there is a fixed number of parameters returned.

The function *protcharge* (coded in listing 6.1) could be modified to return, besides the net charge, the proportion of charged amino acids:

Listing 6.2: Function to calculate two parameters (py3.us/18)

```
 1 def chargeandprop(AAseq):
 2     """ Returns the net charge of a protein sequence
 3     and proportion of charged amino acids
 4     """
 5     protseq = AAseq.upper()
 6     charge = -0.002
 7     cp = 0
 8     AACharge={"C":-.045,"D":-.999,"E":-.998,"H":.091,
 9              "K":1,"R":1,"Y":-.001}
10     for aa in protseq:
11         charge += AACharge.get(aa,0)
12         if aa in AACharge:
13             cp += 1
14     prop = 100.*cp/len(AAseq)
15     return (charge,prop)
```

If we call the function with the same parameters of the last example, we get another result:

```
>>> chargeandprop("QTALLVVLVLLAVALQATEAGPYGA")
(-1.0009999999999999, 8.0)
>>> chargeandprop("EEARGPLRGKGDQKSAVSQKPRSRGILH")
(4.0940000000000003, 39.285714285714285)
```

Use subscripts to get only one value:

```
>>> chargeandprop("QTALLVVLVLLAVALQATEAGPYGA")[0]
-1.0009999999999999
>>> chargeandprop("EEARGPLRGKGDQKSAVSQKPRSRGILH")[1]
39.285714285714285
```

All function returns something. A function can be used to "do something" instead of returning a value. In this case the value returned is None. For example the following function stores the contents of a list into a text file:[4]

Listing 6.3: Converts a list into a text file

```
1 def savelist(L,fname):
2     """ A list (L) is saved in a file (fname) """
```

---

[4]For a way to save all kind of Python data structures, see **Pickle** on page 96.

```
3      fh = open(fname,"w")
4      for x in L:
5            fh.write('%s\n'%x)
6      fh.close()
7      return None
```

The **return None** statement is optional. The function will work without it, but Python coders prefer explicit statements than implicit assumptions.

Note that in Python 3, code 6.3 can be written with the **print** function:

Listing 6.4: Converts a list into a text file, using print

```
1 # Works on Python 3.
2 def savelist2(L,fname):
3      """ A list (L) is saved to a file (fname) """
4      fh = open(fname,"w")
5      for x in L:
6            print(x,file=fh)
7      fh.close()
8      return None
```

The "for loop" in line 5 can be avoided by using a property not seen yet. Code 6.7 on page 111 shows an alternative whitout the loop.

Function Scope

Variables declared inside a function are valid only inside the function. To access the contents of a function variable from outside the function, the variable must be returned to the main program by using the **return** statement. In the following example, the variable z, defined inside the **test** function, is not affected by an assignation of the same variable outside the function. The same code also shows how an internal variable (x) is accessible only from inside the function:

```
>>> def test(x):
...     z=10
...     print("The value of z is "+str(z))
...     return x*2

>>> z=50
>>> test(3)
The value of z is 10
6
>>> z
50
>>> x
```

```
Traceback (most recent call last):
  File "<pyshell#27>", line 1, in <module>
    x
NameError: name 'x' is not defined
```

It can be specified inside a function that a variable is of **global** type, so its life won't be confined to the place it was defined. It is not a good idea to use global variables, since it can be modified at unexpected places. Another problem related to global variables is that Python has to keep track of its value for the entire runtime so it is not memory-efficient. If despite all warnings against the use of global variables, you still want to use them, here is the way:

```
>>> def test(x):
...     global z
...     z = 10
...     print("z=%s"%z)
...     return x*2
...
>>> z=1
>>> test(4)
z=10
8
>>> z
10
```

When a variable is referred, Python first searches locally in the scope it was referred (in a function, module, object), if it is not there, it looks in the upper level, if it still can't be located there, it is searched at a global level:

```
>>> f=1
>>> def test():
...     print f
...     return None
...
>>> test()
1
```

6.1.2   Function Parameter Options

Placement of Arguments

Up to this point the arguments were put in the same order as originally defined. The function `savelist` can be called this way:

`savelist([1,2,3],"temp.txt")`.

If we try to invert the parameters order (`savelist("temp.txt",[1,2,3])`) we get an error message, since this function expects a string as a second

parameter instead of a list. To call the function with the parameters in a
different order as it was originally defined, the parameter must be named
when calling the function:

```
>>> savelist(fname="temp.txt",L=[1,2,3,4])
```

This way the order of parameters is irrelevant.

Arguments with Default Values

Python allows default value in the arguments. It is done by entering the
default value in function definition:

```
def name(arg1=defaultvalue, arg2=defaultvalue, ... )
```

For example the function `savelist`, which saves the contents of a list to a
file, may have a default file name:

Listing 6.5: Function with a default parameter

```
1 def savelist(L,fname="temp.txt"):
2     """ A list (L) is saved to a file (fname) """
3     fh = open(fname,"w")
4     for x in L:
5         fh.write(str(x)+"\n")
6     fh.close()
7     return None
```

In this way the function can be called with only one parameter:

```
mylist = ['MS233','MS772','MS120','MS93','MS912']
>>> savelist(mylist)
```

Undetermined Numbers of Arguments

Functions can have a variable numbers of arguments if the final parameter
is preceded by a "*". Any excess arguments will be assigned to the last
parameter as a tuple:

Listing 6.6: Function to calculate the average of values entered as parameters

```
1 def average(*numbers):
2     if len(numbers)==0:
3         return None
4     else:
5         total = sum(numbers)
6         return float(total)/len(numbers)
```

In this way the **average** function can be called with an undetermined number of arguments:

```
>>> print(average(2,3,4,3,2))
2.8
>>> print(average(2,3,4,3,2,6,1,9))
3.75
```

In Python 3, "*" can be used to assign a variable to multiple arguments.[5] This property is used here (line 5 in code 6.7) to avoid using a loop for walk over all elements of L:

Listing 6.7: Converts a list into a text file, using print and *

```
1 # Works on Python 3.
2 def savelist(L,fname):
3     """ A list (L) is saved to a file (fname) """
4     with open(fname,"w") as fh:
5         print(*L,sep='\n',file=fh)
6     return None
```

Undetermined Number of Keyword Arguments

The functions can also accept an arbitrary number of arguments with keywords. In this case we use the final parameter preceded by "**" (two asterisks). The excess arguments are passed to the function as a dictionary:

Listing 6.8: Function that accepts a variable number of arguments

```
1 def commandline(name, **parameters):
2     line = ""
3     for parname,parvalue in parameters.iteritems():
4         line = line + " -" + parname + " " + parvalue
5     return name+line
```

This function can be called with a variable number of keyword parameters:

```
>>> commandline("formatdb",t="Caseins",i="indata.fas")
'formatdb -i indata.fas -t Caseins'
>>> commandline("formatdb",t="Caseins",i="indata.fas",p="F")
'formatdb -i indata.fas -p F -t Caseins'
```

---

[5]This is explained in detail in PEP-3132 (http://www.python.org/dev/peps/pep-3132).

Tip: A Word about docstrings.

Functions can have a text string immediately following the function definition, this line (or lines) is called "docstring".

These strings are used for online help, automatic documentation generation systems and for anyone who cares to read the source code. You can write anything inside a *docstring*, but there are written guidelines to standardize the structure of a docstring. Please refer to PEP-257 (`http://www.python.org/dev/peps/pep-0257`) for more information on Docstring format conventions.

Not only functions can have docstrings, modules, class and method definition are expected to have its documentation as the first statement.

6.1.3   Generators

Generators are a special kind of function. Functions perform some action and the result is returned to the point where it was called. It is like a process that is executed in batch until completion. If the result of the function is a big object (like a big list, tuple, set, file, etc.), this could cause problems such as system instability.

Take for example a function that reads records from a file and returns a data structure with data from this file. If the file is too big (like several times the available memory), the resulting data structure may not fit in memory. If the idea is to process each record, we would need a function that returns a record at a time. A function can't do that because it doesn't keep a state, so each time it is executed, it has to process all the data again. **Generators** can be defined as functions that can keep their internal state. They introduce a new keyword: **yield**. When a **yield *VALUE*** statement is found, it returns (or yields) *VALUE* back to where it was called (as a function) but keeps track of its internal values, so next time it is called, it resumes operation from where it was before yielding the value.

Before Python 3, there was **range()** and **xrange()** as built-in function. **range()** was a function, but **xrange()** works as a generator:

```
>>> range(5)
[0, 1, 2, 3, 4]
>>> xrange(5)
xrange(5)
```

Both allow iteration over the resulting values, but **xrange()** doesn't return a list, but it returns a value each time it is called. Since Python 3, **range()** works as a generator and **xrange()** is deprecated. If you need the old funcionality of **range()** (that is, a list of numbers), you can apply **list()** function over Python 3 **range()**.

Creating a Generator

Code 6.9 has a function (`putn()`) that returns all prime numbers present up to a given value. It returns them all together in a list:

Listing 6.9: Function that returns all prime numbers up to a given value (py3.us/19)

```
 1 def isprime(n):
 2     for i in range(2,n-1):
 3         if n%i == 0:
 4             return False
 5     return True
 6
 7 def putn(n):
 8     p = []
 9     for i in xrange(1,n):
10         if isprime(i):
11             p.append(i)
12     return p
```

Function `putn()` from code 6.9 can be replaced with generator `gputn()`:

Listing 6.10: Generator that replaces `putn()` in code 6.9. (py3.us/20)

```
 1 def gputn(n):
 2     for i in xrange(1,n):
 3         if isprime(i):
 4             yield i
```

Note that code 6.10 doesn't use a list, since there is no need because it yields one result at a time. Both functions can be used to walk over the results, but `putn()` generates a list, while `gputn()` doesn't:

```
for x in putn(1000):
    print x
for x in gputn(1000):
    print x
```

6.2 Modules

A module is a file with function definitions, constants or any type of object that you can use from other module or from your main program. Modules also provide namespaces, so two functions may be given the same name provided that they are defined in different modules. The name of the module is taken from the name of the file. If the module filename is `my_module.py`, the module name is `my_module`.

6.2.1 Using Modules

To access contents of a module, use **import**. Usually import is issued at the beginning of the program. It is not mandatory to place the imports at the beginning of the file, but it must be placed before calling any of the elements of the module. It is customary, however, to place the **import** statement at the beginning of the program. There are many ways to use import. The most used form is by calling a module by its name. To call the built-in module **os**, use,

```
>>> import os
```

When a module is imported for the first time, its contents are executed. If the module is imported more than once, the successive imports will not have any effect. This gives us the assurance that we can put an import statement inside a function and not worry about if it is called repeatedly.

Once a module is imported, to access a function or a variable, use the name of the module as a prefix:

```
>>> os.getcwd()
'/mnt/hda2'
>>> os.sep
'/'
```

It is also possible to import from a module only a required function. This way we can call it without having to use the name of the module as a prefix.

```
>>> from os import getcwd
>>> getcwd()
'/mnt/hda2'
```

To import all the contents of a module, use the "*" operator (asterisk)

```
>>> from os import *
>>> getcwd()
'/mnt/hda2'
>>> sep
'/'
```

Warning: Don't use the `from module import *` unless you know what you are doing. The problem with importing all the elements of the module is that it may produce conflicts with the names already defined in the main program (or defined in other modules and imported the same way). In Python programming standards, wildcard imports are equivalent to the dark side of the force. They're quicker, easier, more seductive but dangerous.

It is also possible to import a module using a different name than it has:

```
>>> import xml.etree.ElementTree as ET
>>> tree = ET.parse("/home/sb/bioinfo/smallUniprot.xml")
```

Don't worry if you don't know what is `xml.etree.ElementTree`, we will look at this in the XML chapter, but from this moment take into account that this entire name (`xml.etree.ElementTree`) is called "ET."

6.2.2 Installing Modules

Python comes with several modules (so called built-in modules). These modules are installed together with Python so they are ready to use as soon as you have a working Python interpreter[6].

There are also third party modules that extend Python functionality, as mentioned on page 10. Installation can be as easy as copying a single file to a specific location up to executing several programs in a predetermined order. It depends on the complexity of the modules (they range from self contained in one file to several files spanned in multiple directories that interact with other programs) and each particular Python installation. So there is no single way to install every module available to Python.

Copying to PYTHONPATH

This is not the most frequent module installation procedure, but it is mentioned first because it is very simple. Just copy the module where Python seaches for modules. Where does Python search for modules? There are three places:

- In the same directory where the program that will call the module is located.

- In the same directory where the Python executable is located. This directory is different on each operating system.[7]

- In a directory created especially for our modules. In this case, it must be specified in the environment variable **PYTHONPATH** or in the

---

[6]Check `http://docs.python.org/library/index.html` for a complete description of the Python Standard Library including built-in modules.

[7]On Windows it is usually found at `C:\program files\Python` while on Linux it is found at `/usr/bin/python`. To find out the path to the Python executable in *nix, use `which python`.

variable **sys.path**. This final variable lists all the paths where Python
should look for a module. To add a directory to **sys.path**, modify it as
you would do with any list, using the **append** method:

```
>>> import sys
>>> sys.path
['/home/sb', '/usr/local/bin', '/usr/local/lib/python2.5', <=
'/usr/local/lib/python2.5/site-packages', <=
'/usr/local/lib/python2.5/site-packages/Numeric']
>>> sys.path.append("/home/sb/MyPyModules")
```

A module that is installed this way is **PathModule**.[8] This module helps
when dealing with files by adding an interface layer over **os**, **glob** and **shutil**
module. See module documentation (inside the source code) for usage information.

Using System Package Management

Some Python modules are installed as any other application. For example
to install NumPy in Windows or MacOSX (a fundamental package needed
for scientific computing with Python), just download the corresponding Windows or MacOSX installer from NumPy webpage[9] and install it as any other
application.

Under Linux use the package management software that you normally use
in your distribution (`rpm` for Red Hat-based system, and `apt-get` for Debian,
and Ubuntu-based systems).

The advantage of using system package management is that you can keep
track of installed Python modules the same way you keep track of every other
software in your system. Upgrades and uninstallations are easier and without nasty consequences such as orphan files or broken installations. Package
management also has its drawbacks, like a gap between last program version
and the version available in your distro repository. Some modules develop at
a fast pace, sometimes so fast that package managers can't keep up to date.
For example Ubuntu users who want to install Biopython using apt-get, at
time of writing, are limited to version 1.45 when 1.49 is the last version available at Biopython website. Another problem is that you need administration
rights to use package management. The last problem involving package management is that sometimes the required package is not available. For example
Windows users who want to install Biopython 1.49 using provided executable
files are unable to do it since a required package (NumPy for Python 2.6)

---

[8]PathModule is available at `http://wiki.python.org/moin/PathModule`.

[9]NumPy webpage is located at `http://numpy.scipy.org`.

is not available. Therefore Windows users who want to use only their package management system should install Biopython 1.49 under Python 2.5 (or resign using executable files and do a manual install to get the last version).

Easy Install with easy_install

Some modules support an alternative installation method called **easy_install**. The first step is to install the **easy_install** script, that is part of **setuptools** package. Windows installer is available from setuptools homepage (`http://pypi.python.org/pypi/setuptools`) and in Linux it can be installed from all major distro repositories (with `python-setuptools` as package name):

For Debian/Ubuntu:

```
$ sudo apt-get install python-setuptools
```

Once installed, Python modules can be installed with:

```
$ sudo easy_install MODULE_NAME
```

Where *MODULE_NAME* is the name of the module you want to install. When you request a module (like MODULE_NAME), **easy_install** searches if it is available in this URL: `http://pypi.python.org/simple/MODULE_NAME`. If found, it downloads the last version and installs it. Here is a sample installation of **pyexcelerator**, a program to write Excel files:

```
$ sudo easy_install pyexcelerator
[sudo] password for sb:
Searching for pyexcelerator
Reading http://pypi.python.org/simple/pyexcelerator/
Reading http://sourceforge.net/projects/pyexcelerator/
Reading http://www.sourceforge.net/projects/pyexcelerator
Best match: pyexcelerator 0.6.0a
Downloading http://pypi.python.org/packages/source/p/pyExcelerat<=
or/pyexcelerator-0.6.0a.zip#md5=df116f024919e129487366729e619928
Processing pyexcelerator-0.6.0a.zip
Running pyExcelerator-0.6.0a/setup.py -q bdist_egg --dist-dir <=
/tmp/easy_install-23lOQH/pyExcelerator-0.6.0a/egg-dist-tmp-YR3GKH
zip_safe flag not set; analyzing archive contents...
Adding pyExcelerator 0.6.0a to easy-install.pth file

Installed /usr/lib/python2.5/site-packages/pyExcelerator-0.6.0a-<=
py2.5.egg
Processing dependencies for pyexcelerator
Finished processing dependencies for pyexcelerator
```

To see what modules are available to install using easy_install, see the list in `http://pypi.python.org/simple`.

Easy Install without Administrative Rights

When using easy_install with administrative rights, the program is installed system-wide. This is not a problem if we have such administrative rights and we are the only user. Both conditions are usually met on personal computers, but not Web servers and computer clusters.

The first step is to install **virtualenv**. It can be installed with easy_install:

```
$ sudo easy_install virtualenv
```

Virtualenv is a tool to set up isolated Python environments. Using **virtualenv** you can install Python modules without affecting other users (or even yourself, since you can set up unlimited virtual environments for private use). If can't make your System Administrator to install **virtualenv**, download the .tar.gz package from http://pypi.python.org/pypi/virtualenv, unpack it, change to the new directoy and use the virtualenv.py file from it:

```
$ wget http://pypi.python.org/packages/source/v/virtualenv/<=
virtualenv-1.3.2.tar.gz
$ tar xfz virtualenv-1.3.2.tar.gz
$ cd virtualenv-1.3.2
```

Once the program is installed (system wide or by unpacking it in a user directory), run it like this:

```
$ mkdir MY_DIR
$ virtualenv --no-site-packages MY_DIR
New python executable in MY_DIR/bin/python
Also creating executable in MY_DIR/bin/python
Installing setuptools............done.
```

Then cd to MY_DIR (replace MY_DIR with any name you want) and activate the new virtual environment:

```
$ cd MY_DIR
$ . bin/activate
(MY_DIR)$
```

This is valid for *nix systems. In Windows there is a .bat file that must be executed:

```
> \path\to\env\bin\activate.bat
(MY_DIR)>
```

Note that the prompt changed to (MY_DIR)$, this indicate that every program that we install from now on, won't interfere with any other Python installation. What is modified in MY_DIR, remains in MY_DIR. For example:

```
(MY_DIR)$ easy_install Numpy
Searching for Numpy
Reading http://pypi.python.org/simple/Numpy/
(...cut...)
Finished processing dependencies for Numpy
(MY_DIR)$ easy_install biopython
Searching for biopython
Reading http://pypi.python.org/simple/biopython/
(...cut...)
Finished processing dependencies for biopython
```

This NumPy and Biopython installation will not interfere with any other installation in your system (if any).

To use this virtual installation, run the Python interpreter from the `bin` directory inside MY_DIR:

```
(MY_DIR)$ ./bin/python2.4
Python 2.4 (#2, Dec  3 2004, 17:59:05)
[GCC 3.3.5 (Debian 1:3.3.5-2)] on linux2
Type "help", "copyright", "credits" or "license" for more <=
information.
>>> import Bio
>>>
```

Once you are done with working with the virtual environment, you should deactivate it to return to your standard prompt:

```
(MY_DIR)$ deactivate
$
```

In windows:

```
(MY_DIR)> \path\to\env\bin\deactivate.bat
>
```

Standard Build and Install

If you can't use system packages and don't want to (or can't) use easy_install, there is always a manual way to install packages. Download the module files (usually packet and compressed in ".tar.gz"), unpack it and look for a `setup.py` file. In most cases installing it is a matter of running:

```
python setup.py install
```

If there are any problems, see the README file. In fact it is advisable to check the README file before trying to install the program (who does that?). In most cases problem arises from missing dependencies (like you need module X to run module Y), that you will have to fulfill. That is why it is better to install, when possible, Python modules with the package manager and let it handle the dependency hell.[10]

6.2.3   Creating Modules

To create a module you have to create a file and save it with the ".py" extension. It should be saved in a directory where the Python interpreter searches for it, like those in the PYTHONPATH variable (see page 115 for more information).

For example, store the function `savelist` in a module and call it `utils`. For this create the file `utils.py` with the following contents:

```
# utils.py file
def savelist(anylist,fn="temp.txt"):
    """ A list (anylist) is saved in a file (fn) """
    fh = open(fn,"w")
    for x in anylist:
        fh.write("%s\n"%x)
    fh.close()
    return None
```

This way, this function (`savelist`) can be used from any program, provided that this file is saved in a location accessible from Python:

```
>>> import utils
>>> utils.savelist([1,2,3])
```

6.2.4   Testing Our Modules

A good programming practice involves the creation of tests to verify the correct functioning of the module. As the modules are designed to be used from within a program, these tests must be executed only when called from the command line. This has the advantage that the tests will not interfere with the normal operation of the program.

To achieve this, we need to be able to differentiate when code is being executed as a stand alone program and when it is executed as a module from another program. When the code is executed as a program, the variable __name__ takes the value "__main__". As a result, the way to incorporate test code is by doing it after verifying that the program executes independently.

---

[10]See this Wikipedia article if you never found this term before: `http://en.wikipedia.org/wiki/Dependency_hell`.

```
if __name__ == "__main__":
    #Do something
```

This type of test is usually included at the end of a module. In code B.11 (page 443) we can see a test in action.

Python provides a module that facilitates the task of testing that our code works as we expect. This module is called **doctest**.

Doctest, Testing Modules in an Automatic Way

Doctest is a module that searches for pieces of Python code inside a docstring. This code is executed as if it were an interactive Python session. The module tests if this code works exactly as shown in the docstring or in an external file.

In code 6.11 we have `isprime()`, a function that checks if a given number (n) is prime. Let's see how we can incorporate a test unit and run it:

Listing 6.11: Module with doctest (py3.us/21)

```
1  def isprime(n):
2      """ Check if n is a prime number.
3      Sample usage:
4      >>> isprime(0)
5      False
6      >>> isprime(1)
7      True
8      >>> isprime(2)
9      True
10     >>> isprime(3)
11     True
12     >>> isprime(4)
13     False
14     >>> isprime(5)
15     True
16     """
17
18     if n<=0:
19         # This is only for numbers>0.
20         return False
21     for x in range(2,n):
22         if n%x==0:
23             return False
24         else:
25             pass
26     return True
```

```
27
28 def _test():
29      import doctest
30      doctest.testmod()
31
32 if __name__ == "__main__":
33      _test()
```

Code Explanation: The `isprime(n)` function is defined from line 1 to 26, but the actual functionality starts at line 18. Up to this line there are some tests. These tests are not executed if the program is called from another program, this is checked in line 32. If the program is executed as a stand alone program, all test are run:

```
$ python prime.py
$
```

No news, good news. Let's see what happens when I change line 21 in 6.11 to "for x in range(1,n):":

```
>>> isprime(1)
False
```

In this case, the test fails:

```
****************************************************************
File "./doctestSN.py", line 10, in __main__.isprime
Failed example:
    isprime(2)
Expected:
    True
Got:
    False
****************************************************************
File "./doctestSN.py", line 12, in __main__.isprime
Failed example:
    isprime(3)
Expected:
    True
Got:
    False
****************************************************************
File "./doctestSN.py", line 16, in __main__.isprime
Failed example:
    isprime(5)
Expected:
```

```
    True
Got:
    False
****************************************************************
1 items had failures:
   3 of   6 in __main__.isprime
***Test Failed*** 3 failures.
```

Some people believe that testing is so important that they propose to design a test for every function before start writing code. This is called test-driven development. Testing may not be perceived as a primary need for a program, but one cannot be certain that a function works unless one tests it (and testing it does not assure us that it will work). Python has extensive support for software testing (with modules **doctest** and **unittest**), but this is out of the scope of this book. For more information on testing see "Additional Resources."

6.3   Additional Resources

- "Modules, The Python Tutorial."
  http://docs.python.org/tutorial/modules.html

- "Default Parameter Values in Python, by Fredrik Lundh."
  http://effbot.org/zone/default-values.htm

- "Python Library Reference. Unittest API."
  http://www.python.org/doc/2.5.2/lib/doctest-unittest-api.html

- "Installing Python Modules."
  http://docs.python.org/install/index.html

- Easy Install.
  http://peak.telecommunity.com/DevCenter/EasyInstall

- "Recipe 305292: doctest, unittest, and python 2.4's cool doctest.DocFileSuite."
  http://code.activestate.com/recipes/305292

- "Extreme Programming." Wikipedia article.
  http://en.wikipedia.org/wiki/Extreme_Programming

6.4 Self-Evaluation

1. What is a function?

2. How many values can return a function?

3. Can a function be called without any parameters?

4. What is a docstring and why is it related to functions and modules?

5. Does every function need to know in advance how many parameters will receive?

6. Write a code that shows that xrange() is not a generator.

7. Why must all optional arguments in a function be placed at the end in the function call?

8. What is a module?

9. Why are modules invoked at the beginning of the program?

10. How do you import all contents of a module? Is this procedure advisable?

11. How can you test if your code is being executed as a stand alone program or called as a module?

Chapter 7

Error Handling

NAESER'S LAW: You can make it foolproof, but you can't make it damn-foolproof.

7.1   Introduction to Error Handling

A program rarely works as expected, at least on the first try.

Traditionally a programmer would choose between one of these two strategies when faced with runtime program errors. The problem is ignored or each condition is verified where an error may occur and then he or she would write code in consequence. The first option which is very popular is not advisable if we want our program to be used by anyone besides ourselves. The second option which is also known as **LBYL** (**L**ook **B**efore **Y**ou **L**eap), is time consuming and may make code unreadable. Let's have a look at an example of each strategy.

The following program reads a file (`myfile.csv`) separated by tabs and looks for a number that is found in the first column of the first line. This value is multiplied by 0.2 and that result is written to another file (`otherfile.csv`).

This version does not check for any types of errors and limits itself to its core functionality.

Listing 7.1: Program with no error checking

```
1 fh = open('myfile.csv')
2 line = fh.readline()
3 fh.close()
4 value = line.split('\t')[0]
5 fw = open('other',"w")
6 fw.write(str(int(value)*.2))
7 fw.close()
```

This program may do its job provided that there are no unexpected events. What does "unexpected events" mean in this context? The first line is prone to error. For example, it may be trying to open a file that doesn't exist. In this case when the program executes it will immediately stop after executing the first line and the user will face something like the following:

```
Traceback (most recent call last):
  File "wotest.py", line 1, in ?
    fh = open('myfile.csv')
IOError: [Errno 2] No such file or directory: 'myfile.csv'
```

This is a problem because the program stops, and it is not professional to show the end user a system error.

This program can fail on various possible spots. There may be no tabs in the file, there may be letters instead of numbers and we may not have the write permissions in the directory where we intend to write the output file.

That is what happens when the file exists but there is no tabs inside.

```
Traceback (most recent call last):
  File "wotest.py", line 6, in ?
    fw.write(str(int(value)*.2))
ValueError: invalid literal for int():
```

The result is similar to the previous one. It causes the program to stop and the interpreter shows us another error message. This way we may continue with all the blocks of code that are prone to the failure.

Let's look at the strategy of checking each condition likely to generate an error in order to prevent its occurrence. (LBYL)

Listing 7.2: Error handling LBYL version (py3.us/22)

```
1  import os
2  while True:
3      iname = raw_input("Enter input filename: ")
4      oname = raw_input("Enter output filename: ")
5      if os.path.exists(iname):
6          fh = open(iname)
7          line = fh.readline()
8          fh.close()
9          if "\t" in line:
10             value = line.split('\t')[0]
11             if os.access("/home/sb/"+oname,os.W_OK)==0:
12                 fw = open("/home/sb/"+oname,"w")
13                 if value.isdigit():
14                     fw.write(str(int(value)*.2))
15                     fw.close()
16                     break
17                 else:
18                     print("It can't be converted to int")
```

```
19              else:
20                  print("Output file is not writable")
21          else:
23              print("There is no TAB. Check the input file")
24      else:
25          print("The file doesn't exist")
```

This program considers almost all the possible errors. If the file that the user enters does not exist, the program will not have an abnormal termination. Instead it will display an error message designed by the programmer that would allow the user to reenter the name of the input file.

The disadvantage of this option is that the code is both difficult to read and maintain because the error checking is mixed with its processing and with the main objective of the program. It is for this reason that new programming languages have included a specific system for the control of exceptional conditions. Contrary to LBYL, this strategy is known as EAFP (It's easier to ask forgiveness than permission). With Python the statements **try, except, else** y **finally** are used.

7.1.1 Try and Except

try delimits the code that we want to execute, while the **except** delimits the code that will be executed if there is an error in the code under the **try** block. Errors detected during execution are called *exceptions*. Let's look at the general outline:

```
try:
    code block 1
    # ...some error prone code...
except:
    code block 2
    # ...do something with the error...
[else:
    code block 3
    # ...to do when there is no error...
finally:
    code block 4
    #...some clean up code..].
```

This code will first try to execute the code in block 1. If the code is executed without problems, the flow of execution continues through the code in block 3 and finally through block 4. In case the code in block 1 produces an error (or raises an exception according to the jargon), the code in block 2 will be executed and then the code in block 4. The idea behind this mechanism is to put the block of code that we suspect may produce an error (block 1), inside the **try** clause. The code that does should be called in case of a problem is placed in the **except** block. This code (code block 2) deals with the exception, or in another words, it *handles the exception.* Error messages are what the user gets when exceptions are not handled:

```
>>> 0/0
Traceback (most recent call last):
  File "<stdin>", line 1, in <module>
ZeroDivisionError: integer division or modulo by zero
```

Optionally, it is possible to add the statement **else**, which will be executed only if the code inside **try** (code block 1) executes successfully. Note that the code below **else** can be placed in the **try** block because it would have the same effect (it would execute if there are no errors). The actual difference is conceptual, the block inside **try** should contain only the code that is suspected to raise an exception, while we would have to leave for the block below **else** the instructions that should be executed when the instructions below **try** are executed without error. Note that the code inside **finally** is always executed.

This is an oversimplified example:

```
try:
    print 0/0
except:
    print("Houston, we have a problem...")
```

The result is:

```
Houston, we have a problem...
```

The first thing that we take note of is that neither **else** nor **finally** is included as they are optional statements. In this case, the statement **print 0/0** raises an exception. This exception is "caught" by the code that follows **except**. This way we can securely test a block of code as any error that occurs will redirect the program flow in a predictable way.

In this code exception handling is applied to code listing 7.2:

Listing 7.3: Similar to 7.2 but with exception handling (py3.us/23). Python 2.x only.

```
1 import os
2 while True:
3     try:
4         iname = raw_input("Enter input filename: ")
5         oname = raw_input("Enter output filename: ")
6         fh = open(iname)
7         line = fh.readline()
```

```
8          fh.close()
9          value = line.split('\t')[0]
10         fw = open("/home/sb/"+oname,"w")
11         fw.write(str(int(value)*.2))
12         fw.close()
13     except IOError, (errno,errtext):
14         if errno==13:
15             print "Can't write to outfile."
16         elif errno==2:
17             print "File not exist"
18     except ValueError, strerror:
19         if "substring not found" in strerror.message:
20             print "There is no tab"
21         elif "invalid literal for int" in strerror.message:
22             print "The value can't be converted to int"
23     else:
24         print "Thank you!. Everything went OK."
25         break
```

At first look it is noticieable that this code is easier to follow than the previous version (7.2). At least the code logic is separated from the error handling. We can note that from line 4 to 12, the same code is repeated as in the original listing (7.1). It is from line 13 where the exception handling begins. According to the type of exception, it is the code that will be executed below. We will see how to distinguish between the different types of exceptions later.

Listing 7.3 is an introductory example of how to apply exception handling to the listing 7.1, and not a definitive guide of how to handle exceptions.

For example if the integer conversion of line 11 fails, an exception of type ValueError will be raised, a message will be printed and the program flow will return to line 3 (because it is under a **while TRUE**), without releasing the resources used (the filehandle **fw**). One way to solve this problem is to copy the statement where the resource is closed (line 12) to the block where the corresponding exception is managed (after line 22). This way we ensure to release the resource.

Listing 7.4: Another version of 7.3 code (py3.us/24)

```
1 import os
2 while True:
3     try:
4         iname = raw_input("Enter input filename: ")
5         oname = raw_input("Enter output filename: ")
6         fh = open(iname)
7         line = fh.readline()
```

```
8           fh.close()
9           value = line.split('\t')[0]
10          fw = open(os.path.join("/home/sb/",oname),"w")
11          fw.write(str(int(value)*.2))
12          fw.close()
13      except IOError, (errno,errtext):
14          if errno==13:
15              print("Can't write to outfile.")
16          elif errno==2:
17              print("File not exist")
18      except ValueError, strerror:
19          if "substring not found" in strerror.message:
20              print("There is no tab")
21          elif "invalid literal for int" in strerror.message:
22              print("The value can't be converted to int")
23              fw.close()
24      else:
25          print("Thank you!. Everything went OK.")
26          break
```

Even if this code does its job, it is not a good idea to repeat the same statement (`fw.close()`) in two places (line 12 and 23). If we have a block of code that always needs to be executed whether or not an error occurs we can include it in **finally**, which is where the code that is executed independently of what happens with the code in **try**. The problem with including `fw.close()` under **finally** is that there may be an exception before opening `fh` (for example in the integer conversion, line 10 of the listing 7.4) and we are going to try to close a file that was never opened, which will cause another error by itself. This error in turn, can be predicted, for which we can use the exception mechanism and include a try/except clause within **finally**:

Listing 7.5: Code with nested exceptions

```
1  import os
2  while True:
3      try:
4          iname = raw_input("Enter input filename: ")
5          oname = raw_input("Enter output filename: ")
6          fh = open(iname)
7          line = fh.readline()
8          fh.close()
9          value = line[:line.index("\t")]
10         fw = open(os.path.join("/home/sb/",oname),"w")
11         fw.write(str(int(value)*.2))
12     except IOError, (errno,errtext):
```

```
13          if errno==13:
14              print("Can't write to outfile.")
15          elif errno==2:
16              print("File not exist")
17      except ValueError, strerror:
18          if "substring not found" in strerror.message:
19              print("There is no tab")
20          elif "invalid literal for int" in strerror.message:
21              print("The value can't be converted to int")
22      else:
23          print("Thank you!. Everything went OK.")
24          break
25      finally:
26          try:
27              fw.close()
28          except:
29              pass
30          else:
31              print("All resources freed")
```

The code from the listing 7.5 was made to illustrate the use of a nested **try**, but it is not the best way to solve the problem. We may avoid causing an exception while the filehandle is open by making the integer conversion before opening the file. Another change to consider is to remove the **raw_input** statements in the **try** block, because it is convenient to include only the statements that are expected to cause exceptions.

Listing 7.6: Similar to code 7.4 without nested exceptions

```
1 import os, errno
2 while True:
3     iname = raw_input("Enter input filename: ")
4     oname = raw_input("Enter output filename: ")
5     try:
6         fh = open(iname)
7         line = fh.readline()
8         value = str(int(line[:line.index("\t")])*.2)
9         fw = open("/home/sb/"+oname,"w")
10        fw.write(value)
11    except IOError, (errno,errtext):
12        if errno==errno.EACCES:
13            print("Permission denied")
14        elif errno==errno.ENOENT:
15            print("No such file")
16    except ValueError, strerror:
```

```
17        if "substring not found" in strerror.message:
18            print("There is no tab")
19        elif "invalid literal for int" in strerror.message:
20            print("The value can't be converted to int")
21    else:
22        fw.close()
23        fh.close()
24        break
```

We've seen in general terms how the **try/except**, clause works and now we can go a little deeper to discuss the types of exceptions:

7.1.2   Exception Types

Exceptions can be distinguished. A non existent variable and mixing incompatible data types are not the same. The first exception is of the **NameError** type while the second is of the **TypeError** type. A complete list of exceptions can be found in Section D.7 (page 499).

How to Respond to Different Exceptions:

It is possible to handle an error generically using **except** without a parameter:

```
d = {"A":"Adenine","C":"Cisteine","T":"Timine","G":"Guanine"}
try:
    print d[raw_input("Enter letter: ")]
except:
    print "No such nucleotide"
```

Just because we may be able to respond generically to all errors doesn't mean that it is a good idea (in fact it is a bad idea). This makes debugging our code difficult because an unanticipated error can pass unnoticed. This code will return a "No such nucletide" for any type of error. If we introduce an EOF signal, EOF (end of file, CONTROL-D), the program will output "No such nucleotide". It is useful to distinguish between the different types of abnormal events, and react in consequence. For example to differentiate the entrance of an EOF from a nonexistent dictionary key:

```
d={"A":"Adenine", "C":"Cisteine", "T":"Timine", "G":"Guanine"}
try:
    print(d[raw_input("Enter letter: ")])
except EOFError:
    print("Good bye!")
except KeyError:
    print("No such nucleotide")
```

This way, the program prints "No such nucleotide" when the user enters a key that does not exist in **d** dictionary and "Good bye!" when it gets an EOF.

We are now ready to consider the code 7.3 (page 128) in detail. Let's start in line 13, where an exception of type IOError is caught: **except IOError, (errno,errtext)**. Besides catching the exception there are also two values: `errno` and `errtext`. The first value corresponds to a code related to the type of IOError that was produced. The IO error includes various distinct related errors. This is why an error code exists which is used (in the lines 14 and 16) to determine exactly which error was produced. The error type ValueError also includes more than one error. On one hand it is called when a value is searched for in a list that does not exist (it may occur on line 9) and on the other hand, when one wants to perform data type conversions between incompatible types (line 11). The way to distinguish between these conditions is by the error message, as shown on lines 19 and 21.

To get information about the exception that is currently being handled, use **sys.exc_info()**:

Listing 7.7: Using sys.exc_info() (py3.us/25)

```
1 import sys
2
3 try:
4     0/0
5 except:
6     a,b,c=sys.exc_info()
7     print('Error name: %s' % a.__name__)
8     print('Message: %s' % b.message)
9     print('Error in line: %s' % c.tb_lineno)
```

This program prints:

```
Error name: ZeroDivisionError
Message: integer division or modulo by zero
Error in line: 4
```

Listing 7.8: Another use of sys.exc_info()

```
1 import sys
2
3 try:
4     x=open('a_random_filename')
5 except:
6     a,b=sys.exc_info()[:2]
7     print 'Error name: %s' % a.__name__
8     print 'Error code: %s' % b.args[0]
9     print 'Error message: %s' % b.args[1]
```

This program prints:

```
Error name: IOError
Error code: 2
Error message: No such file or directory
```

7.1.3   Provoking Exceptions

Exceptions can be activated manually, without the need to wait for them to occur. The statement used to raise an exception is **raise**. At this point, you may be wondering why you would want to provoke an exception. An appropriately raised exception may be more helpful to the programmer (or to the user if he or she is not the same programmer) than an exception that is fired in an uncontrolled way. This is especially true when debugging programs.

This type of concept is better understood with an example. Below there is a function that calculates the average of a sequence of numbers:

```
def avg(numbers):
    return sum(numbers)/len(numbers)
```

A function of this type will have problems with an empty list:

```
>>> avg([])
```

```
Traceback (most recent call last):
  File "<pyshell#4>", line 1, in <module>
    avg([])
  File "<pyshell#3>", line 2, in avg
    return sum(numbers)/len(numbers)
ZeroDivisionError: integer division or modulo by zero
```

The problem with this error message is that it does not tell us that it was caused by the empty list, but says that it was provoked but trying to divide by zero. Knowing how the function works, one can deduce that an empty list causes this error. However, it would be more interesting if this error points this out, without having to know the internal structure of the function. For this we can raise an error by ourselves.

```
def avg(numbers):
    if not numbers:
        raise ValueError("Please enter at least one element")
    return sum(numbers)/len(numbers)
```

In this case, the error type is closer to the actual problem.

```
>>> avg([])

Traceback (most recent call last):
  File "<pyshell#8>", line 1, in <module>
    avg2([])
  File "<pyshell#6>", line 3, in avg2
    raise ValueError("Please enter at least one element")
ValueError: Please enter at least one element
```

We could have avoided the error if we printed a string without raising the error, but this will be against pythonic principles, especially against "errors should not pass unnoticed". In practice this may cause problems because if a function returns an unanticipated value, the effects can be unpredictable. On raising the exception, we assure ourselves that the error will not pass unnoticed.

In some texts or old code you will find a syntax of the form **raise "This is an error"**. These types of exceptions (called chained exceptions), will not work from Python 2.6 and greater. It is also not recommended using **raise ValueError, 'A message'**, instead of **raise ValueError('A message')**. The latter form is preferred because in large chains the parentheses indicates "continue on the next line". In Python 3.0 the latter form is mandatory.[1]

7.2 Creating Customized Exceptions

An advantage of the exception system is that we don't have to limit ourselves to those provided by Python. We can define new exceptions in function of our needs. In order to create an exception, we need to work with Object Oriented Programming, a topic that has not been covered. As a result, if you're reading this book from the start, my recommendation is that you skip the rest of this chapter and proceed directly to OOP. After reading OOP, return to this section if you still want to create your own exceptions.

All Exceptions Derive from Exception Class

Since all exception derive from the exceptions class, we can make our own exception by subclassing the exception class. Take for example this exception that I called NotDNAException. It should be raised when there is a DNA sequence with a character not belonging to either 'a', 'c', 't' or 'g'. Let's see a custom exception defined:

---

[1]Please see PEP 3109 (http://www.python.org/dev/peps/pep-3109) regarding the rationale for this.

```
class NotDNAException(Exception):
    """ A user-defined exception.
    """
    def __init__(self, dna):
        self.dna = dna
    def __str__(self):
        for nt in self.dna:
            if nt not in 'atcg':
                return nt
```

The programmer should create a code to detect the exception:

```
dnaseq = 'agcttacagt'
if set(dnaseq) != set('atcg'):
    raise NotDNAException(dnaseq)
else:
    print 'OK'
```

If **dnaseq** is an iterable object with either 'a', 'c', 't' and 'g', this code prints
OK. But if **dnaseq** contains a non-DNA character, the exception will be raised.
This is the result of the former code but with a 'w' in **dnaseq**:

```
Traceback (most recent call last):
  File "/home/sb/newexcep.py", line 16, in <module>
    raise NotDNAException(dnaseq)
NotDNAException: w
```

7.3   Additional Resources

- Mark Pilgrim. "Dive into Python. Exceptions and File Handling."
  http://www.diveintopython.org/file_handling/index.html

- Python Documentation. "Built-in Exceptions."
  http://www.python.org/doc/current/lib/module-exceptions.html

- Python Documentation. "Standard errno system symbols."
  http://docs.python.org/library/errno.html

- C H. Swaroop. "Python en:Exceptions."
  http://www.swaroopch.com/notes/Python_en:Exceptions

- Ian Bicking. "Re-Raising Exceptions."
  http://blog.ianbicking.org/2007/09/12/re-raising-exceptions

7.4   Self-Evaluation

1. What is the meaning of LBYL and EAFP? Which one is used in Python?

2. What is an **exception**?

3. What is an "unhandled exception"?

4. When do you use **finally** and when do you use **else**?

5. Exceptions are often associated with file handling. Why?

6. How do you sort an error derived from a disk full condition from trying to write to a read-only file system?

7. Why is it not advisable to use **except:** to catch all kind of exceptions, instead of using, for example, **except IOError:**?

8. Exceptions can be raised at will. Why would you do that?

9. What is the purpose of **sys.exc_info()**?

10. Explain the purpose of this function:

```python
def formatExceptionInfo():
    """ Author: Arturo 'Buanzo' Busleiman """
    cla, exc = sys.exc_info()[:2]
    excName = cla.__name__
    try:
        excArgs = exc.__dict__["args"]
    except KeyError:
        excArgs = str(exc)
    return (excName, excArgs)
```

Chapter 8

Introduction to Object Orienting Programming (OOP)

8.1 Object Paradigm and Python

As mentioned in the introduction of the book, Python is an object oriented language. Unlike other languages that handle objects, Python allows us to program in a classic procedural way, without considering the objects paradigm. Sometimes this is called "multi-paradigm language." Although this makes it easy to start programming, in the long run it can tempt the programmer not to take advantage of all the possibilities that Python offers.

We already used objects, even without stating it in an explicit way. Data types included in Python are objects. A string, a dictionary and a list, are implementation of objects. Each of them has its associated functions (methods in the jargon) and its attributes. We have seen that **lower()** returns a string in lower case, this is because all the objects of the class string have the method **lower()** associated with them. The same is true for other types of data that are included in Python.

A class can be used to define a data type. Although data types included in Python are many and varied, its capacity to include all our information modelling needs is limited. One of the goals of programming is to represent the real world. We can use a dictionary to represent a translation table between nucleotides and amino acids, a string to represent a DNA sequence or a tuple to represent the space coordinates of an atom in a protein. But, what data type do we use to represent a metabolic state of a cell? The different domains in a protein? The result of a BLAST run? What about an ecosystem?

There is a need for the ability to define our own data types, to be able to model any system, either biological or of any other type. Although the functions are useful to modularize the code, they are not designed to fulfill this role. The functions cannot store states, since the values of variables only have life while the function is being executed. Other languages have their personalized data types, like "structs" in C or "record" in Pascal, but they do not have the same flexibility as the objects of languages based on OOP (like Java, C++ or Python). Only objects have enough ductility to be able to model any type of system and its possible relations with other systems.

8.2 Exploring the Jargon

The world of OOP has its own vocabulary. In this section I will try to clarify a few of the many new words such as class, method, instance, attributes, polymorphism, inheritance, etc. The definitions will not be exhaustive. Some of them will not even be exact, but the priority will be the understanding of the subject rather than being overly formal. Let's remember that the objective of this book is to provide programming tools to solve biological problems. Keeping this in mind, the following definitions and their respective examples have been written.

Classes: Object Generators

We can see a class as a user defined data type. However this definition is incomplete, it does not consider that a class has associated functions and is not just a data container. That is why sometimes it is presented as a mold to generate objects, since the general characteristics of objects are defined in a class. A class can be genome, people, sequences, etc. Any object capable of being abstracted can be a class.

Instance: Particular Implementation of Class

An instance is the implementation of a class. For instance, if we have a class orca, an instance can be Willy. Several instances from the same class can be created (for example, Shamu) and all are independent of each other.

Attributes or Instance Variables: Characteristic of Objects

Each object will have its own characteristics (or attributes), for example weight. Willy surely will have a weight different from Shamu, but in spite of having variations in their attributes, both instances, belong to the same class orca. They share at least the "type of attributes." We could create a class dog, with instances Lassie, Laika and Rin-tin-tin. This class can have the attribute hair_color, which is not going to be shared by instances of the Orca class.

Methods: Behavior of Objects

A method is a function that belongs to a class. Methods define how the objects derived from that class "behave." For example the DNA class can have the **method translate** that allows translating an amino acid sequence into a protein. This method is nothing but a function associated to a class. It could require as parameters a string with the DNA sequence and a dictionary including its related translation table.

Class Variables: The Characteristics of Classes

They are variables associated to all the objects of a class. Whenever an object is created from a class, this object inherits the variable of the class.

Inheritance: Properties Are Transmitted between the Related Classes

Classes can be related between themselves and are not isolated entities. It is possible to have a *mammal class* with common properties with the *orca class* and the *dog class*. For example, the method *reproduction* can be defined for the *mammal class*. When we create the classes *dog* and *orca*, we can define them as "children" of *mammal*. It won't be necessary to create for them the method *reproduction*. Child classes will be able to have their own methods themselves that make them unique, like *immersion* and *race*.

Polymorphism

Is the ability of different types of objects to respond to the same method with a different behaviour. For example you can iterate over a list, a set, a dictionary, a file, a database, and more in the same way.

Encapsulation

Is the ability to hide the internal operation of an object and leave access for the programmers only through their public methods. The term **encapsulation** is not associated with Python because this language does not have a **true encapsulation**. It is possible to make the access to certain methods difficult, but not to prevent it. It is not in the philosophy of Python to be in the way of the programmer. What it is possible to do in Python is to make clear what methods and properties are owned by a class and those thought to be shared. Sometimes this behavior is referred to as pseudo-encapsulation or translucent encapsulation. It is up to the programmer to make a rational use of this option. That is called in Python: Protection by convention, not by legislation. See the section "Making our code private" on page 154 for using this property.

8.3   Creating Classes

Remember that classes are the template of the objects. The syntax to create classes in Python is very simple:

```
class NAME:
    [body]
```

Let's see a class that actually does something:

```
class Square:
    def __init__(self):
        self.side=1
```

This class (`Square`) has a method called `__init__`. It is a special method that has as a characteristic the fact that it does not return anything and that it is executed whenever an instance of `Square` is created. In this case it sets the value of the attribute side. Another peculiarity to consider is the word self that is repeated as parameter of the method and as part of the name of the attribute. `Self` is a variable that is used to represent the instance of `Square`. It is possible to use another name instead of self, but self is used by convention. It is advisable to follow the convention because it makes our program easier to understand by other programmers who may want to read our code in the future.[1]

Instantiation uses function notation. It is like a function without parameters that returns a new instance of the class. Let's see an example, the use of the `Square` class, with the creation of the instance `Bob`:

```
>>> Bob=Square() # Bob is an instance of Square.
>>> Bob.side #Let's see the value of side
1
```

It is possible to change the value of the attribute `side` of the instance Bob:

```
>>> Bob.side=5 #Assing a new value to side
>>> Bob.side #Let's see the new value of side
5
```

Although this change is specific for this instance, when new instances are created the method **__init__** is executed again to assign the `side` value to the new instance:

---

[1]There are also code analyzers that depend on this convention for working.

```
>>> Krusty=Square()
>>> Krusty.side
1
```

In the case that the variable `side` is a variable that must be accessible from all the instances of the class, it is advisable to use a **class variable**. These variables are shared by all the objects of the same class.

```
class Square:
    side=1
```

This way, the value of *side* will be defined even before we create an instance of `Square`:

```
>>> Square.side
1
```

Of course if we created instances of `Square`, they will also have this value of `side`:

```
>>> Crab=Square()
>>> Crab.side
1
```

The class variables can have information on the instances. For example it is possible to use them to control how many instances of a class have been created.

```
class Square:
    CountObjects=0
    def __init__(self):
        Square.CountObjects=Square.CountObjects+1
        print "Object created successfully"
```

This version of `Square` can count the number of instances that have been created. Note that the `CountObjects` variable is acceded within the class as `Square.CountObjects` to distinguish itself from an instance variable, which is noted with the prefix *self.VARIABLENAME*. Let's see how this object is used:

```
>>> Bob=Square()
Object created successfully
>>> Patrick=Square()
Object created successfully
>>> Square.CountObjects
2
```

Let's see a more useful, though simple class:

```python
class Sequence:
    TranscriptionTable = {"A":"U","T":"A","C":"G","G":"C"}
    def __init__(self, seqstring):
        self.seqstring=seqstring.upper()
    def transcription(self):
        tt = ""
        for x in self.seqstring:
            if x in 'ATCG':
                tt += self.TranscriptionTable[x]
        return tt
```

This class has two methods and one attribute. The method **__init__** is used to set the value of *seqstring* in each instance:

```python
>>> DangerousVirus=Sequence('atggagagccttgttcttggtgtcaa')
>>> DangerousVirus.seqstring
'ATGGAGAGCCTTGTTCTTGGTGTCAA'
>>> HarmlessVirus=Sequence('aatgctactactattagtagaattgatgcca')
>>> HarmlessVirus.seqstring
'AATGCTACTACTATTAGTAGAATTGATGCCA'
```

The Sequence class also has a method called *transcription* that has as its only parameter the instance itself (represented by *self*). This parameter does not appear when the function is called, because it is implicit. Notice that the function *transcription* uses the class variable of the *TranscriptionTable* (that is a dictionary) to convert the sequence *seqstring* to its transcript equivalent:

```python
>>> DangerousVirus.transcription()
'GCUAAGAGCUCGCGUCCUCAGAGUUUAGGA'
```

The methods can also have parameters. In order to show this, here is a new mehod (*restriction*) in the *Sequence* class. This method calculates how many restriction sites has a sequence for a given enzyme.[2] Therefore, this method will require as parameter a restriction enzyme. Another difference is that this class will contain a dictionary that relates the name of the enzyme to the recognition sequence:

Listing 8.1: Sequence class (py3.us/26)

```python
class Sequence:
    TranscriptionTable = {"A":"U","T":"A","C":"G","G":"C"}
```

---

[2]Remember that a restriction enzyme is a protein that recognizes a specific DNA sequence and produces a cut within the recognition zone.

```
# Dictionary with the name of the restriction enzyme and
# the recognition sequence.
EnzDict = {"EcoRI":"GAATTC","EcoRV":"GATATC"}
def __init__(self, seqstring):
    self.seqstring = seqstring.upper()
def restriction(self,enz):
    try:
        return self.seqstring.count(Sequence.EnzDict[enz])
    except:
        return 0
def transcription(self):
    tt = ""
    for x in self.seqstring:
        if x in 'ATCG':
            tt += Sequence.TranscriptionTable[x]
    return tt
```

Using the Sequence class:

```
>>> other_virus=Sequence('atgatatcggagaggatatcggtgtcaa')
>>> other_virus.transcription()
'UACUAUAGCCUCUCCUAUAGCCACAGUU'
>>> other_virus.restriction("EcoRV")
2
```

Tip: Not All Classes Are Created Equal: Classic Versus New-Style Classes.

The type of class that is available from Python version 2.2 is called "new-style" classes, in contrast to "traditional" classes (known also as "classic"). The difference between them is that the new-style classes derive from "object," inheriting its methods and properties. This was done because before Python 2.2, built-in types and user-defined classes were different. There was no way to create classes derived from the built-in types (lists, dictionaries, sets, and so on). Inherit from types is not its unique advantage. New-style classes provides new methods, as described in PEP 252[3] and the "Unifying types and classes in Python 2.2" document.[4] From Python 3, all classes are "new-style" classes, so there is no need to distinguish between classic and new-style.

---

[3]http://www.python.org/dev/peps/pep-0252/

[4]http://www.python.org/download/releases/2.2.3/descrintro

8.4 Inheritance in Action

Remember that the inheritance of classes implies that the new class "inherits" the methods and attributes of the base class. The following is the syntax used to create a class that inherits from other class:

```
class DerivedClass(BaseClass):
    [body]
```

Let's see as an example a class called `Plasmid`[5] that is based on the `Sequence` class. Because plasmid is a type of DNA sequence, we created the `Plasmid` class that inherits methods and properties from `Sequence`. We also defined methods and attributes that are exclusive to this new class, like `AbResDict` and `ABres`. The method `ABres` is used to know if our plasmid has resistance to a particular antibiotic, whereas the `AbResDict` attribute has the information of the regions that characterize the different antibiotic resistances.

Listing 8.2: Plasmid class (py3.us/82)

```
class Plasmid(Sequence):
    AbResDict = {"Tet":"ctagcat","Amp":"CACTACTG"}
    def __init__(self,seqstring):
        Sequence.__init__(self,seqstring)
    def ABres(self,ab):
        if self.AbResDict[ab] in self.seqstring:
            return True
        else:
            return False
```

Notice that within the method _init_ of Plasmid we called to the method _init_ of `Sequence`. This is the way that our class inherits the attributes and methods of the "father" class. Let's see how the `Plasmid` class uses its own methods and those of its father (`Sequence`). The method `ABres` works in a way similar to `Restriction` with the difference that instead of giving back the position which we are looking for, it simply informs us if it is present or absent.

Introducing Some Biopython Objects

While there is a special section of Biopython ahead in this book, we will see some Biopython structures to get familiar with them.

---

[5]A plasmid is a DNA molecule that is independent of the chromosomal DNA of a microorganism.

The IUPAC nucleic acid notation			
	Symbol	Meaning	Mnemonic
DNA Bases	G	Guanine	Guanine
	T	Thymine	Thymine
	A	Adenine	Adenine
	C	Cytosine	Cytosine
Ambiguity Characters	R	G + A	puRine
	Y	T + C	pYrimidine
	S	G + C	Strong interactions (3 H bonds)
	W	T + A	Weak interactions (2 H bonds)
	K	G + T	Keto
	M	A + C	aMino
	D	G + T + A	Not-C (D follows C in alphabet)
	H	T + A + C	Not-G (H follows G)
	B	G + T +C	Not-A (B follows A)
	V	G + A + C	Not-T or U (V follows U)
	N	G + A + T + C	aNy

FIGURE 8.1: IUPAC nucleic acid notation table.

Class IUPACAmbiguousDNA: The class **IUPACAmbiguousDNA** is in the module *IUPAC*. It is a class that derives from **alphabet** and holds the information regarding the IUPAC[6] approved letters for DNA sequences. In this case (*AmbiguousDNA*) ambiguity is taken into account, that is, we can denote when a position is not fully determined. For example, if a nucleotide in a specific position can be A or G, it is indicated with an R (See figure 8.1 for the complete IUPAC nucleic acid notation table). For this reason IUPACAmbiguousDNA has a class variable `letters` that holds the string 'GATCRYWSMKHBVDN'. At first sight it doesn't seem a very useful class, but in the class Seq its usefulness will be shown.

Class IUPACUnambiguousDNA: Like **IUPACAmbiguousDNA**, there is **IUPACUnambiguousDNA**. This class derives from the former, so it keeps its properties. The only difference is that this class defines again the **letters** attribute, with 'GATC' as the content.

---

[6]IUPAC stands for **I**nternational **U**nion for **P**ure and **A**pplied **C**hemistry; it is the body that regulates the nomenclature used in chemistry.

Class Seq: In the module **Seq** there is a class called **Seq**. The purpose of the objects derived from this class is to store sequence information. Up until now we have represented the sequences as strings. The problem with this approach is that the string just holds sequence information, and we have to guess about what kind of sequence it is (DNA, RNA, amino acids). In the **Seq** class, there are two parameters: **data** and **alphabet**. **Data** is a string with the sequence and **alphabet** is an object of the **alphabet** type. It contains information about the type of sequence alphabet. Another feature of this class, is that it is "immutable", that is, once a sequence is defined, it can't be modified (just as a string). This way we are sure the sequence remains the same even after several manipulations. In order to change the sequence, we have to use a **MutableSeq** kind of object.

The **Seq** class defines several methods, being the most important: *complement* (Returns the complement sequence), *reverse_complement* (Returns the reverse complement sequence), *tomutable* (returns a **MutableSeq** object) and *tostring* (returns the sequence as a string). Let's see it in action:

```
>>> from Bio.Alphabet import IUPAC
>>> from Bio.Seq import Seq
>>> first_seq = Seq('GCTATGCAGC', IUPAC.unambiguous_dna)
>>> first_seq
Seq('GCTATGCAGC', IUPACUnambiguousDNA())
>>> first_seq.complement()
Seq('CGATACGTCG', IUPACUnambiguousDNA())
>>> first_seq.tostring()
'GCTATGCAGC'
```

This object has special methods that allows the programmer to work with a **Seq** type object as if it were a string:

```
>>> first_seq[:10] #slice a sequence
Seq('GCTAT', IUPACUnambiguousDNA())
>>> len(first_seq) #get the length of the sequence
10
>>> first_seq[0] #get one character
'G'
```

Class MutableSeq: It is an object very similar to **Seq**, with the main difference that its sequence can be modified. It has the same methods as **Seq**, with some methods tailored to handle mutable sequences.

We can create it from scratch or it can be made from a **Seq** object using the *tomutable* method:

```
>>> first_seq
Seq('GCTATGCAGC', IUPACUnambiguousDNA())
>>> AnotherSeq=first_seq.tomutable()
>>> AnotherSeq.extend("TTTTTTT")
>>> print(AnotherSeq)
MutableSeq('GCTATGCAGCTTTTTTT', IUPACUnambiguousDNA())
>>> AnotherSeq.pop()
'T'
>>> AnotherSeq.pop()
'T'
>>> print(AnotherSeq)
MutableSeq('GCTATGCAGCTTTTT', IUPACUnambiguousDNA())
```

8.5   Special Methods Attributes

Some methods have a special meaning. We have already seen the __init__ method that is executed each time a new instance is created (or a new object is instantiated). Other methods are executed under other conditions. Each special method is executed under a pre-established condition. What can be modified is how the object responds to a particular condition.

Take for example the __len__ method. This method is activated in an object each time the function **len(instance)** is called. What this method returns is the responsibility of the programmer. Recall the *Sequence* class (listing 8.1) and see what happens when you want to find out the length of a sequence:

```
>>> len(Sequence("ACGACTCTCGACGGCATCCACCCTCTCTGAGA"))
Traceback (most recent call last):
  File "<stdin>", line 1, in <module>
AttributeError: Sequence instance has no attribute '__len__'
```

This was somewhat expected. We didn't define what is the meaning of the length of *Sequence*. This object has several attributes, the interpreter has no way to know which attribute returns when *len(Sequence)* is required. The error message gives us a clue about the problem: "Sequence instance has no attribute '__len__' ". Hence if we want to set a behaivor for **len()** function, we have to define the special method attribute __len__:

```
def __len__(self):
    return len(self.seqstring)
```

This method must be included in the class definition (8.1). Now that we have defined the __len__ method, we can apply the function **len** to the *Sequence* objects:

```
>>> M13=Sequence("ACGACTCTCGACGGCATCCACCCTCTCTGAGA")
>>> len(M13)
32
```

In the same way that we can control what is returned by **len()**, we can do it with other methods that can be programmed in our class. Let's see some of them:[7]

- __str__ This method is invoked when the string representation of an object is required. This representation is obtained with **str(object)** or with **print object**. This way the programmer can choose how its objects "looks." For example, the translation table provided by Biopython Bio.Data.CodonTable is stored as a dictionary, but its representation appears as a table:

```
>>> import Bio.Data.CodonTable
>>> print(Bio.Data.CodonTable.standard_dna_table)
Table 1 Standard, SGC0

  |  T      |  C      |  A      |  G      |
--+---------+---------+---------+---------+--
T | TTT F   | TCT S   | TAT Y   | TGT C   | T
T | TTC F   | TCC S   | TAC Y   | TGC C   | C
T | TTA L   | TCA S   | TAA Stop| TGA Stop| A
T | TTG L(s)| TCG S   | TAG Stop| TGG W   | G
--+---------+---------+---------+---------+--
C | CTT L   | CCT P   | CAT H   | CGT R   | T
C | CTC L   | CCC P   | CAC H   | CGC R   | C
C | CTA L   | CCA P   | CAA Q   | CGA R   | A
C | CTG L(s)| CCG P   | CAG Q   | CGG R   | G
--+---------+---------+---------+---------+--
A | ATT I   | ACT T   | AAT N   | AGT S   | T
A | ATC I   | ACC T   | AAC N   | AGC S   | C
A | ATA I   | ACA T   | AAA K   | AGA R   | A
A | ATG M(s)| ACG T   | AAG K   | AGG R   | G
--+---------+---------+---------+---------+--
G | GTT V   | GCT A   | GAT D   | GGT G   | T
G | GTC V   | GCC A   | GAC D   | GGC G   | C
G | GTA V   | GCA A   | GAA E   | GGA G   | A
G | GTG V   | GCG A   | GAG E   | GGG G   | G
--+---------+---------+---------+---------+--
```

---

[7]In Table D.21 (page 503) there is a list of **Special methods**.

- __repr__ Invoked with **repr()** built-in function and when the object is entered into the interative shell. It should look like a valid Python expression that could be used to recreate an object with the same value, when not possible, a string of the form <...some useful description...>. It is used mostly on debugging. See the same object as above but with **repr()** instead of **print()**:

```
>>> repr(Bio.Data.CodonTable.standard_dna_table)
'<Bio.Data.CodonTable.NCBICodonTableDNA instance at 0xb7da0c>'
```

- __getitem__ Is used to access an object sequentially or by using a subscript like *object[n]*. Each time you try to access an object as *object[n]*, *object.__getitem__(n)* is executed. This method requires two parameters: The object (usually *self*) and the index. There is a usage sample in listing 8.4.

- __iter__ Allows walking over a sequence. With **__iter__** we can iterate the same way over many different objects such as dictionaries, lists, files, strings and so on. The **for** statement calls the build-in function **iter** on the object being iterated over. **__iter__** defines how the items are returned when using the **next** method. It is easy to understand them with a couple of examples. In the first example we create the *Straight* class, where its elements are returned in the same order as they are stored, while the *Reverse* class returns its elements using an inverted order:

Listing 8.3: Straight and Reverse classes (py3.us/28)

```
class Straight:
    def __init__(self, data):
        self.data = data
        self.index = 0
    def __iter__(self):
        return self
    def next(self):
        if self.index == len(self.data):
            raise StopIteration
        answer = self.data[self.index]
        self.index = self.index + 1
        return answer

class Reverse:
    def __init__(self, data):
        self.data = data
        self.index = len(data)
```

```
    def __iter__(self):
        return self
    def next(self):
        if self.index == 0:
            raise StopIteration
        self.index = self.index - 1
        return self.data[self.index]
```

Let's see them in action:

```
>>> A=Straight("123")
>>> for x in A:
    print x

1
2
3
>>> B=reverse("123")
>>> for x in B:
    print x

3
2
1
```

- __setitem__ Is used to assign a value to a key (with the form *self[key]=value*). Normally we use it to change the value of a dictionary key, but remember that we are the ones who define what the method actually does. For example, we could use it to replace a character in a string:

```
def __setitem__(self, key, value):
    if len(value)==1:
        self.seq=self.seq[:key]+value+self.seq[key+1:]
        return None
    else:
        raise ValueError
```

- __delitem__ Implements the deletion of objects of the form *self[key]*. It can be used with any object that supports the deletion of its elements.

Sequence class with some special methods attributes:

Listing 8.4: Sequence class with special methods attributes (py3.us/29)

```python
class Sequence:
    TranscriptionTable = {"A":"U","T":"A","C":"G","G":"C"}
    CompTable = {"A":"T","T":"A","C":"G","G":"C"}
    def __init__(self, seqstring):
        self.seqstring=seqstring.upper()
    def restriction(self,enz):
        EnzDict={"EcoRI":"ACTGG","EcoRV":"AGTGC"}
        if EnzDict.get("EcoRI") in self.seqstring:
            return self.seqstring.count(EnzDict[enz])
        else:
            return 0
    def __getitem__(self,index):
        return self.seqstring[index]
    def __getslice__(self,low,high):
        return self.seqstring[low:high]
    def __len__(self):
        return len(self.seqstring)
    def __str__(self):
        if len(self.seqstring)>=28:
            return self.seqstring[:25]+"..."+self.seqstring[-3:]
        else:
            return self.seqstring
    def transcription(self):
        tt = ""
        for x in self.seqstring:
            if x in 'ATCG':
                tt += self.TranscriptionTable[x]
        return tt
    def complement(self):
        tt=""
        for x in self.seqstring:
            if x in 'ATCG':
                tt += self.CompTable[x]
        return tt
```

8.5.1   Create a New Data Type Out of a Built-in Data Type

"New style" allows us to create our own classes derived from built-in data types. To illustrate this point, see how to create a variant of the *dict type*. Zdict is a dictionary-like object, it behaves like a dictionary with one difference: Instead of raising an exception when trying to retrieve a value with a nonexistent key, it returns 0 (zero).

Listing 8.5: Sequence class with special methods attributes (py3.us/30)

```
1 class Zdic(dict):
2     """ A dictionary-like object that return 0 when a user
3         request a non-existent key.
4
5     """
6
7     def __missing__(self,x):
8         return 0
```

Code explanation: In line 1 we name the class and pass a data type as argument (`dict`). This means that the resulting class (`Zdic`) inherits from the `dict` type. From line 7 to 8 there is the definition of a special method: **__missing__**. This method is triggered when the user tries to retrieve a value with a nonexistent key. It takes as argument the value of the key, but in this case the program don't use such value (`x`) since it returns 0 disregarding of the key value:

```
>>> a=Zdic()
>>> a['blue']='azul'
>>> a['red']
0
```

8.6 Making Our Code Private

At the beginning of this chapter it was highlighted that one of the characteristics of the OOP is encapsulation. Encapsulation is about programmers ignoring the internal operation of objects and only being able to see their available methods. Some of these methods that we will create will not be for "external consumption," but they will serve as support for other methods of the class, which we do want to be used from other sections of the program. Some languages allow the hiding of methods and properties. In the jargon, this is called "to make a method private." Python does not allow the hiding of a method, because it is one of its premises not to be in the way of the programmer. But it has a syntax that makes it difficult to access a method or a property from outside of a class. This is called **mangling** and its syntax consists of adding two underscores at the beginning (but not at the end) of the name of the method or attribute that we want to be private. Let's see an example of a class that defines 2 methods, `a` and `__b`:

```
class TestClass:
    def a(self):
```

```
      pass
   def __b(self):
      # mangled to _TestClass__b
      pass
```

Trying to access to __b() raises an error:

```
>>> MyObject = TestClass()
>>> MyObject.a()
>>> MyObject.__b()

Traceback (most recent call last):
  File "<pyshell#14>", line 1, in <module>
    MyObject.__b()
AttributeError: TestClass instance has no attribute '__b'
```

It is possible to access the method **a**, but not __**b**, at least not directly. The notation *object._Class_method* should be used. For example:

```
>>> MyObject._TestClass__b()
```

You may be wondering what the point is of a privacy method that is not really private. On one hand the methods that have this "semi-protection" are inherited/associated with the child classes (and the name space is not contaminated). On the other, when an object is explored using **dir**, this class of objects won't be seen. An important thing to consider is that the protection offered by this notation is a convention to know how to proceed more than a real protection.

8.7 Additional Resources

- Python Programming/OOP:
  http://en.wikibooks.org/wiki/Python_Programming/OOP

- Introduction to OOP with Python:
  http://www.voidspace.org.uk/python/articles/OOP.shtml

- Dive into Python (Mark Pilgrim). Chapter 5. "Objects and Object-Orientation":
  http://diveintopython.org/object_oriented_framework

- Python Objects (Fredrik Lundh):
  http://www.effbot.org/zone/python-objects.htm

- Java Tutorial: Lesson: Object-Oriented Programming Concepts:
  `http://java.sun.com/docs/books/tutorial/java/concepts/`

- Introduction to New-Style Classes in Python:
  `http://www.geocities.com/foetsch/python/new_style_classes.htm`

8.8   Self-Evaluation

1. Why is Python often characterized as a multi-paradigm language?

2. Name the main characteristics of Object-Oriented Programming (OOP).

3. Explain the following concepts: Inheritance, Encapsulation and Polymorphism.

4. What is the difference between class attributes and instance attributes?

5. What is a special method attribute? Name at least four.

6. What is the difference between __str__ and __repr__

7. What is a private method? Are they really private in Python?

8. Why were "new style" classes introduced?

9. Define a class that keeps track of how many instances have instantiated.

10. Define a new type based in a built-in type.

Chapter 9

Regular Expressions

9.1 Introduction to Regular Expressions (REGEX)

A common feature of every scripting language is support of regular expression (REGEX in programming jargon). What are regular expressions? They are expressions that sumarize a text pattern. A known case of regular expression is the abreviations used in most operating systems, like using "ls *.py" (or "dir *.py") to list all files ended in ".py". These are known as wildchars.

When doing text processing it is often necessary to give special treatment to strings containing a specific condition. For example, you may want to extract everything that is between <pre> and </pre> in an html file, or remove from a file any character that is not A, T, C, or G.

Biological applications of this feature are straightforward. Regular expressions can be used to locate domains in proteins, sequence patterns in DNA like CpG islands, repeats, restriction enzyme, nuclease recognition sites and so on. There are even biological databases devoted to protein domains, like PROSITE.[1]

Nevertheless, your programming needs may not include the use of regular expressions. In this case, you can skip this chapter and read it when you need it. The rest of this book can be read without knowledge of regular expressions.

Each language has its own REGEX syntax. In Python, this syntax is close to the one used in Perl. So if you know Perl, learning Python REGEX is easy. If you never heard of REGEX before, don't worry, basic REGEX syntax is not so hard to learn. Some REGEX could turn into obscure and complex expressions in specific cases. Due to this potential complexity, there are even whole books on this subject.[2]

---

[1] http://www.expasy.ch/prosite/
[2] Please see **Additional Resources** for book recommendations.

9.1.1   REGEX Syntax

In general the letters and characters match with themselves. "Python" is going to match with "Python" (but not with "python"). The exceptions to this rule are metacharacters, which are characters that have a special meaning in the context of the REGEX:

. ˆ $ * + ? { [ ] \ | ( )

Let's see the meaning of most commonly used special characters:

. **(dot)**: Matches any character, except new line: "ATT.T" will match "ATTCT", "ATTFT" but not "ATTTCT".

ˆ(carat): Matches the beginning of the chain: "ˆAUG" will match "AU-GAGC" but not "AAUGC". Using inside a group means "opposite".

$(dollar): Matches the end of the chain or just before a new line at the end of the chain: "UAA$" will match "AGCUAA" but not "ACUAAG".

* **(star)**: Matches 0 or more repetitions of the preceding token: "AT*" will match "AAT", "A", but not "TT".

+ **(plus)**: The resulting REGEX will match 1 or more repetitions of the preceding REGEX: "AT+" will match "ATT", but not "A".

? **(question mark)**: The resulting REGEX matches 0 or 1 repetitions of the preceding RE. "AT?" will match either "A" or "AT".

(...): Matches whatever regular expression is inside the parentheses, and indicates the start and end of a group. To match the literals "(" or ")", use \(or \), or enclose them inside a character class: [(] [)].

(?:...): A non-grouping version of regular parentheses. The substring matched by the group cannot be retrieved after performing a match.

{n}: Exactly n copies of the previous REGEX will match: "(ATTG){3}" will match "ATTGATTGATTG" but not "ATTGATTG".

{m,n}: The resulting REGEX will match from m to n repetitions of the preceding REGEX: "(AT){3,5}" will match "ATATT**ATATAT**" but not "ATATTATAT". Without m, it will match from 0 repetitions. Without n, it will match all repetitions.

[] (square brackets): Indicates a set of characters. "[A-Z]" will match any uppercase letter and "[a-z0-9]" will match any lowercase letter or digit. Meta characters are not active inside REGEX sets. "[AT*]" will match "A", "T" or "*". The ˆinside a set will math the complement of a set. "[ˆR]" will match any character but "R".

"\" (backslash): Used to escape reserved characters (to match characters like "?", "*"). Since Python also uses backslash as the escape character, you should pass a raw string to express the pattern.

| (vertical bar): As in logic, it reads as "or". Any number of REGEX can be separated by "|". "A|T" will match "A", "T" or "AT".

There are also special sequences with "\" and a character. They are listed in Table 9.1.

TABLE 9.1: REGEX Special Sequences

Name	Description
number	The contents of the group of the same number, starting from 1
\A	Only at the start of the string
\b	The empty string, only at the beginning or end of a word
\B	The empty string, only when it is not at the beginning or end of a word
\d	Any decimal digit (as [0-9])
\D	Any non-digit (as [^0-9])
\s	Any whitespace character (as [\t\n\r\f\v])
\S	Any non-whitespace character (as [^\t\n\r\f\v])
\w	Any alphanumeric character (as [a-zA-Z0-9_])
\W	Any non-alphanumeric character (as [^a-zA-Z0-9_]
\Z	Only the end of the string

9.2   The re Module

The **re** module provides methods like compile, search, findall, match, and other. These functions are used to process a text using a pattern built with the REGEX syntax.

A basic search works like this:

```
>>> import re
>>> mo = re.search("hello","Hello world, hello Python!")
```

The **search** from **re** method requires a pattern as a first argument and as a second argument, a string where the patter will be searched. In this case the pattern can be translated as "H or h, followed by ello". When a match is found, this function returns a **match object** (called mo in this case) with information about the first match. If there is no match, it returns **None**. A **match object** can be queried with methods shown here:

```
>>> mo.group()
'hello'
>>> mo.span()
(13, 18)
```

group() returns the string matched by the REGEX, while **span()** returns a tuple containing the (start, end) positions of the match (that is the (0, 5) returned by mo.span()).

This result is very similar to what the index method returns:

```
>>> "Hello world, hello Python!".index("hello")
13
```

The difference lies in the chance of using REGEX instead of plain strings. For example we would like to match "Hello" and "hello":

```
>>> import re
>>> mo = re.search("[Hh]ello","Hello world, hello Python!")
```

The first match now is,

```
>>> mo.group()
'Hello'
```

re.findall

To find all the matches, and not just the first one, use **findall**:

```
>>> re.findall("[Hh]ello","Hello world, hello Python,!")
['Hello', 'hello']
```

Note that **findall** returns a list with the actual matches instead of match objects.

re.finditer

If we want to have a match object for each match, there is the **finditer** method. As an additional bonus, it doesn't return a list, but an iterator. This means that each time finditer is invoked it returns the next element without having to calculate them all at once. As with any iterator, this optimizes memory usage:

```
>>> re.finditer("[Hh]ello","Hello world, hello Python,!")
<callable-iterator object at 0xb6f43d8c>
```

Walking on the results:

```
>>> mos = re.finditer("[Hh]ello","Hello world, hello Python,!")
>>> for x in mos:
    print x.group()
    print x.span()

Hello
(0, 5)
hello
(13, 18)
```

re.match

There is a **match** method that works like **search** but it looks only at the start of a string. When the pattern is not found, it returns **None**:

```
>>> mo = re.match("hello", "Hello world, hello Python!")
>>> print mo
None
```

Ah **search**, when the pattern is found, it returns a match object:

```
>>> mo = re.match("Hello", "Hello world, hello Python!")
>>> mo
<_sre.SRE_Match object at 0xb7b5eb80>
```

This match object can be queried as before:

```
>>> mo.group()
'Hello'
>>> mo.span()
(0, 5)
```

9.2.1 Compiling a Pattern

A pattern can be compiled (converted to an internal representation) to speed up the search. This step is not mandatory but recommended for large amounts of text. Let's see **findall** with a regular pattern and then with a "compiled" pattern (**rgx**):

```
>>> re.findall("[Hh]ello","Hello world, hello Python,!")
['Hello', 'hello']
>>> rgx = re.compile("[Hh]ello")
>>> rgx.findall("Hello world, hello Python,!")
['Hello', 'hello']
```

Compiled patterns have all methods available in the **re** module:

```
>>> rgx = re.compile("[Hh]ello")
>>> rgx.search("Hello world, hello Python,!")
<_sre.SRE_Match object at 0xb6f494f0>
>>> rgx.match("Hello world, hello Python,!")
<_sre.SRE_Match object at 0xb6f493d8>
>>> rgx.findall("Hello world, hello Python,!")
['Hello', 'hello']
```

Program 9.1 shows how to compile a pattern in the context of a search:

Listing 9.1: Find the first "TAT" repeat (py3.us/31)

```
1 import re
2 seq = "ATATAAGATGCGCGCGCTTATGCGCGCA"
3 rgx = re.compile("TAT")
4 i = 1
5 for mo in rgx.finditer(seq):
6     print('Ocurrence %s: %s'%(i,mo.group()))
7     print('Position: From %s to %s'%(mo.start(),mo.end()))
8     i += 1
```

Code explanation: In line 3 the pattern (TAT) is compiled. The compiled object returned in line 3 (rgx) has the methods found in the **re** module, like **finditer**. This operation returns a "match" type object (mo). From this object, in lines 6 and 7, the **group** and **span** methods are invoked. Note that mo.start() and mo.end() are equivalent to mo.span()[0] and mo.span()[1].

This is the result of running the program:

```
Ocurrence 1: TAT
Position: From 1 to 4
Ocurrence 2: TAT
Position: From 18 to 21
```

Groups

Sometimes you need to match more than one pattern; this can be done by grouping. Groups are marked by a set of parentheses ("()"). Groups can be "capturing" ("named" or "unnamed") and "non-capturing." The difference between them will be clear later.

A "capturing" group is used when you need to retrieve the contents of a group. Groups are captured with **groups**. Don't confuse **group** with **groups**. As seen on page 161, **group** returns the string matched by the REGEX.

```
>>> import re
>>> seq = "ATATAAGATGCGCGCGCTTATGCGCGCA"
>>> rgx = re.compile("(GC){3,}")
>>> result = rgx.search(seq)
>>> result.group()
'GCGCGCGC'
```

This case is just like code snipet shown in page 162. Instead, **groups** return a tuple with all the subgroups of the match. In this case, since **search** returns one match and there is one group in the pattern, the result is a tuple with one group:

```
>>> result.groups()
('GC',)
```

There is a "CG" group, like in the pattern. If you want to whole pattern returned by groups, you need to declare another group like in this example:

```
>>> rgx = re.compile("((GC){3,})")
>>> result = rgx.search(seq)
>>> result.groups()
('GCGCGCGC', 'GC')
```

Both groups present in the pattern are retrieved (counting from left to right). This is this way because by default every group is "capturing." If you don't need the internal subgroup (the "CG" group), you can label as "non-capturing." This is done by adding "?:" at the beginning of the group:

```
>>> # Only the inner group is non-capturing
>>> rgx = re.compile("((?:GC){3,})")
>>> result = rgx.search(seq)
>>> result.groups()
('GCGCGCGC',)
```

findall also behaves differently if there is a group in the pattern. Without a group it returns a list of matching strings (as seen on page 162). If there is one group in the pattern, it returns a list with the group. If there is more than one group, it return a list of tuples:

```
>>> rgx = re.compile("TAT") # No group at all.
>>> rgx.findall(seq) # This returns a list of matching strings.
['TAT', 'TAT']
>>> rgx = re.compile("(GC){3,}") # One group. Return a list
>>> rgx.findall(seq)            # with the group for each match.
['GC', 'GC']
>>> rgx = re.compile("((GC){3,})") # Two groups. Return a
>>> rgx.findall(seq)  # list with tuples for each match.
[('GCGCGCGC', 'GC'), ('GCGCGC', 'GC')]
>>> rgx = re.compile("((?:GC){3,})") # Using a non-capturing
>>> rgx.findall(seq) # group to get only the matches.
['GCGCGCGC', 'GCGCGC']
```

Groups can be labeled to refer to them later. To give a name to a group, use: **?P<*name*>**. Code 9.2 shows how to use this feature:

Listing 9.2: Find multiple sub-patterns (py3.us/32)

```
1 import re
2 rgx = re.compile("(?P<TBX>TATA..).*(?P<CGislands>(?:GC){3,})")
```

```
3 seq = "ATATAAGATGCGCGCGCTTATGCGCGCA"
4 result = rgx.search(seq)
5 print(result.group('CGislands'))
6 print(result.group('TBX'))
```

This program returns:

```
GCGCGC
TATAGA
```

9.2.2   REGEX Examples

As a REGEX example, code 9.3 shows how many lines in a given file have a
pattern entered from the command line.[3] The program is executed like this:

   program_name.py file_name pattern

Where file_name is the name of the file where **pattern** is searched.

Listing 9.3: Count lines with a user-supplied pattern on it

```
1 import re, sys
2 myregex = re.compile(sys.argv[2])
3 counter = 0
4 fh = open(sys.argv[1])
5 for line in fh:
6     if myregex.search(line):
7         counter += 1
8 fh.close()
9 print(counter)
```

Code explained: The re module is imported and the expression to search
is "compiled" (line 1 and 2). This "compilation" is optional but recommended.
It accelerates the search by compiling the REGEX into an internal structure
that is later used by the interpreter. **sys.argv** is a list of strings. Each
string is an argument taken from the command line. If the command line is
program.py word myfile.txt, the contents of sys.argv is ['program.py',
'word', 'myfile.txt']. In line 4 the program opens the file entered as the
first argument. In line 5 it parses the open file and in 5 and 6 it does the regular
expression search "Python" within each line. If the expression is found, the
counter variable (**counter**) is incremented (line 7).

This script doesn't count how many occurrences of your word are in the
file. If a word is repeated more than once in the same line, it is counted as
one. The following script counts all the occurrences of a given pattern:

---

[3]There are more efficient ways to accomplish this, like using the Unix **grep** command, but
it is shown here for a didactic purpose.

Listing 9.4: Count the occurrence of a pattern in a file (py3.us/33)

```
1 import re, sys
2 myregex = re.compile(sys.argv[2])
3 i = 0
4 fh = open(sys.argv[1])
5 for line in fh:
6     i += len(myregex.findall(line))
7 fh.close()
8 print(i)
```

Tip: Testing a REGEX with Kodos.

Kodos is a nice GUI utility (made in Python) that allows you to test and debug your regular expressions. It has a window where you enter your REGEX pattern and another window where you enter a string to test your REGEX pattern against. As a result you will have the matching group information (if applicable), the match of the REGEX pattern in relation to the text string by using colors and several variations of using the REGEX pattern in a Python application.

The program is released under the GNU Public License (GPL) and it is available at http://kodos.sourceforge.net.

9.2.3 Pattern Replace

The **re** module, can be used to replace patterns, with the **sub** function:

re.sub

sub(rpl,str[,count=0]): Replace *rpl* with the portion of the string (*str*) that coincides with the REGEX to which it applies. The third parameter, which is optional, indicates how many replacements we want made. By default the value is zero and means that it replaces all of the occurrences. It is very similar to the string method called **replace**, just that instead of replacing one text for another, the replaced text is located by a REGEX.

Listing 9.5: Delete GC repeats (more than 3 GC in a row) (py3.us/34)

```
1 import re
2 regex = re.compile("(?:GC){3,}")
3 seq="ATGATCGTACTGCGCGCTTCATGTGATGCGCGCGCGCAGACTATAAG"
4 print "Before:",seq
5 print "After:",regex.sub("",seq)
```

The product of this program is

```
Before: ATGATCGTACTGCGCGCTTCATGTGATGCGCGCGCGCAGACTATAAG
After: ATGATCGTACTTTCATGTGATAGACTATAAG
```

re.subn

subn(*rpl,str[,count=0]*): It has the same function as **sub**, differing in that instead of returning the new string, it returns a tuple with two elements: the new string and the number of replacements made. This function is used when, in addition to replacing a pattern in a string, it's required to know how many replacements have been made.

With this we have a very general vision of the possibilities that Regular Expressions open for us. The idea was to give an introduction to the subject and tools to start making our own REGEX. Next, we will see an example use of what has been learned so far.

9.3 REGEX in Bioinformatics

As I mentioned at the beginning of the chapter, the REGEX can be used to search PROSITE style patterns.[4] The patterns are sequences of characters that describe a group of sequences in a condensed form. For example, the following is the pattern for the active site of the enzyme isocitrate lyase:

```
K-[KR]-C-G-H-[LMQR]
```

This pattern is interpreted as: a K in the first position, a K or R in the second, then the sequence CGH and finally, one of the following amino acids: L ,M, Q or R. If we want to search for this pattern in this sequence, as a first measure one must convert the pattern from PROSITE to a Python REGEX. The conversion in this case is immediate:

```
"K[KR]CGH[LMQR]"
```

To change a PROSITE profile to REGEX basically consists of removing the hyphens (-), replacing the numbers between parentheses with numbers between braces and replacing the "x" with a period. Let's see an example of the adenylyl cyclase associated protein 2:

PROSITE version:

---

[4]If you are not familiar with protein patterns, please take a look at the PROSITE user manual, located at: http://www.expasy.org/prosite/prosuser.html.

```
[LIVM](2)-x-R-L-[DE]-x(4)-R-L-E
```

REGEX version:

```
"[LIVM]{2}.RL[DE].{4}RLE"
```

Let's suppose that we want to find a pattern of this type in a sequence in FASTA format. Besides finding the pattern, we may need to retrieve it in a context, that is, 10 amino acids before and after the pattern. Here is a sample FASTA file:

```
>Q5R5X8|CAP2_PONPY CAP 2 - Pongo pygmaeus (Orangutan).
MANMQGLVERLERAVSRLESLSAESHRPPGNCGEVNGVIGGVAPSVEAFDKLMDSMVAEF
LKNSRILAGDVETHAEMVHSAFQAQRAFLLMASQYQQPHENDVAALLKPISEKIQEIQTF
RERNRGSNMFNHLSAVSESIPALGWIAVSPKPGPYVKEMNDAATFYTNRVLKDYKHSDLR
HVDWVKSYLNIWSELQAYIKEHHTTGLTWSKTGPVASTVSAFSVLSSGPGLPPPPPPPPP
PGPPPLLENEGKKEESSPSRSALFAQLNQGEAITKGLRHVTDDQKTYKNPSLRAQGGQTR
SPTKSHTPSPTSPKSYPSQKHAPVLELEGKKWRVEYQEDRNDLVISETELKQVAYIFKCE
KSTLQIKGKVNSIIIDNCKKLGLVFDNVVGIVEVINSQDIQIQVMGRVPTISINKTEGCH
IYLSEDALDCEIVSAKSSEMNILIPQDGDYREFPIPEQFKTAWDGSKLITEPAEIMA
```

The program in code 9.6 reads the FASTA file.

Listing 9.6: Search a pattern in a FASTA file (py3.us/89)

```
1 import re
2 pattern = "[LIVM]{2}.RL[DE].{4}RLE"
3 fh = open('/home/sb/bioinfo/prot.fas')
4 fh.readline() # Discard the first line.
5 seq = ""
6 for line in fh:
7     seq += line.strip()
8 rgx = re.compile(pattern)
9 result = rgx.search(seq)
10 patternfound = result.group()
11 span = result.span()
12 leftpos = span[0]-10
13 if leftpos<0:
14     leftpos = 0
15 print(seq[leftpos:span[0]].lower()+patternfound+
16     seq[span[1]:span[1]+10].lower())
17 fh.close()
```

The result of this program is

```
lrsyrrdewaLLTRLDAQWERLElwmdrfatki
```

Code explanation: Up to line 7 the program reads the FASTA file and stores the protein sequence (`seq`). In line 8 the pattern defined in line 2 is compiled. The search is done at line 9. From line 10 onwards the program works on displaying the result. As requested, the resulting pattern is shown in a context of 10 amino acids on each side.

9.3.1 Cleaning Up a Sequence

It's more than common to find a file with sequences in a nonstandard format, such as the following sequence:

```
  1   ATGACCATGA TTACGCCAAG CTCTAATACG ACTCACTATA GGGAAAGCTT GCATGCCTGC

 61   AGGTCGACTC TAGAGGATCT ACTAGTCATA TGGATATCGG ATCCCCGGGT ACCGAGCTCC

121   AATTCACTGG CCGTCGTTTT
```

The following code reads a text file with the sequence in this format and returns only the sequence, without any strange (number or whitespace) character:

Listing 9.7: Cleans a DNA sequence

```
1 import re
2 regex = re.compile(' |\d|\n|\t')
3 seq = ''
4 for line in open('pMOSBlue.txt'):
5     seq += regex.sub('',line)
6 print seq
```

This program prints:

```
ATGACCATGATTACGCCAAGCTCTAATACGACTCACTATAGGGAAAGCTTGCATGCCTGC<=
AGGTCGACTCTAGAGGATCTACTAGTCATATGGATATCGGATCCCCGGGTACCGAGCTCG<=
AATTCACTGGCCGTCGTTTT
```

Code explained: Line 2 defines the characters we are going to search for removal. In this case the characters are: whitespaces, numbers, carriage return and tabs. In lines 4 and 5 the program parses all the lines of the file (`pMOSBlue.txt`) and removes the pattern each time it's found.

9.4 Additional Resources

- Jeffrey EF Friedl, "Mastering Regular Expressions," Third Edition, 2006, O'Reilly Media.

http://www.oreilly.com/catalog/regex2

- Tony Stubblebine, "Regular Expression Pocket Reference," Second Edition, 2007. O'Reilly Media.
  http://www.oreilly.com/catalog/9780596514273/

- "The Premier Web site about Regular Expressions."
  http://www.regular-expressions.info.

- "The Python Regular Expression Debugger." Local application to test your regular expressions. See **Tip** on page 167 for more information.
  http://kodos.sourceforge.net

- "Regular Expressions in Java." Test your regular expressions online.
  http://www.javaregex.com/test.html

- "Python Regular Expression Builder." Pyreb is a wxPython GUI to the re python module; it will speed up the development of Python regular expression (similar to PCRE).
  http://savannah.nongnu.org/projects/pyreb

- Harry J Mangalam. "tacg - a grep for DNA." BMC Bioinformatics 2002, 3:8.
  http://www.biomedcentral.com/1471-2105/3/8

9.5 Self-Evaluation

1. What is a REGEX?

2. What is the difference between a "capturing" and a "non-capturing" group?

3. How text patterns search can be applied to biology?

4. Line 13 of Code 9.6 (page 169) is checked if the value `leftpos` is less than 0. Why?

5. In Code 9.7, the pattern used was "|\d|\n|\t". What other alternative could have been employed?

6. Make a program that retrieves all phone number found in a file. The numbers must be in the format **nnn-nnn-nnnn**, where **n** is a number.

7. Make a program to retrieve every e-mail ending in .com present in every file in a given directory.

8. Make a program to sort if a sequence is made out of DNA or amino acids. Hint: DNA sequences can only have these characters: "ATCGN."

9. Write a REGEX pattern to detect a HindII restriction site. This enzyme recognizes the DNA sequence **GTYRAC** (where "Y" means "C" or "T" and "R" means "G" or "A").

10. What is the meaning of the following REGEX and write a string that match with it.

    `"[0-9]{1,4}/[0-9]{1,2}/[0-9]{1,2}"`

Part II

Biopython

Chapter 10

Introduction to Biopython

10.1 What Is Biopython?

Biopython[1] is a package of useful modules to develop bioinformatics applications. Although each bioinformatics analysis is unique, there are some tasks that are repeated, constants shared between programs and standard file formats. This situation suggests the need for a package to deal with biological problems.

Biopython started as an idea in August of 1999, it was an initiative by Jeff Chang and Andrew Dalke. Although they came up with the idea, collaborators soon joined the project. Among the most active developers, Brad Chapman, Peter Cock, Michiel de Hoon and Iddo Friedberg stand out. The project began to take code form in February 2000 and in July of the same year the first release was made. The original idea was to build a package equivalent to BioPerl which back then was the principal bioinformatics package. Although BioPerl may have been Biopython's inspiration, the conceptual differences between Perl and Python have given Biopython a particular way of doing things. Biopython is part of the family of open-bio projects (also known as Bio*), for which institutionally it is a member of the Open Bioinformatics Foundation.[2]

10.1.1 Project Organization

It is an open source community project. Although the Open Bioinformatics Foundation takes care of administrative, economic and legal aspects, its content is managed by the programmers and users.

Anyone can participate in the project. The code is public domain and is available in CVS form through the Web.[3] The procedure that you have to

---

[1] Available from http://www.biopython.org.

[2] http://www.open-bio.org

[3] The following address is likely to change as Biopython moves from CVS to Git or another version control system: http://cvs.biopython.org/cgi-bin/viewcvs/viewcvs.cgi/?cvsroot=biopython Please see http://biopython.org/wiki/GitMigration for more information on the migration to Git distributed version control.

follow to collaborate on Biopython is similar to other open source projects. You have to use the software and then determine if it needs any additional features or if you want to modify any of the existing features. Before writing any code my recommendation is to discuss your ideas on the development mailing list.[4] first There you will find out if that feature had already been discussed and was rejected or if it was not included because no one needed it until that time. In the case of a bug fix, you don't need to ask, just report it in the bug tracking software,[5] and if possible, add a solution proposal. If what we have is a proposal for improvement and there weren't any objections from the list, we can send the code using the bug tracking system, although it has to be marked as "enhancement" in the "Severity" drop-down menu.

Due to the open nature of the project tens of people have contributed code from diverse fields within Bioinformatics, from information theory to population genetics.

I was involved in Biopython as a user since 2002 and submited my first contribution on 2003 with `lcc.py`, a function to calculate the local compositional complexity of a sequence. In 2004 I submitted code for melting point calculation of oligonucleotides. My last submission was some functions for the CheckSum module in 2007.[6] In every case I found a supportive community, especially in the first submission when my coding skills were at a beginner level.

For more information concerning how to participate in the Biopython project, see the specific instructions at `http://biopython.org/wiki/Contributing`.

The Biopython code is developed under the "Biopython License.[7]" It is very liberal and there are virtually no restrictions to its use.[8]

10.2 Biopython Components

Biopython has various modules. Some facilitate tasks that are undertaken on a daily basis in a molecular biology laboratory while others have very specific objectives. What is "commonly used" will depend on the work environment of the reader, but after having worked giving IT support to molecular

---

[4]`http://lists.open-bio.org/mailman/listinfo/biopython-dev`

[5]`http://bugzilla.open-bio.org`

[6]Bassi, Sebastian and Gonzalez, Virginia. New checksum functions for Biopython. Available from Nature Precedings <http://dx.doi.org/10.1038/npre.2007.278.1> (2007).

[7]The license is included in the biopython package and available online at `http://www.biopython.org/DIST/LICENSE`.

[8]The only condition imposed for using Biopython are related to publishing the copyright notice and not to use the name of the contributors in advertising.

biologists at a biotech research center, reading the mailing list for Biopython for a few years and doing consulting work, I think I can identify key modules.

As with all enumerations, it is arbitrary and it is possible that it would not reflect the interests of all readers. It's sorted in didactic fashion with the intention that the first items will help you to understand the rest.

10.2.1 Alphabet

In bioinformatics we constantly deal with alphabets. DNA has a 4 letter alphabet (A,C,T,G) while proteins have their 20 amino acids, each one represented by a letter of the alphabet. There are also special "alphabets" like the ones that contemplate ambiguity positions, these are, positions where more than one nucleotide may be present. For example the letter S may represent the nucleic acids C or G, the letter H represents A, C, or T. This ambiguous alphabet in Python is called **ambiguous_dna**. Concerning the proteins, there is also an extended dictionary, which is, the dictionary that contains amino acids that are not normally found in proteins[9] (**ExtendedIUPACProtein**). Similarly, there is an extended alphabet for nucleotides (**ExtendedIUPACDNA**) that allows letters with modified bases. Going back to proteins, there is also a reduced alphabet that, taking into account common physicochemical properties, lumps together several amino acids into one letter.

There is even one alphabet that is not DNA or amino-acid based: **SecondaryStructure**. This alphabet represents domains like **H**elix, **T**urn, **S**trand and **C**oil.

Alphabets defined by IUPAC are stored in Biopython as classes of the IUPAC module. Parent module (**Bio.Alphabet**) includes more general/generic cases. Here are some attributes of the alphabets:

```
>>> import Bio.Alphabet
>>> Bio.Alphabet.ThreeLetterProtein.letters
['Ala', 'Asx', 'Cys', 'Asp', 'Glu', 'Phe', 'Gly', 'His', <=
'Ile', 'Lys', 'Leu', 'Met', 'Asn', 'Pro', 'Gln', 'Arg', <=
'Ser', 'Thr', 'Sec', 'Val', 'Trp', 'Xaa', 'Tyr', 'Glx']
>>> from Bio.Alphabet import IUPAC
>>> IUPAC.IUPACProtein.letters
'ACDEFGHIKLMNPQRSTVWY'
>>> IUPAC.unambiguous_dna.letters
'GATC'
>>> IUPAC.ambiguous_dna.letters
'GATCRYWSMKHBVDN'
>>> IUPAC.ExtendedIUPACProtein.letters
'ACDEFGHIKLMNPQRSTVWYBXZ'
```

---

[9]Selenocysteine and pyrrolysine are typical examples.

```
>>> IUPAC.ExtendedIUPACDNA.letters
'GATCBDSW'
```

Alphabets are used to define the content of a sequence. How do you know that sequence made of "CCGGGTT" is a small peptide with several cysteine, glycine and threonine or it is a DNA fragment of cytosine, guanine and thymine? If sequences were stored as strings, there would be no way to know what kind of sequence it is. This is why Biopython introduces **Seq** objects.

10.2.2 Seq

This object is composed of the sequence itself and an **alphabet** that defines the nature of the sequence.

Let's create a sequence object as a DNA fragment:

```
>>> from Bio.Seq import Seq
>>> import Bio.Alphabet
>>> seq = Seq('CCGGGTT',Bio.Alphabet.IUPAC.unambiguous_dna)
```

Since this sequence (`seq`) is defined as DNA, you can apply operations that are permitted to DNA sequences. Seq objects have the **transcribe** and **translate** methods:

```
>>> seq.transcribe()
Seq('CCGGGUU', IUPACUnambiguousRNA())
>>> seq.translate()
Seq('PG', IUPACProtein())
```

An RNA sequence can't be transcribed, but it can be translated:

```
>>> rna_seq = Seq('CCGGGUU',Bio.Alphabet.IUPAC.unambiguous_rna)
>>> rna_seq.transcribe()
Traceback (most recent call last):
  File "<stdin>", line 1, in <module>
  File "/home/sb/Seq.py", line 520, in transcribe
    raise ValueError("RNA cannot be transcribed!")
ValueError: RNA cannot be transcribed!
>>> rna_seq.translate()
Seq('PG', IUPACProtein())
```

You can go back from RNA to DNA using the **back_transcribe** method

```
>>> rna_seq.back_transcribe()
Seq('CCGGGTT', IUPACUnambiguousDNA())
```

Tip: The Transcribe Function in Biopython.

Note that the **transcribe** function may not work as expected by most biologists. This function replaces each occurrence of "T" in the sequence with a "U". In biology, a transcription means replace each DNA nucleotide with its complementary nucleotide and reverse the resulting string. **transcribe** function works this way because all biological publications show the non-template strand. Biopython is assuming that you are giving to the function the non-template strand. The Bio.Seq module also has transcribe, back transcribe and translate functions that can be used on Seq objects or strings:

```
>>> from Bio.Seq import translate, transcribe, back_transcribe
>>> dnaseq = "ATGGTATAA"
>>> translate(dnaseq)
'MV*'
>>> transcribe(dnaseq)
'AUGGUAUAA'
>>> rnaseq = transcribe(dnaseq)
>>> translate(rnaseq)
'MV*'
>>> back_transcribe(rnaseq)
'ATGGTATAA'
```

Seq Objects as a String

Seq objects behave almost like a string, so many string operations are allowed:

```
>>> seq = Seq('CCGGGTTAACGTA',Bio.Alphabet.IUPAC.unambiguous_dna)
>>> seq[:5]
Seq('CCGGG', IUPACUnambiguousDNA())
>>> len(seq)
13
>>> print seq
CCGGGTTAACGTA
```

This behaivor is constantly evolving, so expect more string-like features in the next Biopython releases.[10]

If you need a string representation of a Seq object, since Biopython 1.45 you can use the Python built-in **str()** function. There is also a **tostring()**

---

[10]Biopython Bug# 2351 deals with this feature (`http://bugzilla.open-bio.org/show_bug.cgi?id=2351`).

method that still works but it is recommended only if you want to make your code compatible with older Biopython versions.

10.2.3   MutableSeq

Seq objects are not mutable. This is intended since you may want to keep your data without changes. This way immutable seq matches Python string behavior. Attempting to modify it raises an exception:

```
>>> seq[0]='T'
Traceback (most recent call last):
  File "<stdin>", line 1, in ?
AttributeError: 'Seq' instance has no attribute '__setitem__'
```

This problem can be solved by generating a **MutableSeq** with the **to-mutable()** method:

```
>>> mut_seq = seq.tomutable()
>>> mut_seq
MutableSeq('CCGGGTT', IUPACUnambiguousDNA())
```

Introduce a change to test that it is mutable:

```
>>> mut_seq[0]='T'
>>> mut_seq
MutableSeq('TCGGGTT', IUPACUnambiguousDNA())
```

You can change the sequence as if it were a list, with **append()**, **insert()**, **pop()** and **remove()**. There are also some methods specific for changing a DNA sequence:

```
>>> mut_seq.reverse()
>>> mut_seq
MutableSeq('TTGGGCT', IUPACUnambiguousDNA())
>>> mut_seq.complement()
>>> mut_seq
MutableSeq('AACCCGA', IUPACUnambiguousDNA())
>>> mut_seq.reverse_complement()
>>> mut_seq
MutableSeq('TCGGGTT', IUPACUnambiguousDNA())
```

10.2.4   SeqRecord

The **Seq** class is important because it stores the main subject of study in bioinformatics: The sequence. Sometimes we need more information than the plain sequences, like the name, id, description, and cross references to external databases and annotations. For all this information related with the sequence, there is the **SeqRecord** class. In other words, a **SeqRecord** is a **Seq** object with associated metadata:

```
>>> SeqRecord(seq, id='001', name='My Sequence')
SeqRecord(seq=Seq('CCGGGTTAACGTA', IUPACUnambiguousDNA()), <=
id='001', name='My Sequence', description='<unknown descrip<=
tion>', dbxrefs=[])
```

SeqRecord has two main attributes:

id A string with an identifier. This attribute is optional but highly recommended.

seq A Seq object. This attribute is required.

There are some additional attributes:

name A string with the name of the sequence.

description A string with more information.

dbxrefs A list of strings, each string is a database cross reference id.

features A list of SeqFeature objects. This represents those Sequence Feature found in Genbank records. This attribute is usually populated when we retrieve a sequence from a Genbank file (using for example the **SeqIO** parser). It contains the sequence location, type, strand and other variables.

annotations A dictionary with further information about the whole sequence. This attribute can't be set when initializing a **SeqRecord** object.

Creating a **SeqRecord** object from scratch:

```
>>> from Bio.SeqRecord import SeqRecord
>>> from Bio.Seq import Seq
>>> from Bio.Alphabet import generic_protein
>>> rec = SeqRecord(Seq("mdstnvrsgmksrkkkpkttviddddddcmtcsacqs"\
            + "klvkisditkvsldyintmrgntlacaacgsslkllndfas",
            generic_protein),
            id="P20994.1", name="P20994",
            description="Protein A19",
            dbxrefs=["Pfam:PF05077", "InterPro:IPR007769",
            "DIP:2186N"])
>>> rec.annotations["note"] = "A simple note"
```

To create a SeqRecord from a Genbank file, please see page 187.

10.2.5 Align

The Align module contains code for dealing with alignments. The central object of this module is the **Alignment** class. This object stores sequence alignments. It is not meant for making alignments, it is supposed that the sequences are already aligned before trying to store it in.

Here is a simple two small peptide sequence alignment:

```
MHQAIFIYQIGYPLKSGYIQSIRSPEYDNW
||   |||||||||*||||||||||||| ||
MH--IFIYQIGYALKSGYIQSIRSPEY-NW
```

This alignment can be stored in one object by using Biopython as in code 10.1:

Listing 10.1: Using Align module (py3.us/36)

```
1 # Import all required classes
2 from Bio import Alphabet
3 from Bio.Alphabet import IUPAC
4 from Bio.Align.Generic import Alignment
5 from Bio.Seq import Seq
6 # Create and name our two sequences
7 seq1 = 'MHQAIFIYQIGYPLKSGYIQSIRSPEYDNW'
8 seq2 = 'MH--IFIYQIGYALKSGYIQSIRSPEY-NW'
9 # Initialize an alignment object
10 a = Alignment(Alphabet.Gapped(IUPAC.protein))
11 # Add the sequences to this alignment object
12 a.add_sequence("asp",seq1)
13 a.add_sequence("unk",seq2)
```

Code explanation: The **Alignment** class is instantiated in line 10. a is the name of the Alignment object. Both sequences are added in line 12 and 13.

Let's see the contents of this object:

```
>>> print a
ProteinAlphabet() alignment with 2 rows and 30 columns
Seq('MHQAIFIYQIGYPLKSGYIQSIRSPEYDNW', ProteinAlphabet()) asp
Seq('MH--IFIYQIGYALKSGYIQSIRSPEY-NW', ProteinAlphabet()) unk
```

Here are the methods associated with Alignment:

- **get_all_seqs**: Retrieves all sequences stored in the alignment. It returns a list with a SeqRecord object for each sequence. It can be used to iterate over all the alignment sequences. The following code calculates the isoelectric point of each sequence in the alignment:

```
from Bio.SeqUtils.ProtParam import ProteinAnalysis as PA
for s in a.get_all_seqs():
    print PA(str(s.seq)).isoelectric_point()
```

Since the alignment object allows iteration over the records directly, this code can be rewritten as,

```
from Bio.SeqUtils.ProtParam import ProteinAnalysis as PA
for s in a:
    print PA(str(s.seq)).isoelectric_point()
```

- **get_seq_by_num(n)**: Retrieves only the selected sequence:

```
>>> str(a.get_seq_by_num(0))
'MHQAIFIYQIGYPLKSGYIQSIRSPEYDNW'
>>> str(a.get_seq_by_num(1))
'MH--IFIYQIGYALKSGYIQSIRSPEY-NW'
```

Since Biopython 1.48 **SeqRecord** supports subindexes:

```
>>> a[0]
SeqRecord(seq=Seq('MHQAIFIYQIGYPLKSGYIQSIRSPEYDNW', Gapped(<=
IUPACProtein(), '-')), id='asp', name='<unknown name>', des<=
cription='asp', dbxrefs=[])
```

Note that the subindexes return a **SeqRecord** object while **get_seq_by_num** returns a **Seq** object. To get the Seq object using subindexes, get it from the resulting **SeqRecord**:

```
>>> a[0].seq
Seq('MHQAIFIYQIGYPLKSGYIQSIRSPEYDNW', Gapped(IUPACProtein(), '-'))
>>> str(a[0].seq)
'MHQAIFIYQIGYPLKSGYIQSIRSPEYDNW'
```

- **get_alignment_length()**: Get the length of the alignment:

```
>>> a.get_alignment_length()
30
```

- **get_column(n)**: Returns a string with all the letters in the n column:

```
>>> a.get_column(0)
'MM'
>>> a.get_column(2)
'Q-'
```

This may change in the future since there are plans to also offer array like access to the data.

The usefulness of the Align objects will be clear after reading the Clustalw section in this chapter.

AlignInfo

The AlignInfo module is used to extract information from alignment objects. It provides the **print_info_content** function, the **SummaryInfo** and **PSSM** class:

- **print_info_content()**:

  Let's see them in action:

  ```
  >>> from Bio.Align import AlignInfo
  >>> from Bio.Align.AlignInfo import SummaryInfo
  >>> summary = SummaryInfo(a)
  >>> print(summary.information_content())
  120.674950704
  >>> summary.dumb_consensus()
  Seq('MHQAIFFIYQIGYPLKSGYIQSIRSPEYDNW', ProteinAlphabet())
  >>> summary.gap_consensus()
  Seq('MHQAI-FIYQIGYPLKSGYIQSIRSPEYDNW', ProteinAlphabet())
  >>> summary.get_column(2)
  'Q-'
  >>> summary.get_column(1)
  'HH'
  >>> summary.dumb_consensus()
  Seq('MHQAIFIYQIGYXLKSGYIQSIRSPEYDNW', ProteinAlphabet())
  >>> print summary.alignment
  ProteinAlphabet() alignment with 2 rows and 30 columns
  MHQAIFIYQIGYPLKSGYIQSIRSPEYDNW asp
  MH--IFIYQIGYALKSGYIQSIRSPEY-NW unk
  ```

10.2.6 ClustalW

This module has clasess and functions to interact with **ClustalW**.[11] You may know **ClustalX**, a popular graphical multiple alignment program authored by Julie Thompson and Francois Jeanmougin. ClustalX is a graphical front-end for ClustalW, a command line multiple alignment program.

The main object in **ClustalW** is **MultipleAlignCL**. It is used to build the ClustalW command line:

---

[11]This program is available from `ftp://ftp-igbmc.u-strasbg.fr/pub/ClustalW`.

```
>>> from Bio.Clustalw import MultipleAlignCL
>>> cl = MultipleAlignCL('inputfile.fasta')
>>> cl.set_output('cltest.txt')
>>> print("Command line: %s"%cl)
Command line:  clustalw inputfile.fasta -OUTFILE=cltest.txt
```

If the clustalw program is not in your system path, you have to specify its location when initializing the object. For example, if clustalw is in c:\windows\program file\clustal\clustalw.exe, **MultipleAlignCL** is initialized as:

```
>>> clpath='c:\\windows\\program file\\clustal\\clustalw.exe'
>>> cl = MultipleAlignCL('inputfile.fasta',command=clpath)
```

To run the program use the **do_alignment(*command_line*)** function:

```
>>> from Bio.Clustalw import do_alignment
>>> align = do_alignment(cl)
```

The function returns an Alignment object (the same object already seen on page 182):

```
>>> for seq in align.get_all_seqs():
      print seq.description
      print seq.seq

40|1tetH|gi|13278069
QVQLQQSDAELVKPGASVKISCKVSGYTFTDHT----IHWVKQRPE
139|1tetH|gi|19343851
QVQLLQSGPELVKPGASVKISCRASGYAFSKSW----MNWVKRRPG
84|8fabA|gi|18044241
SVLTQPP-SVSGAPGQRVTISCTGSSSNIG---AGYDVHWYQQLPG
```

Passing Parameters to ClustalW

Some parameters can be set as attributes:

```
>>> from Bio.Clustalw import MultipleAlignCL
>>> cl = MultipleAlignCL('inputfile.fasta')
>>> cl.gap_open_pen=5
>>> cl.gap_ext_pen=3
>>> cl.new_tree='outtree.txt'
>>> print(cl)
clustalw inputfile.fasta -NEWTREE=outtree.txt -align -GAPOPEN=5<=
  -GAPEXT=3
```

TABLE 10.1:   Parameters for ClustalW

Attribute Name	Description
gap_open_pen	Gap opening penalty
gap_ext_pen	Gap extension penalty
is_no_end_pen	A flag as to whether or not there should be a gap separation penalty for the ends
gap_sep_range	The "pairs" and "simple" alignment format from the EMBOSS tools
is_no_pgap	A flag to turn off residue specific gaps
is_no_hgap	A flag to turn off hydrophilic gaps
h_gap_residues	A list of residues to count a hydrophilic
max_div	A percent identity to use for delay
trans_weight	The weight to use for transitions

MultipleAlignCL parameters are in Table 10.1.

Some parameters are set via functions:

- output parameters: set_output

    - output_file

    - output_type

    - output_order

    - change_case

    - add_seqnos

- a guide tree to use: set_guide_tree

    - guide_tree

    - new_tree

- matrices: set_protein_matrix and set_dna_matrix

    - protein_matrix

    - dna_matrix

- type of residues: set_type

    - type

Code 10.2 shows how to set up a ClustalW command line:

Listing 10.2: Set up a ClustalW command line (py3.us/37)

```
 1 from Bio.Clustalw import MultipleAlignCL
 2 from Bio.Clustalw import do_alignment
 3
 4 cl = MultipleAlignCL('myseqs.fasta')
 5 cl.gap_open_pen = 5
 6 cl.gap_ext_pen = 3
 7 cl.type='protein'
 8 cl.set_output('outfile.aln', output_type='PHYLIP',
 9                 output_order='ALIGNED')
10 cl.set_protein_matrix('PAM')
11 # cl.set_guide_tree('tree1.txt')
12 cl.set_new_guide_tree('newtree.txt')
13 print cl
```

10.2.7 SeqIO

Bio.SeqIO is a common interface to input and output sequence file formats. Sequences retrieved with this interface are passed to your program as SeqRecord objects. Bio.SeqIO can also read alignment file formats, and it will return each record as a SeqRecord object. To retrieve an alignment as an Alignment object, use the Bio.AlignIO module.

Reading Sequence Files

The method used for reading sequences is **parse(file_handle, format)**. Where **format** can be "fasta", "genbank" or any other present in table 11.1. This parser returns a generator. The elements returned by this generator are of the **SeqRecord** type:

```
>>> from Bio import SeqIO
>>> f_in = open('/home/sb/bioinfo/a19.gbk')
>>> SeqIO.parse(f_in,'genbank').next()
SeqRecord(seq=Seq('MDSTNVRSGMKSRKKKPKTTVIDDDDDCMTCSACQSKLVKISDIT<=
KVSLDYINT...FAS', IUPACProtein()), id='P20994.1', name='P20994',<=
description='Protein A19.', dbxrefs=[])
```

Where there is only one sequence in the file, use **SeqIO.read()** instead of **SeqIO.parse()**:

```
>>> f_in = open('/home/sb/bioinfo/a19.gbk')
>>> SeqIO.read(f_in,'genbank')
SeqRecord(seq=Seq('MDSTNVRSGMKSRKKKPKTTVIDDDDDCMTCSACQSKLVKISDIT<=
KVSLDYINT...FAS', IUPACProtein()), id='P20994.1', name='P20994',<=
description='Protein A19.', dbxrefs=[])
```

In Listing 10.3 there is a script that reads a file full of sequences in FASTA format and displays the title and the length of each entry.

Listing 10.3: Read a FASTA file

```
1 from Bio import SeqIO
2 fh = open("myprots.fas")
3 for record in SeqIO.parse(fh, "fasta"):
4     id = record.id
5     seq = record.seq
6     print("Name: %s, size: %s"%(id,len(seq)))
7 fh.close()
```

Content of the input file:

Listing 10.4: A file with protein sequences in FASTA format

```
>Protein-X [Simian immunodeficiency virus]
NYLNNLTVDPDHNKCDNTTGRKGNAPGPCVQRTYVACH
>Protein-Y [Homo sapiens]
MEEPQSDPSVEPPLSQETFSDLWKLLPENNVLSPLPSQAMDDLMLSPDDIEQWFTEDPGPDA
>Protein-Z [Rattus norvegicus]
MKAAVLAVALVFLTGCQAWEFWQQDEPQSQWDRVKDFATVYVDAVKDSGRDYVSQFESST
```

Code 10.3 parses the file 10.4 and generates the following output:

```
Name: Protein-X, size: 38
Name: Protein-Y, size: 62
Name: Protein-Z, size: 60
```

Writing Sequence Files

SeqIO has a method for writing sequences: **write(iterable, file_handle, format)**. The first parameter that this function takes is an iterable object with SeqRecord objects (e.g. a list of SeqRecord objects). The second parameter is the file handle that will be used to write the sequences. The *format* argument works as in **parse**.

Code 10.5 shows how to read a file with a sequence as a plain text and write it as a FASTA sequence:

Listing 10.5: Read a file and write it as a FASTA sequence (py3.us/38)

```
1 from Bio import SeqIO
2 from Bio.Seq import Seq
3 from Bio.SeqRecord import SeqRecord
4 fh = open('NC2033.txt')
```

TABLE 10.2: Sequence and Alignment Formats

Format name	Description	Alignment - Sequence
ace	Reads the contig sequences from an ACE assembly file.	S
clustal	Ouput from Clustal W or X	A
embl	The EMBL flat file format.	S
emboss	The "pairs" and "simple" alignment format from the EMBOSS tools.	A
fasta	A simple format where each record starts with an identifer line starting with a ">" character, followed by lines of sequence.	A/S
fasta-m10	Alignments output by Bill Pearson's FASTA tools when used with the -m 10 command line option.	A
genbank	The GenBank or GenPept flat file format.	S
ig	IntelliGenetics file format, also used by MASE.	A/S
nexus	Used by MrBayes and PAUP. See also the module Bio.Nexus which can read any phylogenetic trees in these files.	A
phd	Output from PHRED.	S
phylip	Used by the PHYLIP tools.	A
stockholm	Used by PFAM.	A
swiss	Swiss-Prot (UniProt) format.	S
tab	Simple two column tab separated sequence files.	S

```
 5 f_out = open('NC2033.fasta','w')
 6 rawseq = fh.read().replace('\n','')
 7 #record = [SeqRecord(Seq(rawseq),'NC2033.txt','','')]
 8 record = (SeqRecord(Seq(rawseq),'NC2033.txt','',''),)
 9 SeqIO.write(record, f_out,'fasta')
10 f_out.close()
11 fh.close()
```

Knowing how to read and write most biological file formats allows one to read a file with sequences in one format and write them into another format:

```
from Bio import SeqIO
fo_handle = open('myseqs.fasta','w')
readseq = SeqIO.parse(open('myseqs.gbk'), "genbank")
SeqIO.write(readseq, fo_handle, "fasta")
fo_handle.close()
```

There are more examples of SeqIO usage in Chapter 15.

10.2.8 AlignIO

AlignIO is the Input/Output interface for alignments. It works mostly as SeqIO, but instead of returning a SeqRecord object, it returns an Alignment object. It has three main methods: **read, parse** and **write**. The first two methods are used for input and the last one for output.

- **read(handle,format[,sec_count])**: Take the file handle and the alignment format as arguments and return an Alignment object.

```
>>> from Bio import AlignIO
>>> fn = open("secu3.aln")
>>> align = AlignIO.read(fn, "clustal")
>>> print align
SingleLetterAlphabet() alignment with 3 rows and 1098 columns
----------------------------------------...--- secu3
----------------------------------------...--- AT1G14990.1-CDS
GCTTTGCTATGCTATATGTTTATTACATTGTGCCTCTG...CAC AT1G14990.1-SEQ
```

The *sec_count* argument can be used with any file format although it is used mostly with FASTA alignments. It indicates the number of sequences per alignment, useful to sort out if a file is only one alignment with 15 sequences or three alignments of 5 sequences.

- **parse(handle,format)**: This method is used for parsing alignments from a file with more than one alignment. Taking the same arguments as **read**, it returns an iterator with all the alignments present in this file. It is meant to be used in a loop:

```
>>> from Bio import AlignIO
>>> fn = open("example.aln")
>>> for alignment in AlignIO.parse(fn, "clustal") :
    alignment.get_alignment_length()

1098
233
```

- **write(iterable,handle,format)**: Take a set of Alignment objects, a file handle and a file format, to write them into a file. You are expected to call this function with all alignments in `iterable` and close the file handle. The following code reads an alignment in Clustal format and writes it in Phylip format.

Listing 10.6: Alignments

```
fi = open('/home/sb/bioinfo/example.aln')
fo = open('/home/sb/bioinfo/example.phy','w')
align = AlignIO.read(fi,"clustal")
AlignIO.write([alig],fo,"phylip")
fo.close()
```

10.2.9 BLAST

Basic Local Alignment Search Tool (BLAST) is a sequence similarity search program used to compare a user's query to a database of sequences. Given a DNA or amino acid sequence, the BLAST heuristic algorithm finds short matches between the two sequences and attempts to start alignments from these "hot spots." BLAST also provides statistical information about an alignment such as the "expect" value.[12]

BLAST is one of the most widely used bioinformatics research tools, since it has several applications. Here is a list of typical BLAST applications:

- Following the discovery of a previously unknown gene in one species, search other genomes to see if other species carry a similar gene.

- Finding functional and evolutionary relationships between sequences.

- Search for consensus regulatory patterns such as promoter signals, splicing sites and transcription factor binding sites.

- Infer protein structure based on previously crystallized proteins.

- Help identify members of gene families.

If you work in bioinformatics, chances are that you will need to run some BLAST queries or face the need to process BLAST queries generated by you or by another person. Biopython provides tools for both tasks:

BLAST Running and Processing with Biopython

BLAST can be run online on the NCBI webserver or locally on your own computer. Running BLAST over the Internet is a good option for small jobs involving few sequences. Larger jobs tend to get aborted by the remote server with the message "CPU usage limit was exceeded." Since NCBI BLAST is a public service, they have to put quotas on CPU usage to avoid overloading their servers. Another compelling reason to use a local version of BLAST is

---

[12]The expect value (E) is a parameter that describes the number of hits one can "expect" to see by chance when searching a database of a particular size.

when you need to query a custom database. There is some flexibility regarding the database(s) you could use in the NCBI BLAST server, but it can't accomodate custom data.[13]

For all these reasons, it is not uncommon for most research laboratories to run in-house BLAST searches.

Starting a BLAST Job

Biopython has a wrapper to the BLAST executable, so you can run the BLAST program from inside your script. This wrapper is a function called *blastall*, inside the **Bio.Blast.NCBIStandalone** module.

Here is the *blastall* syntax:

```
blastall(blast executable, program name, database, input file, <=
[align_view=7], [parameters])
```

This function returns a tuple with two file objects. The first one is the actual result while the second one is the BLAST error message (if any). Most parameters are self explanatory. Code in listing 10.7 will make it clear:

Listing 10.7: Running a local NCBI BLAST

```
1 from Bio.Blast import NCBIStandalone as BLAST
2 b_exe = '/home/sb/blast-2.2.20/bin/blastall'
3 f_in = 'seq3.txt'
4 b_db = '/home/sb/blast-2.2.20/data/TAIR8cds'
5 rh, eh = BLAST.blastall(b_exe, "blastn", b_db, f_in)
```

The BLAST program is run in line 5. To retrieve the result, you have to read the returned file like object **rh** as already seen in chapter 5:

```
>>> rh.readline()
<?xml version="1.0"?>
>>> rh.readline()
'<!DOCTYPE BlastOutput PUBLIC "-//NCBI//NCBI BlastOutput/EN"<=
 "http://www.ncbi.nlm.nih.gov/dtd/NCBI_BlastOutput.dtd">\n'
```

The output is in XML format. This information can be parsed using the tools learned in chapter 12 or with the tools provided by Biopython (more on this in the next section). There is also a way to avoid dealing with the XML output by forcing blastall to use plain text as output. This is done by using −m 1 as an optional parameter in the command line or align_view=1 in the blastall Biopython function. This will result in an easier to read (by a human) but hard to parse (by a computer) output. If the last sentence seems strange,

---

[13]You can't query your private database in the public NCBI server.

bear with me for a few paragraphs to understand why a "human readable" format may not be suitable for automated processing.

The `eh` filehandle stores the error message returned by `blastall`. In this case it is empty (since there was no error):

```
>>> eh.readline()
''
```

Function call present in line 5 is the equivalent of entering the following statement in the command line:

```
$ ./blastall -p blastn -i seq3.txt -d TAIR8cds -m 7
```

Most parameters in this command line can be matched up with a parameter in the Biopython blastall function, except **-m 7**, the last parameter. This argument is used to force the output to the XML format. Biopython blastall function defaults its output to XML since this is the more reliable format for parsing. Other BLAST output formats like plain text and HTML tend to change from version to version, making keeping an up to date parser very difficult.[14] This is why an easy to read output ends up being harder to parse.

There are many aspects of a blast query that can be controlled via optional parameters that are appended at the end of the function call. In table 10.3 there is a list of all accepted parameters.

Once you have the BLAST result as a file object, you may need to process it. If you plan to store the result for later processing, you need to save it:

```
>>> fh = open('testblast.xml','w')
>>> fh.write(rh.read())
>>> fh.close()
```

Most of the time you will need to extract some information from the BLAST output. For this purpose the NCBIXML parser, featured in the next subsection, comes in handy.

Reading the BLAST Output

Parsing the contents of a BLAST file is something any bioinformatician has to deal with. Biopython provides a useful parser in the `Bio.Blast.NCBIXML` module (called **parse**). With this parser the programmer can extract any significant bit from a BLAST output file. This parser takes as input a file object with the BLAST result and returns an iterator for each record inside

---

[14]There is an official statement from NCBI about this: "NCBI does not advocate the use of the plain text or HTML of BLAST output as a means of accurately parsing the data." For more information on this please see this letter: `http://www.bioperl.org/w/index.php?title=NCBI_Blast_email&oldid=5114`.

TABLE 10.3: Parameters for blastall

Variable	Class Effect
	Scoring
matrix	Matrix to use.
gap_open	Gap open penalty.
gap_extend	Gap extension penalty.
window_size	Multiple hits window size.
npasses	Number of passes.
passes	Hits/passes. Integer 0-2.
	Algorithm
gapped	Whether to do a gapped alignment. T/F
expectation	Expectation value cutoff.
wordsize	Word size.
keep_hits	Number of best hits from a region to keep.
xdrop	Dropoff value (bits) for gapped alignments.
hit_extend	Threshold for extending hits.
region_length	Length of region used to judge hits.
db_length	Effective database length.
search_length	Effective length of search space.
nbits_gapping	Number of bits to trigger gapping.
pseudocounts	Pseudocounts constants for multiple passes.
xdrop_final	X dropoff for final gapped alignment.
xdrop_extension	Dropoff for blast extensions.
model_threshold	E-value threshold to include in multipass model.
required_start	Start of required region in query.
required_end	End of required region in query.
	Processing
program	The blast program to use. (PHI-BLAST)
filter	Filter query sequence for low complexity (with SEG) T/F
believe_query	Believe the query defline T/F
nprocessors	Number of processors to use.
	Formatting
html	Produce HTML output T/F
descriptions	Number of one-line descriptions.
alignments	Number of alignments.
align_view	Alignment view. Int or str 0-11
show_gi	Show GI's in deflines T/F
seqalign_file	seqalign file to output.
align_outfile	Output file for alignment.
checkpoint_outfile	Output file for PSI-BLAST checkpointing.
restart_infile	Input file for PSI-BLAST restart.
hit_infile	Hit file for PHI-BLAST.
matrix_outfile	Output file for PSI-BLAST matrix in ASCII.

the file. In this context, `record` represents an object with all the information of each BLAST result (assuming that the BLAST file has inside the result

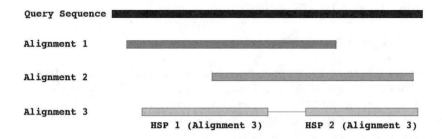

FIGURE 10.1: Anatomy of a BLAST result. This query sequence has three alignments. Each alignment has at least one HSP. Note that an alignment (or hit) can have more than one HSP like the "Alignment 3".

of several BLAST queries[15]). Since it returns an iterator, you can retrieve BLAST records one by one using a *for loop*:

```
from Bio.Blast import NCBIXML
for blast_record in NCBIXML.parse(rh):
    # Do something with blast_record
```

What's in a BLAST Record Object?

Every bit of information present in a BLAST file can be retrieved from the *blast record object*. Here is the big picture: A BLAST record contains the information of the BLAST run. This information is divided in two groups. First there are fixed features such as the characteristics of the program, query sequence and database (like program name, program version, query name, database length, name). The other group of data is related with the alignments (or hits). Each hit is the alignment between the query sequence and the target found. In turn, each alignment may have more than one HSP (High-scoring Segment Pairs). An HSP is a segment of an alignment. Figure 10.1 should make these concepts more accessible.

The BLAST record object mirrors this structure. It has an *alignments* property which is a list of (BLAST) alignment objects. Each alignment object has the information of the hit (hit_id, hit_definition, title) and a list (**hsps**) with the information of each HSP. The data associated with each HSP is usually the most requested information from a BLAST record (like bit score, E value, position). Let's see a plain text BLAST output in listing 10.8:

---

[15]This is a bug in BLAST versions prior to 2.2.14 with the way it formats the XML results for multiple queries, so you must use newer NCBI BLAST versions.

Listing 10.8: A BLAST output

```
BLASTN 2.2.18 [Mar-02-2008]

Reference: Altschul, Stephen F., Thomas L. Madden, Alejandro
A. Schaffer, Jinghui Zhang, Zheng Zhang, Webb Miller, and
David J. Lipman (1997), "Gapped BLAST and PSI-BLAST: a new
generation of protein database search programs",
Nucleic Acids Res. 25:3389-3402.


Query= sample-5
        (73 letters)

Database: NCBI genome chromosomes - other
          6369 sequences; 2,950,119,688 total letters

Searching............................................done

                                                Score    E
Sequences producing significant alignments:    (bits) Value

ref|NC_008258.1| Shigella flexneri 5 str. 8401    76   2e-12
ref|AC_000091.1| Escherichia coli W3110 DNA       76   2e-12
ref|NC_008563.1| Escherichia coli APEC 01         60   1e-07

>ref|NC_008258.1| Shigella flexneri 5 str. 8401, complete genome
 gb|CP000266.1| Shigella flexneri 5 str. 8401, complete genome
          Length = 4574284

 Score = 75.8 bits (38), Expect = 2e-12
 Identities = 41/42 (97%)
 Strand = Plus / Plus

Query: 1     taataagcggggttaccggttgggatagcgagaagagccagt 42
             |||||||||||||||||||||||||| ||||||||||||||||
Sbjct: 67778 taataagcggggttaccggttgggttagcgagaagagccagt 67819

>ref|AC_000091.1| Escherichia coli W3110 DNA, complete genome
 dbj|AP009048.1| Escherichia coli W3110 DNA, complete genome
          Length = 4646332

 Score = 75.8 bits (38), Expect = 2e-12
 Identities = 41/42 (97%)
 Strand = Plus / Plus
```

```
Query: 1     taataagcggggttaccggttgggatagcgagaagagccagt 42
             |||||||||||||||||||||||||| ||||||||||||||||
Sbjct: 70268 taataagcggggttaccggttgggttagcgagaagagccagt 70309
```

```
 Score = 56.0 bits (28), Expect = 2e-06
 Identities = 31/32 (96%)
 Strand = Plus / Plus
```

```
Query: 42    tgcttgatcggcgactggatttctttctggct 73
             ||||||||| ||||||||||||||||||||||||
Sbjct: 71549 tgcttgatgggcgactggatttctttctggct 71580
```

```
>ref|NC_008563.1| Escherichia coli APEC O1, complete genome
 gb|CP000468.1| Escherichia coli APEC O1, complete genome
          Length = 5082025
```

```
 Score = 60.0 bits (30), Expect = 1e-07
 Identities = 39/42 (92%)
 Strand = Plus / Plus
```

```
Query: 1     taataagcggggttaccggttgggatagcgagaagagccagt 42
             |||||||||||||||||||||||||   ||| ||||||||||||
Sbjct: 72088 taataagcggggttaccggttggattagtgagaagagccagt 72129
```

```
   Database: NCBI genome chromosomes - other
     Posted date: Aug 29, 2007  6:02 PM
   Number of letters in database: 2,950,119,688
   Number of sequences in database:  6369
```

```
Lambda     K       H
    1.37    0.711    1.31
Gapped
Lambda     K       H
    1.37    0.711    1.31
```

```
Matrix: blastn matrix:1 -3
Gap Penalties: Existence: 5, Extension: 2
Number of Sequences: 6369
Number of Hits to DB: 1,810,978
Number of extensions: 84567
Number of successful extensions: 129
Number of sequences better than 10.0: 40
Number of HSP's gapped: 129
Number of HSP's successfully gapped: 42
Length of query: 73
```

```
Length of database: 2,950,119,688
Length adjustment: 19
Effective length of query: 54
Effective length of database: 2,949,998,677
Effective search space: 159299928558
Effective search space used: 159299928558
X1: 11 (21.8 bits)
X2: 15 (29.7 bits)
X3: 50 (99.1 bits)
S1: 16 (32.2 bits)
S2: 17 (34.2 bits)
```

Listing 10.8 is the product of a *blastn* of a DNA query sequence against the NCBI genome chromosomes database, using default program settings.[16] Note that there are three alignments in this result. The first and last alignments have only one HSP, while the second one, has two HSP's.

See in code 10.9 how to retrieve the name of all the target sequence names:

Listing 10.9: Extract alignments title from a BLAST output (py3.us/39)

```
1 from Bio.Blast import NCBIXML
2 xmlfh = open('/home/sb/bioinfo/other.xml') # BLAST output file.
3 for record in NCBIXML.parse(xmlfh):
4     for align in record.alignments:
5         print align.title
```

Code 10.9 produces an output like this:

```
gi|110804074|ref|NC_008258.1| Shigella flexneri 5 str. 8401
gi|89106884|ref|AC_000091.1| Escherichia coli W3110 DNA
gi|117622295|ref|NC_008563.1| Escherichia coli APEC O1
```

You can get more information from each alignment like the length of the target sequence, and other related information:

```
>>> alig.length
3630528
>>> alig.hit_id
u'gi|23097455|ref|NC_004193.1|'
>>> alig.accession
u'NC_004193'
>>> alig.hit_def
```

---

[16]This listing is a reduced version of the actual output, some results were intentionally left out to avoid showing redundant data and facilitate the reader focusing on how the parser works.

```
u'Oceanobacillus iheyensis HTE831, complete genome'
>>> alig.hsps
[<Bio.Blast.Record.HSP instance at 0xb65eb8cc>]
```

hsps contain a list of *HSP*. Each *HSP* instance, as already mentioned, has the information most users want to extract from a BLAST output. Look at an HSP:

```
>ref|NC_008258.1| Shigella flexneri 5 str. 8401, complete genome
 gb|CP000266.1| Shigella flexneri 5 str. 8401, complete genome
            Length = 4574284

 Score = 75.8 bits (38), Expect = 2e-12
 Identities = 41/42 (97%)
 Strand = Plus / Plus

Query: 1      taataagcggggttaccggttgggatagcgagaagagccagt 42
              |||||||||||||||||||||||||| ||||||||||||||||
Sbjct: 67778 taataagcggggttaccggttgggttagcgagaagagccagt 67819
```

This is how this information can be retrieved with the BLAST parser:

```
>>> blast_record = NCBIXML.parse(open(xmlfile)).next()
>>> align = blast_record.alignments[0]
>>> hsp = align.hsps[0]
>>> hsp.bits
75.822299999999998
>>> hsp.score
38.0
>>> hsp.expect
2.3846099999999999e-12
>>> hsp.identities
41
>>> hsp.align_length
42
>>> hsp.frame
(1, 1)
>>> hsp.query_start
1
>>> hsp.query_end
42
>>> hsp.sbjct_start
67778
>>> hsp.sbjct_end
67819
>>> hsp.query
```

```
u'TAATAAGCGGGGTTACCGGTTGGGATAGCGAGAAGAGCCAGT'
>>> hsp.match
u'|||||||||||||||||||||||||| |||||||||||||||||'
>>> hsp.sbjct
u'TAATAAGCGGGGTTACCGGTTGGGTTAGCGAGAAGAGCCAGT'
```

Having this in mind, we can answer questions like: What are accession numbers of the alignments with **E value** lesser than a threshold value? (listing 10.10) and other questions involving any parameter in the BLAST output. This parser can be used in more complex programs like 20.2 (page 349).

Listing 10.10: Extract accession numbers of sequences that have an E value less than a specific threshold (py3.us/40)

```
1 from Bio.Blast import NCBIXML
2 threshold = 0.0001
3 xmlfh = open('/home/sb/bioinfo/other.xml')
4 blast_record = NCBIXML.parse(open(xmlfh)).next()
5 for align in blast_record.alignments:
6     if align.hsps[0].expect < threshold:
7         print align.accession
```

Code explained: This program is very similar to code 10.9. It retrieves the first BLAST record in the xml file (note the **next()** method in line 4). This method is used because older Biopython version lacks the NCBIXML.read() method. If you are using Biopython 1.50, and there is only one BLAST record in the xml file, use NCBIXML.read(open(xmlfh)). The program walks through all *alignments* in blast_record (from line 5). For each alignment (align in line 5), it checks the expect value of the first *HSP* (line 6), if the E value is less than the threshold defined in line 2, the program prints the accession number of the alignment.

Note that when doing a BLAST search you can set an E value either from command line or from Biopython blastall wrapper. Once the output is generated you can apply a filter like in code 10.10.

Detailing how to extract every possible data from a BLAST record can result in a dull reading; if you need to extract a bit of data that is not present in the examples above, I recommend reading listing 20.2 on page 349 that converts an XML BLAST output into HTML.

Tip: BLAST Running and Processing without Biopython.

Although Biopython can be used to run and parse BLAST searches, we can get by without Biopython if necessary.

BLAST can be executed as any external program with **os.system, os.-popen3** or better yet, with **subprocess.Popen**. Remember to set up the "m" option according to how you plan to process the output.

There are two ways to process the BLAST output. If the BLAST was set produce the output in XML (with command line the option "-m 7"), the result can be parsed with the tools shown in chapter 12. Another easier way to parse BLAST results is to use the CSV module (seen on page 92). To do this, the BLAST output should be formated in a compatible way (with the command line option "-m 8").

10.2.10 Data

Biopython is not just a collection of tools. It has some biological related data. This data is included in Biopython for internal usage, like translation tables (**CodonTable.unambiguous_dna_by_name**) for the **Translate** function, amino acid weights (**protein_weights**) for **molecular_weight** function.

Your code can also accesses these data. For example the code in this interactive session access to the dictionary that converts an "ambiguous dna value" to its possible values:

```
>>> from Bio.Data import IUPACData
>>> IUPACData.ambiguous_dna_values['M']
'AC'
>>> IUPACData.ambiguous_dna_values['H']
'ACT'
>>> IUPACData.ambiguous_dna_values['X']
'GATC'
```

Remember the protein weight calculator from code 4.8 on page 77? With Biopython there is no need to define a dictionary with amino acid weights since such a dictionary is already included:

Listing 10.11: Protein weight calculator with Biopython

```
1 from Bio.Data.IUPACData import protein_weights as protweight
2 protseq = raw_input("Enter your protein sequence: ")
3 totalW = 0
4 for aa in protseq:
5     totalW += protweight.get(aa.upper(),0)
6 totalW -= 18*(len(protseq)-1)
7 print("The net weight is: %s" % totalW)
```

The resulting program is shorter than the original version and there is no need to define a dictionary with values taken from a reference table, let Biopython programmers handle this for you.

Most data available from **Bio.Data.IUPACData** and **Bio.Data.Codon-Table** is presented in listing 10.12 and 10.13, respectively.

Listing 10.12: Data from Bio.Data.IUPACData

```
protein_letters
extended_protein_letters
ambiguous_dna_letters
unambiguous_dna_letters
ambiguous_rna_letters
unambiguous_rna_letters
ambiguous_dna_complement
ambiguous_dna_values
ambiguous_dna_weight_ranges
ambiguous_rna_complement
ambiguous_rna_values
ambiguous_rna_weight_ranges
avg_ambiguous_dna_weights
avg_ambiguous_rna_weights
avg_extended_protein_weights
extended_protein_values
extended_protein_weight_ranges
protein_weight_ranges
protein_weights
unambiguous_dna_weight_ranges
unambiguous_dna_weights
unambiguous_rna_weight_ranges
unambiguous_rna_weights
```

Listing 10.13: Data from Bio.Data.CodonTable

```
ambiguous_dna_by_id
ambiguous_dna_by_name
ambiguous_generic_by_id
ambiguous_generic_by_name
ambiguous_rna_by_id
ambiguous_rna_by_name
generic_by_id
generic_by_name
standard_dna_table
standard_rna_table
unambiguous_dna_by_id
unambiguous_dna_by_name
unambiguous_rna_by_id
unambiguous_rna_by_name
```

To get the bacterial DNA translation table,

```
>>> from Bio.Data.CodonTable import unambiguous_dna_by_id
>>> bact_trans=unambiguous_dna_by_id[11]
>>> bact_trans.forward_table['GTC']
'V'
>>> bact_trans.back_table['R']
'CGT'
```

To have a graphical representation of a translation table:

```
>>> from Bio.Data import CodonTable
>>> print CodonTable.generic_by_id[2]
Table 2 Vertebrate Mitochondrial, SGC1

  | U         | C         | A         | G         |
--+-----------+-----------+-----------+-----------+--
U | UUU F     | UCU S     | UAU Y     | UGU C     | U
U | UUC F     | UCC S     | UAC Y     | UGC C     | C
U | UUA L     | UCA S     | UAA Stop| UGA W     | A
U | UUG L     | UCG S     | UAG Stop| UGG W     | G
--+-----------+-----------+-----------+-----------+--
C | CUU L     | CCU P     | CAU H     | CGU R     | U
C | CUC L     | CCC P     | CAC H     | CGC R     | C
C | CUA L     | CCA P     | CAA Q     | CGA R     | A
C | CUG L     | CCG P     | CAG Q     | CGG R     | G
--+-----------+-----------+-----------+-----------+--
A | AUU I(s)| ACU T     | AAU N     | AGU S     | U
A | AUC I(s)| ACC T     | AAC N     | AGC S     | C
A | AUA M(s)| ACA T     | AAA K     | AGA Stop| A
A | AUG M(s)| ACG T     | AAG K     | AGG Stop| G
--+-----------+-----------+-----------+-----------+--
G | GUU V     | GCU A     | GAU D     | GGU G     | U
G | GUC V     | GCC A     | GAC D     | GGC G     | C
G | GUA V     | GCA A     | GAA E     | GGA G     | A
G | GUG V(s)| GCG A     | GAG E     | GGG G     | G
--+-----------+-----------+-----------+-----------+--
```

10.2.11 Entrez

Entrez is a search engine that integrates several health sciences databases at the National Center for Biotechnology Information (NCBI) website. From a single webpage you can search on diverse datasets such as "scientific literature, DNA and protein sequence databases, 3D protein structure and protein domain data, expression data, assemblies of complete genomes, and taxonomic information.[17"]

This search engine is available at `http://www.ncbi.nlm.nih.gov/sites/gquery`. You can use it online as any standard search engine, but using it from a browser is not useful for incorporating data to your scripts. That is why the NCBI created the "Entrez Programming Utilities" (eUtils). This is a server side set of tools for querying the Entrez database without a web browser and can be used for retrieving search results to include them in your own programs.

eUtils at a Glance

The user must construct a specially crafted URL. This URL should contain the name of the program to use in the NCBI web server and all required parameters (like database name and search terms). Once this URL is posted, the NCBI sends the resulting data back to the user. This data is sent, in most cases, in XML format.

The rationale behind this procedure is that the program must build the URL automatically, post it, retrieve and process the results. The URL is not supposed to be built by hand, neither parsing the resulting XML file.

It is possible to combine eUtils components to form customized data pipelines within these applications.

Biopython and eUtils

Python has tools to fetch a URL (urllib2) and to parse XML files (like miniDOM), so it could be used to access eUtils. Even using the relevant Python modules to interact with the eUtils involves a lot of work. For this reason Biopython includes the **Entrez** module. The **Bio.Entrez** module provides functions to call every **eUtils** program without having to know how to build a URL or how to parse an XML file.

There are two ways to interact with the Entrez database: Query the database and retrieving actual data. The first action can be performed with **esearch** and **egquery Bio.Entrez** functions, while the **efetch** and **esummary** functions are used for data retrieval. Table 10.4 summarizes all functions available in the **eUtils** module.

eUtils: Retrieving Bibliography

The following script queries Pubmed through Entrez. Pubmed is a search engine for MEDLINE, a literature database of life sciences and biomedical information.

---

[17]`http://www.ncbi.nlm.nih.gov/books/bv.fcgi?rid=helpbook.TOC`

TABLE 10.4: eUtils

Name	Description
efetch	Retrieves records in the requested format from a list of one or more primary IDs or from the user's environment.
einfo	Provides field index term counts, last update, and available links for each database.
egquery	Provides Entrez database counts in XML for a single search using Global Query.
elink	Checks for the existence of an external or Related Articles link from a list of one or more primary IDs.
epost	Posts a file containing a list of primary IDs for future use in the user's environment to use with subsequent search strategies.
esearch	Searches and retrieves primary IDs (for use in EFetch, ELink, and ESummary).
espell	Retrieves spelling suggestions.
esummary	Retrieves document summaries from a list of primary IDs or from the user's environment.
read	Parses the XML results returned by any of the above functions.

Listing 10.14: Retrieve and display data from Pubmed (py3.us/41)

```
1 from Bio import Entrez
2 my_em = 'user@example.com'
3 db = "pubmed"
4 # Search de Entrez website using esearch from eUtils
5 # esearch returns a handle (called h_search)
6 h_search = Entrez.esearch(db=db, email=my_em,
7                           term="python and bioinformatics")
8 # Parse the result with Entrez.read()
9 record = Entrez.read(h_search)
10 # Get the list of Ids returned by previous search
11 res_ids = record["IdList"]
12 # For each id in the list
13 for r_id in res_ids:
14     # Get summary information for each id
15     h_summ = Entrez.esummary(db=db, id=r_id, email=my_em)
16     # Parse the result with Entrez.read()
17     summ = Entrez.read(h_summ)
18     print(summ[0]['Title'])
19     print(summ[0]['DOI'])
20     print('===============================================')
```

Provided that there is a working Internet connection when running code 10.14, it outputs something like this:

```
Optimal spliced alignments of short sequence reads.
10.1093/bioinformatics/btn300
==========================================
Mixture models for protein structure ensembles.
10.1093/bioinformatics/btn396
==========================================
Contact replacement for NMR resonance assignment.
10.1093/bioinformatics/btn167
==========================================
```

eUtils: Retrieving Gene Information

Since eUtils is an interface for several databases, the same program that is used to retrieve bibliographic data (code 10.14) can be used to retrieve gene information. The key change in code 10.15 is the database field (line 3).

Listing 10.15: Retrieve and display data from pubmed (py3.us/42)

```
1 from Bio import Entrez
2 my_em = 'user@example.com'
3 db = "gene"
4 term = 'cobalamin synthase homo sapiens'
5 h_search = Entrez.esearch(db=db, email=my_em, term=term)
6 record = Entrez.read(h_search)
7 res_ids = record["IdList"]
8 for r_id in res_ids:
9     h_summ = Entrez.esummary(db=db, id=r_id, email=my_em)
10    summ = Entrez.read(h_summ)
11    print(r_id)
12    print(summ[0]['Description'])
13    print(summ[0]['Summary'])
14    print('==========================================')
```

Code 10.15 produces a result like this:

```
326625
methylmalonic aciduria (cobalamin deficiency) cblB type
This gene encodes a protein that catalyzes the final step in <=
the conversion of vitamin B(12) into adenosylcobalamin (AdoCb<=
l), a vitamin B12-containing coenzyme for methylmalonyl-CoA m<=
utase. Mutations in the gene are the cause of vitamin B12-dep<=
endent methylmalonic aciduria linked to the cblB complementat<=
ion group. [provided by RefSeq]
```

```
=================================================
4524
5,10-methylenetetrahydrofolate reductase (NADPH)
Methylenetetrahydrofolate reductase (EC 1.5.1.20) catalyzes t<=
he conversion of 5,10-methylenetetrahydrofolate to 5-methylte<=
trahydrofolate, a cosubstrate for homocysteine remethylation <=
to methionine.[supplied by OMIM]
=================================================
(...)
```

Note that there is a number in this output that was not present in the result of code 10.14. This number is the ID returned by **esearch** function. This ID was used to retrieve the summary with the **esummary** function. The next code uses this ID to retrieve an actual DNA sequence:

```
>>> n = "nucleotide"
>>> handle = Entrez.efetch(db=n, id="326625", rettype='fasta')
>>> print handle.read()
>gi|326625|gb|M77599.1|HIVED82FO Human immunodeficiency virus<=
 type 1 gp120 (env) gene sequence
TTAATAGTACTTGGAATTCAACATGGGATTTAACACAACTTAATAGTACTCAGAATAAAGA
AGAAAATATCACACTCCCATGTAGAATAAAACAAATTATAAACATGTGGCAGGAAGTAGGA
AAAGCAATGTATGCCCCTCCCATCAAAGGACAAATTAAATGTTCATCAAATATTACAGGGC
TACTATTAACAAGAGATGGTGGTAATAGTGGTAACAAAAGCAACGACACCACCGAGACCTT
CAGACC
```

By changing the **rettype** parameter to *"genbank"* you can get the genbank record instead of the plain sequence. Once the sequence is in genbank format, it can parse it with the **SeqIO** module as seen on page 187. An alternative way to parse the results is to retrieve them in XML format and then parsing them with the **Entrez.read()** function:

```
>>> handle = Entrez.efetch(db=n, id="326625", retmode='xml')
>>> record[0]['GBSeq_moltype']
'RNA'
>>> record[0]['GBSeq_sequence']
'ttaatagtacttggaattcaacatgggatttaacacaacttaatagtactcagaataaaga<=
agaaaatatcacactcccatgtagaataaaacaaattataaacatgtggcaggaagtaggaa<=
aagcaatgtatgcccctcccatcaaaggacaaattaaatgttcatcaaatattacagggcta<=
ctattaacaagagatggtggtaatagtggtaacaaaagcaacgacaccaccgagaccttcag<=
acc'
>>> record[0]['GBSeq_organism']
'Human immunodeficiency virus 1'
```

10.2.12   PDB

PDB files store information regarding three dimensional structures of molecules held at the Protein Data Bank.

This database, with more than fifty thousand records, is the reference repository of protein structural data. A PDB file stores spacial positions of atoms obtained by X-ray crystallography, NMR spectroscopy and other experimental techniques.

This data is used by several programs, like molecule structure viewers like Deep View,[18] Cn3D[19] and PyMol,[20] protein analysis and structure prediction software such as MakeMultimer[21] and Modeller.[22]

Records of this database can be accessed through the RCSB webpage at `http://www.rcsb.org/pdb/home/home.do`.[23] If you want to make your own application to analyze protein structure data, your program will have to be able to parse the data from PDB files. This is the role of the **Bio.pdb** module.

To effectively use the **Bio.PDB** module, you have first to understand the PDB file structure. A protein structure is modeled with a top-down hierarchy. It begins with the **structure** class, down to the **atom** subclass. Intermediate orders are **model**, **chain** and **residue**. This hierarchy is also known as **SMCRA**. Some proteins don't follow this pattern, but PDB files do.[24]

Bio.PDB Module

The PDB module provides the PDBParser class.[25]   This class has the **get_structure** method. This method needs as input an id and a file name, and it returns a **structure** object. This **SMCRA** hierarchy can be accessed by an identifier as a key:

```
>>> from Bio.PDB.PDBParser import PDBParser
>>> pdbfn = '/home/sb/bioinfo/1FAT.pdb'
>>> parser = PDBParser(PERMISSIVE=1)
>>> structure = parser.get_structure("1fat", pdbfn)
WARNING: Chain A is discontinuous at line 7808.
```

---

[18]`http://spdbv.vital-it.ch`

[19]`http://www.ncbi.nlm.nih.gov/Structure/CN3D/cn3d.shtml`

[20]`http://pymol.sourceforge.net`

[21]`http://watcut.uwaterloo.ca/cgi-bin/makemultimer`

[22]`http://www.salilab.org/modeller`

[23]RCSB is the **R**esearch **Co**llaboratory for **S**tructural **B**ioinformatics, the consortium in charge of the management of the PDB.

[24]There are some malformed PDB out there, when the **Bio.PDB** module finds a problem it can generate an exception or a warning, depending on the *PERMISSIVE* argument (0 for no tolerance and 1 for the parser to issue warnings).

[25]In some Linux installations you have to install the *python-numeric-ext* package for this module to run.

```
(... some warnings removed ...)
WARNING: Chain D is discontinuous at line 7870.
>>> structure.child_list
[<Model id=0>]
>>> model = structure[0]
>>> model.child_list
[<Chain id=A>, <Chain id=B>, <Chain id=C>, <Chain id=D>, <=
<Chain id= >]
>>> chain = model['B']
>>> chain.child_list[:5]
[<Residue SER het=  resseq=1 icode= >, <Residue ASN het= <=
 resseq=2 icode= >, <Residue ASP het=  resseq=3 icode= >,<=
 <Residue ILE het=  resseq=4 icode= >, <Residue TYR het= <=
 resseq=5 icode= >]
>>> residue = chain[4]
>>> residue.child_list
[<Atom N>, <Atom CA>, <Atom C>, <Atom O>, <Atom CB>, <=
<Atom CG1>, <Atom CG2>, <Atom CD1>]
>>> atom = residue['CB']
>>> atom.bfactor
14.130000000000001
>>> atom.coord
array([ 34.30699921,  -1.57500005,  29.06800079],'f')
```

The following program opens a PDB file that is compressed with **gzip**.[26] It scans through all chains of the protein, in each chain it walks through all the atoms in each residue, to print the residue and atom name when there is a disordered atom:

Listing 10.16: Parse a gzipped PDB file (py3.us/43)

```python
1 import gzip
2 from Bio.PDB.PDBParser import PDBParser
3
4 def disorder(structure):
5     for chain in structure[0].get_list():
6         for residue in chain.get_list():
7             for atom in residue.get_list():
8                 if atom.is_disordered():
9                     print residue, atom
10    return None
```

---

[26]gzip is the "standard" application used in *nix systems for file compression but it is also available in most common platforms. It is shown in this example because most of the public accessible molecular data files are compressed in this format.

```
11
12 pdbfn = '/home/sb/bioinfo/pdb1apk.ent.gz'
13 handle = gzip.GzipFile(pdbfn)
14 parser = PDBParser()
15 structure = parser.get_structure("test", handle)
16 disorder(structure)
```

10.2.13 PROSITE

PROSITE is a database of documentation entries describing protein domains, families and functional sites as well as associated patterns and profiles used to identify them.

This database is accessed through the PROSITE site at http://www.expasy.org/prosite or distributed as a single plain text file.[27] This file can be parsed with the *parse* function in the *Prosite* module:

```
>>> from Bio import Prosite
>>> handle = open("prosite.dat")
>>> records = Prosite.parse(handle)
>>> for r in records:
    print(r.accession)
    print(r.name)
    print(r.description)
    print(r.pattern)
    print(r.created)
    print(r.pdoc)
    print("===================================")
```

```
PS00001
ASN_GLYCOSYLATION
N-glycosylation site.
N-{P}-[ST]-{P}.
APR-1990
PDOC00001
===================================
PS00004
CAMP_PHOSPHO_SITE
cAMP- and cGMP-dependent protein kinase phosphorylation site.
[RK](2)-x-[ST].
APR-1990
```

---

[27]Release 20.36, of 02-Sep-2008 is a 22 Mb file available at ftp://ftp.expasy.org/databases/prosite/prosite.dat.

```
PDOC00004
====================================
PS00005
PKC_PHOSPHO_SITE
Protein kinase C phosphorylation site.
[ST]-x-[RK].
APR-1990
PDOC00005
====================================
```

10.2.14 Restriction

Recombinant DNA technology is based on the possibility of combining DNA sequences (usually from different organisms) that would not normally occur together. This kind of biological cut and paste is accomplished by a restriction endonuclease, a special group of enzymes that works as a specific molecular scissors.

The main characteristic of these enzymes is that they recognize a specific sequence of nucleotides and cut both DNA strands. When a researcher wants to introduce a cut in a known DNA sequence, he or she must first check which enzyme has a specificity for a site inside the sequence. All available restriction enzymes are stored in a database called REBASE.[28]

A well known restriction enzyme is EcoRI, this enzyme recognizes the "GAATTC" sequence. So this enzyme cuts any doublestranded DNA having this sequence, like

```
CGCGAATTCGCG
GCGCTTAAGCGC
```

In this case the restriction site is found in the middle of the top strand (marked with '-'): CGC-GAATTC-GCG. The separated pieces look like this:

```
CGC        GAATTCGCG
GCGCTTAA        GCGC
```

Bio.Restriction Module

Biopython provides tools for dealing with restriction enzymes, including enzyme information retrieved from REBASE. All restriction enzymes are available from **Restriction module**:

```
>>> from Bio import Restriction
>>> Restriction.EcoRI
EcoRI
```

---

[28]REBASE is available at http://rebase.neb.com/rebase/rebase.html.

Restriction enzyme objects have several methods, like **search**, that can be used to search for restriction sites in a DNA sequence:

```
>>> from Bio.Seq import Seq
>>> from Bio.Alphabet.IUPAC import IUPACAmbiguousDNA
>>> alfa = IUPACAmbiguousDNA()
>>> gi1942535 = Seq('CGCGAATTCGCG', alfa)
>>> Restriction.EcoRI.search(gi1942535)
[5]
```

Note that the search function returns a list with all positions where the enzyme cuts. The position is the first nucleotide after the cut, beginning in 1 instead of 0 (as usual in other parts of Python). Another parameter in **search** is *linear*. It is defaulted to **False** and should be set as **True** when the sequence is circular.

Segments produced after a restriction can be seen with the **catalyze** function:

```
>>> Restriction.EcoRI.catalyse(gi1942535)
(Seq('CGCG', IUPACAmbiguousDNA()), Seq('AATTCGCG', <=
IUPACAmbiguousDNA()))
```

To analyze several enzymes at the same time there is the **Restriction-Batch class**:

```
>>> enz1 = Restriction.EcoRI
>>> enz2 = Restriction.HindIII
>>> batch1 = Restriction.RestrictionBatch([enz1, enz2])
>>> batch1.search(gi1942535)
{EcoRI: [5], HindIII: []}
```

The **search** function applied over a set of enzymes returns a dictionary:

```
>>> dd = batch1.search(gi1942535)
>>> dd.get(Restriction.EcoRI)
[5]
>>> dd.get(Restriction.HindIII)
[]
```

Enzymes can be added or removed as if the **RestrictionBatch** instance were a set:

```
>>> batch1.add(Restriction.EarI)
>>> batch1
RestrictionBatch(['EarI', 'EcoRI', 'HindIII'])
>>> batch1.remove(Restriction.EarI)
>>> batch1
RestrictionBatch(['EcoRI', 'HindIII'])
```

There are also some predefined sets in the *Restriction* module, like **AllEnzymes**, **CommOnly** and **NonComm**:

```
>>> batch2 = Restriction.CommOnly
```

Analysis Class: All in One

Analysis class simplifies dealing with multiple enzymes:

```
>>> an1 = Restriction.Analysis(batch1,gi1942535)
>>> an1.full()
{HindIII: [], EcoRI: [5]}
```

Up to this point, the result of **full()** method in the **Analysis** object is the same as a **search** over a **RestrictionBatch**. Analysis provides:

```
>>> an1.print_that()

EcoRI      :  5.

   Enzymes which do not cut the sequence.

HindIII

>>> an1.print_as('map')
>>> an1.print_that()

    5 EcoRI
    |
CGCGAATTCGCG
||||||||||||
GCGCTTAAGCGC
1                        12


   Enzymes which do not cut the sequence.

HindIII

>>> an1.only_between(1,8)
{EcoRI: [5]}
```

This covers most of the functions available in **Restriction** module. For more information please refer to the Biopython tutorial at `http://biopython. org/DIST/docs/cookbook/Restriction.html` and see the code 22.1 on page 367.

10.2.15   SeqUtils

This module has several functions to deal with DNA and protein sequences, such as: CG, GC skew, molecular weight, checksum algorithms, Codon Usage, Melting Temperature and others. All functions are properly documented, so I will explain only a few functions to get the idea of how to use them.

DNA Utils

SeqUtils has plenty of functions that can be applied to DNA sequences. Let's see some of them:

GC content: The percentage of bases which are either guanine or cytosine is a parameter that affects some physical properties of the DNA molecule. It is calculated with the GC function:

```
>>> from Bio.SeqUtils import GC
>>> GC('gacgatcggtattcgtag')
50.0
```

DNA Melting Temperature: It can be calculated with the **Melting-Temp.Tm_staluc** function. This function implements the "nearest neighbor method"[29] and can be use for both DNA and RNA sequences:

```
>>> from Bio.SeqUtils import MeltingTemp
>>> MeltingTemp.Tm_staluc('tgcagtacgtatcgt')
42.211472744873447
>>> print '%.2f'%MeltingTemp.Tm_staluc('tgcagtacgtatcgt')
42.21
```

CheckSum functions: A checksum is a usually short alphanumeric string based in an input file mostly used to test data integrity. From any kind of data (like a text file, a DNA sequence), using an algorithm you can generate a small string (usually called "signature") that can represent the original data. Some programs attach a checksum information to sequence information to ensure data integrity. A simple checksum is implemented by the GCG program.

This is a sequence in the gcg format:

```
ID   AB000263 standard; RNA; PRI; 368 BP.
XX
AC   AB000263;
XX
DE   Homo sapiens mRNA for prepro cortistatin like peptide.
XX
```

---

[29]For more information on nearest neighbor method, see the work of "Santalucia, et al. (1996) Biochemistry 35, 3555-3562."

```
SQ    Sequence 37 BP;
AB000263   Length: 37   Check: 1149  ..
       1  acaagatgcc attgtccccc ggcctcctgc tgctgct
```

The **Check** number (1149 in this case) is derived from the sequence. If the sequence is changed, the number is (hopefully) changed. There is always a chance of a random collision, that is, when two different sequences generate the same signature. The "gcg checksum" is weak in the sense it allows only 10000 different signatures. This is why there are some other stronger checksums like the crc32, crc64 and seguid.[30]

All these checksums are available from the **CheckSum** module. They are shown in order from the weaker to the strongest checksum algorithm.

```
>>> from Bio.SeqUtils import CheckSum
>>> myseq = 'acaagatgccattgtcccccggcctcctgctgctgct'
>>> CheckSum.gcg(myseq)
1149
>>> CheckSum.crc32(myseq)
-2106438743
>>> CheckSum.crc64(myseq)
'CRC-A2CFDBE6AB3F7CFF'
>>> CheckSum.seguid(myseq)
'9V7Kf19tfPA5TntEP75YiZEm/9U'
```

Protein Utils

Protein related functions are accessible from the **ProtParam class**. Available protein properties are: *Molecular weight, aromaticity, instability index, flexibility, isoelectric point* and *secondary structure fraction*. Function names are straightforward. See them in code 10.17:

Listing 10.17: Apply PropParam functions to a group of proteins (py3.us/44)

```
1 from Bio.SeqUtils.ProtParam import ProteinAnalysis
2 from Bio.SeqUtils import ProtParamData
3 from Bio import SeqIO
4
5 fh = open('/home/sb/bioinfo/pdbaa')
6 for rec in SeqIO.parse(fh,'fasta'):
7     myprot = ProteinAnalysis(str(rec.seq))
8     print(myprot.count_amino_acids())
9     print(myprot.get_amino_acids_percent())
```

---

[30]For more information on the checksums, refer to "Bassi, Sebastian and Gonzalez, Virginia. New checksum functions for Biopython. Available from Nature Precedings <http://dx.doi.org/10.1038/npre.2007.278.1> (2007)."

```
10      print(myprot.molecular_weight())
11      print(myprot.aromaticity())
12      print(myprot.instability_index())
13      print(myprot.flexibility())
14      print(myprot.isoelectric_point())
15      print(myprot.secondary_structure_fraction())
16      print(myprot.protein_scale(ProtParamData.kd, 9, .4))
17 fh.close()
```

10.2.16   Sequencing

Sequencing projects usually generate *.ace* and *.phd.1* files.[31]

Phd Files

The DNA sequencer trace data is read by the Phred program. This program calls bases, assigns quality values to the bases, and writes the base calls and quality values to output files (with *.phd.1* extension).

The following code (listing 10.18) shows how to extract the data from the *.phd.1* files:

Listing 10.18: Extract data from a .phd.1 file (py3.us/45)

```
1 import pprint
2 from Bio.Sequencing import Phd
3
4 fn = '/home/sb/bt/biopython-1.50/Tests/Phd/phd1'
5 fh = open(fn)
6 rp = Phd.RecordParser()
7 # Create an iterator
8 it = Phd.Iterator(fh,rp)
9 for r in it:
10     # All the comments are in a dictionary
11     pprint.pprint(r.comments)
12     # Sequence information
13     print('Sequence: %s' % r.seq)
14     # Quality information for each base
15     print('Quality: %s' % r.sites)
16 fh.close()
```

If you only want to extract the sequence, it is easier to use SeqIO:

---

[31]This depends on sequencing technology; these files are generated by processing sequence trace chromatogram with popular sequencing processing software such as Phred, Phrap, CAP3, and Consed.

```
>>> from Bio import SeqIO
>>> fn = '/home/sb/bioinfo/biopython-1.43/Tests/Phd/phd1'
>>> fh = open(fn)
>>> seqs = SeqIO.parse(fh,'phd')
>>> seqs = SeqIO.parse(fh,'phd')
>>> for s in seqs:
    print(s.seq)
```

```
ctccgtcggaacatcatcggatcctatcacagagtttttgaacgagttctcg
(...)
```

Ace Files

In a typical sequencing strategy, several overlapping sequences (or "reads") are assembled electronically into one long contiguous sequence. This contiguous sequence is called "contig" and is made with specialized programs like CAP3 and Phrap. Contig files are used for viewing or further analysis. Biopython has the **ACEParser** in the **Ace module**. For each .ace file you can get the number of contigs, number of reads and some file information:

```
>>> from Bio.Sequencing import Ace
>>> fn='836CLEAN-100.fasta.cap.ace'
>>> acefilerecord=Ace.read(open(fn))
>>> acefilerecord.ncontigs
87
>>> acefilerecord.nreads
277
>>> acefilerecord.wa[0].info
['phrap 304_nuclsu.fasta.screen -new_ace -retain_duplicates', <=
'phrap version 0.990329']
>>> acefilerecord.wa[0].date
'040203:114710'
```

The **Ace.read** also retrieves relevant information of each contig as shown in code 10.19.

Listing 10.19: Retrieve data from an ".ace" file (py3.us/46)

```
1 from Bio.Sequencing import Ace
2
3 fn = '/home/sb/bt/biopython-1.50/Tests/Ace/contig1.ace'
4 acefilerecord = Ace.read(open(fn))
5
6 # For each contig:
7 for ctg in acefilerecord.contigs:
```

```
8       print '=========================================='
9       print 'Contig name: %s'%ctg.name
10      print 'Bases: %s'%ctg.nbases
11      print 'Reads: %s'%ctg.nreads
12      print 'Segments: %s'%ctg.nsegments
13      print 'Sequence: %s'%ctg.sequence
14      print 'Quality: %s'%ctg.quality
15      # For each read in contig:
16      for read in ctg.reads:
17          print 'Read name: %s'%read.rd.name
18          print 'Align start: %s'%read.qa.align_clipping_start
19          print 'Align end: %s'%read.qa.align_clipping_end
20          print 'Qual start: %s'%read.qa.qual_clipping_start
21          print 'Qual end: %s'%read.qa.qual_clipping_end
22          print 'Read sequence: %s'%read.rd.sequence
23          print '========================================'
```

10.2.17   SwissProt

SwissProt[32] is a hand annotated protein sequence database. It is maintained collaboratively by the Swiss Institute for Bioinformatics (SIB) and the European Bioinformatics Institute (EBI), forming the UniProt consortium. It is known for its reliable protein sequences associated with a high level of annotation, being the reference database for proteins. As of September 2008 it has almost 400,000 entries while the whole UniProt database has more than 6,000,000 records. Its reduced size is due to its hand curation process.

Swissprot files are text file structured so as to be usable by human readers as well as by computer programs. Specifications for this file format are available at http://www.expasy.org/sprot/userman.html, but there is no need to know it internals to parse it with Biopython.

A sample SwissProt file is shown below:[33]

```
ID  6PGL_ECOLC              Reviewed;         331 AA.
AC  B1IXL9;
DT  20-MAY-2008, integrated into UniProtKB/Swiss-Prot.
DT  29-APR-2008, sequence version 1.
DT  02-SEP-2008, entry version 5.
DE  RecName: Full=6-phosphogluconolactonase;
DE           Short=6-P-gluconolactonase;
DE           EC=3.1.1.31;
```

---

[32]http://www.expasy.org/sprot

[33]This file is slighted modified to fit in this page, the original file can be retrieved from http://www.expasy.org/uniprot/B1IXL9.txt.

```
GN   Name=pgl; OrderedLocusNames=EcolC_2895;
OS   Escherichia coli (strain ATCC 8739 / DSM 1576 / Crooks).
OC   Bacteria; Proteobacteria; Gammaproteobacteria; Enterobacteriales;
OC   Enterobacteriaceae; Escherichia.
OX   NCBI_TaxID=481805;
RN   [1]
RP   NUCLEOTIDE SEQUENCE [LARGE SCALE GENOMIC DNA].
RA   Copeland A., Lucas S., Lapidus A., Glavina del Rio T., Dalin E.,
RA   Tice H., Bruce D., Goodwin L., Pitluck S., Kiss H., Brettin T.;
RT   "Complete sequence of Escherichia coli C str. ATCC 8739.";
RL   Submitted (FEB-2008) to the EMBL/GenBank/DDBJ databases.
CC   -!- FUNCTION: Catalyzes the hydrolysis of 6-phosphogluconolactone
CC       to 6-phosphogluconate (By similarity).
CC   -!- CATALYTIC ACTIVITY: 6-phospho-D-glucono-1,5-lactone + H(2)O
CC       = 6-phospho-D-gluconate.
CC   -!- PATHWAY: Carbohydrate degradation; pentose phosphate pathway;
CC       D-ribulose 5-phosphate from D-glucose 6-phosphate (oxidative
CC       stage): step 2/3.
CC   -!- SIMILARITY: Belongs to the cycloisomerase 2 family.
CC   -----------------------------------------------------------------
CC   Copyrighted by the UniProt Consortium, see
CC   http://www.uniprot.org/terms Distributed under the Creative
CC   Commons Attribution-NoDerivs License
CC   -----------------------------------------------------------------
DR   EMBL; CP000946; ACA78522.1; -; Genomic_DNA.
DR   RefSeq; YP_001725849.1; -.
DR   GeneID; 6065358; -.
DR   GenomeReviews; CP000946_GR; EcolC_2895.
DR   KEGG; ecl:EcolC_2895; -.
DR   GO; GO:0017057; F:6-phosphogluconolactonase activity; IEA:HAMAP.
DR   GO; GO:0006006; P:glucose metabolic process; IEA:HAMAP.
DR   HAMAP; MF_01605; -; 1.
DR   InterPro; IPR015943; WD40/YVTN_repeat-like.
DR   Gene3D; G3DSA:2.130.10.10; WD40/YVTN_repeat-like; 1.
PE   3: Inferred from homology;
KW   Carbohydrate metabolism; Complete proteome; Glucose metabolism;
KW   Hydrolase.
FT   CHAIN         1    331       6-phosphogluconolactonase.
FT                                /FTId=PRO_1000088029.
SQ   SEQUENCE   331 AA;  36308 MW;  D731044CFCF31A8F CRC64;
     MKQTVYIASP ESQQIHVWNL NHEGALTLTQ VVDVPGQVQP MVVSPDKRYL YVGVRPEFRV
     LAYRIAPDDG ALTFAAESAL PGSPTHISTD HQGQFVFVGS YNAGNVSVTR LEDGLPVGVV
     DVVEGLDGCH SANISPDNRT LWVPALKQDR ICLFTVSDDG HLVAQDPAEV TTVEGAGPRH
     MVFHPNEQYA YCVNELNSSV DVWELKDPHG NIECVQTLDM MPENFSDTRW AADIHITPDG
     RHLYACDRTA SLITVFSVSE DGSVLSKEGF QPTETQPRGF NVDHSGKYLI AAGQKSHHIS
```

```
VYEIVGEQGL LHEKGRYAVG QGPMWVVVNA H
//
```

The code 10.20 shows how to retrieve data from a SwissProt file with multiple records:

Listing 10.20: Retrieve data from a SwissProt file (py3.us/47)

```
1 from Bio import SwissProt
2 fh = open('spfile.txt')
3 records = SwissProt.parse(fh)
4 for record in records:
5     print('Entry name: %s' % record.entry_name)
6     print('Accession(s): %s' % ','.join(record.accessions))
7     print('Keywords: %s' % ','.join(record.keywords))
8     print('Sequence: %s' % record.sequence)
9 fh.close()
```

The code 10.21 shows all atributes in record parsed by **SwissProt module**:

Listing 10.21: Atributes of a SwissProt record (py3.us/48)

```
1 from Bio import SwissProt
2 fh = open('/home/sb/bioinfo/spfile.txt')
3 record = SwissProt.parse(fh).next()
4 for att in dir(record):
5     if not att.startswith('__'):
6         print(att,getattr(record,att))
```

10.3 Conclusion

Most used Biopython features had been covered in this chapter. Following the code samples presented here and the full programs in Section IV should give you an insight on how to use Biopython. You should also learn how to use the Python built-in help since online documentation tends to be more up to date than anything printing. Biopython development happens at a fast pace. So fast that this chapter was rewritten several times while I was working on it. The best way to keep updated with Biopython development is to subscribe to the Biopython development mailing list and receive the RSS feed from the code repository.

10.4   Additional Resources

- Chang J, Chapman B, Friedberg I, Hamelryck T, de Hoon M, Cock P, Antão, T. "Biopython Tutorial and Cookbook."
  http://www.biopython.org/DIST/docs/tutorial/Tutorial.html or
  http://www.biopython.org/DIST/docs/tutorial/Tutorial.pdf.

- Hamelryck T, Manderick B., "PDB file parser and structure class implemented in Python." Bioinformatics. 2003 Nov 22;19(17):2308-10.
  http://bioinformatics.oxfordjournals.org/cgi/screenpdf/19/17/2308.pdf

- Sohm, F. "Manual in cookbook style on using the Restriction module."
  http://biopython.org/DIST/docs/cookbook/Restriction.html

- Wu C.H., Apweiler R., Bairoch A., Natale D.A, Barker W.C., Boeckmann B., Ferro S., Gasteiger E., Huang H., Lopez R., Magrane M., Martin M.J., Mazumder R., O'Donovan C., Redaschi N. and Suzek B. (2006). "The Universal Protein Resource (UniProt): an expanding universe of protein information." Nucleic Acids Research 34: D187-D191.

- Magrane M., Apweiler R. (2002). "Organisation and standardisation of information in Swiss-Prot and TrEMBL." Data Science Journal 1(1): 13-18.
  http://journals.eecs.qub.ac.uk/codata/Journal/Contents/1_1/1_1pdf/DS101.Pdf

- Dennis A. Benson, Ilene Karsch-Mizrachi, David J. Lipman, James Ostell, and David L. Wheeler. "GenBank." Nucleic Acids Res. 2008 January; 36(Database issue): D25-D30.
  http://dx.doi.org/10.1093/nar/gkm929

- Larkin MA, Blackshields G, Brown NP, Chenna R, McGettigan PA, McWilliam H, Valentin F, Wallace IM, Wilm A, Lopez R, Thompson JD, Gibson TJ, Higgins DG. "Clustal W and Clustal X version 2.0. Bioinformatics." 2007 Nov 1;23(21):2947-8. Epub 2007 Sep 10.

- Wikipedia contributors. "Restriction enzyme." Wikipedia, The Free Encyclopedia. February 13, 2009, 16:44 UTC.
  http://en.wikipedia.org/wiki/Restriction_enzyme.

- EFetch for Sequence and other Molecular Biology Databases
  http://www.ncbi.nlm.nih.gov/entrez/query/static/efetchseq_help.html

- Cock P. "Clever tricks with NCBI Entrez EInfo (& Biopython)"
  http://news.open-bio.org/news/2009/06/ncbi-einfo-biopython

10.5   Self-Evaluation

1. What is an Alphabet in Biopython? Name at least four.

2. Describe Seq and SeqRecord objects.

3. What advantage provides a Seq object over a string?

4. Seq object provides some string operations. Why?

5. What is a MutableSeq object?

6. What is the relation between the Align and ClustalW module?

7. Name the methods of the SeqIO module.

8. What is the difference between line 7 and 8 in code 10.5?

9. Name five functions found in SeqUtils.

10. What kind of sequence files can be read with **Sequencing** module?

11. What module would you use to retrieve data from the NCBI Web server?

12. Make a program to count all ordered atoms in a PDB file. The PDB file must passed to the program on the command line, in the form: `program.py file.pdb`.

Part III

Advanced Topics

Chapter 11

Web Applications

We have just seen how to run programs on our own computer. This chapter shows how to port your programs to the web using what you have learned up to this point in addition to some new techniques.

The main advantage of making a program available on the Web is that it can reach more users without the need for them to install a copy of the program and to have a Python installation. It also helps users who do not have the permissions necessary to install software on a particular machine. Sometimes the program accesses huge databases that can't be installed on the end user's hard drive.

In order to make web programming it is not enough to know Python. It is also necessary to have a basic understanding of Web servers and Web page design using HTML. Both of these topics are beyond the scope of this book, for which reason I recommend that you read up on them if you have never designed a WEB page before. Knowing the basics of HTML has special importance as most IT Labs have staff dedicated to the setup and maintenance of the WEB server but the HTML design is something that they will rarely do for you. For more information on HTML, please see the "Additional Resources" section. Concerning the Web server, besides using the one provided by the institution where you work you can use for learning purposes the one included with DNA Virtual Desktop Edition that is included with this book.

There are several ways to use Python on a Web server, **CGI** (**C**ommon **G**ateway Interface), **mod_python** and **WSGI** (**W**eb **S**erver **G**ateway Interface). CGI is the most used method, as it is the easiest to configure and is available on almost all Web servers without having to install additional software. It is essentially a protocol to connect an application, written in any language with a Web server. **mod_python** in particular consists of an Apache Module that integrates Python with the Web server. The advantage of this approach is the fast execution time of our scripts, since the Python interpreter is loaded with the Web server.[1] The disadvantage is that it works "only" with the Apache Web server, which is a minor disadvantage as we are probably already using Apache as it is the most used Web server on the Internet.

---

[1]This will depend on a lot of factors, but there are benchmarks that show speed increases of up to 100%. In some cases the difference will be a determining factor, although the processing speed of modern computer systems makes the difference unimportant most of the time.

WSGI, in turn, is a "specification for Web servers and application servers to communicate with Web applications." Since it is a specification, there are several implementations. The main advantage of WSGI is that once you have made a WSGI application, it can be deployed in any WSGI compatible server (or even using a Python provided Web server). As in mod_python, the execution speed is better than CGI, because there is no overhead for starting the Python interpreter on each request.

11.1 CGI in Python

11.1.1 Configuring a Web Server for CGI

Although I have said that the web server configuration is beyond the scope of this book, I think that it is reasonable to offer a few recommendations to help you configure a server for CGI.

In the server configuration file[2] there should be specifications that scripts can be executed via CGI, in which directories and how they will be named.

If we want our scripts to be located at **/var/www/apache2-default/cgi-bin**, we have to include the following lines in the server's configuration file.

```
<Directory /var/www/apache2-default/cgi-bin>
  Options +ExecCGI
</Directory>
```

To specify that the executable scripts are those that have the file extension .py, the following line is added

```
AddHandler cgi-script .py
```

If the configuration file already has a line with the file extensions registered, you only need to add .py to it.

```
AddHandler cgi-script .cgi .pl .py
```

Finally we have to configure the `ScriptAlias` variable to where scripts are stored.

```
ScriptAlias /cgi-bin/ /var/www/apache2-default/cgi-bin/
```

This is all there is to the server configuration file. The only thing that is left to do is make sure that the script has the Apache user permissions. If you have access to the server's terminal you can enter:

---

[2]In Apache Web server, in most cases configuration file is `httpd.conf` or `apache2.conf` and it is located at `/etc/apache2` directory. This can change on each installation.

```
chmod a+x MyScript.py
```

If you only have FTP access, use a FTP client to set the permissions.

11.1.2 Testing the Server with Our Script

To confirm that our server is ready to execute CGI programs, we can do that with the following code:

Listing 11.1: First CGI script

```
1 #!/usr/bin/env python
2 print("Content-Type: text/html\n")
3 print("<html><head><title>Test page</title></head><body>")
4 print("<h1>HELLO WORLD!</h1>")
5 print("</body></html>")
```

Code explanation: The first line indicates the location of the Python interpreter. Usually, this line is optional and we add it only when we want to run the script directly without first having to start the Python interpreter but for CGI programs this line is mandatory.[3] The second line is important for the web server to know that it is going to be sent an HTML page. We have to send the string Content-Type/html followed by two carriage returns. Although on line two there is only one implicit carriage return ($\backslash$n), the other one is added by the print command. The rest of the program is similar to the others that we've done up to this point, the difference being that we print HTML code to be read by the browser.

If we upload this program to a web server and then access the page with our browser, the results we will see will be similar to Figure 11.1. What we have to take special note of from this example is that we are not seeing the content of our file but the product of its execution on the server. Neither do we see directly the results of its execution but what we see is the rendering by a client (Web browser) of the HTML produced by the execution of the code 11.1.

As a result of this it is important to take into account that in order to test our pages we need them to be processed by a web server, and not open them directly from our hard drive. In this case we will have as a result what you see in Figure 11.2, which is not what we want.

---

[3]If you don't know where the Python interpreter is, ask the system administrator to install your script. Another option, if you have access to the server command line is to execute whereis python.

FIGURE 11.1:   Our first CGI.

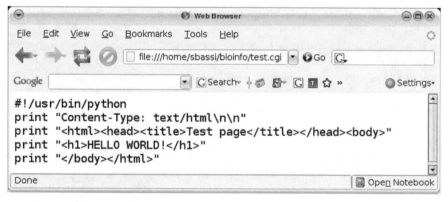

FIGURE 11.2:   CGI accessed from local disk instead of a web server.

FIGURE 11.3: A very simple form.

Sending Data to a CGI Program

The previous program is mostly useless, as it doesn't accept any parameters from the user. Let's see an example of minimalist HTML form that sends data though CGI to a Python program that will run using this data.

The first step is to design the form. In this case we will create a simple form with one field and it will be saved as `greeting.html`:

Listing 11.2: HTML front end send data to a CGI program (py3.us/49)

```
1 <html><head><title>Very Simple Form</title></head>
2 <body>
3 <form action='greeting.cgi' method='post'>
4 Your name: <input type='text' name='username'> <p>
5 <input type='submit' value='Send'>
6 </form></body></html>
```

Code explained: There are two important features to note on this small form. Line 3 is specified where the program that is going to process the data is located (`greeting.cgi`). On line 4 there is the field that the user has to fill ("text" type), with an associated variable (`username`). You have to take note of this variable name because the information entered by the user will be bound to this name. The form looks like the one in Figure 11.3

Let's see how to write the code the will accept that data sent by the form and from this data it builds a Web page "on the fly."

Listing 11.3: CGI program (py3.us/50)

```
1 #!/usr/bin/env python
2 import cgi
3 print "Content-Type: text/html\n"
4 form = cgi.FieldStorage()
```

FIGURE 11.4: CGI output.

```
5 name = form.getvalue("username","NN")
6 print("<html><head><title>A CGI script</title></head>")
7 print("<body><h2>Hello %s</h2></body></html>"%name)
```

Code explained: On line four we create an instance (form) from the class cgi.FieldStorage. This class takes the values sent by the form and is responsible for taking the values sent by the form and making them accessible in a dictionary like fashion. On the next line (5), we see a way to access the data sent by the form. The get-value method takes as a necessary argument, the name of the field whose content we want to access. The second argument is optional and indicates which value will be returned in case the wanted field is blank. Take note that this is similar to the get dictionary function. From line 6 forward the program doesn't have anything new, except that instead of printing ordinary text, we display the HTML code that will be rendered in our browser.

In summary, we used the form 11.2 to enter a name and press "Send." This sends the data, it is then read by the program thanks to the cgi.FieldStorage class and referenced as a variable name that is used in the program to generate a Web page. See the ouput in Figure 11.4.

11.1.3 Web Program to Calculate the Net Charge of a Protein (CGI version)

Using the code from 6.2, we can easily adapt it to use from a Web page. As a first step we need to design a form where a user can enter the data. This is a proposed form:

Listing 11.4: HTML front end to send data to a CGI program (py3.us/51)

```
<html><head><title>Protein Charge Calculator</title></head>
<body><h2>Protein Charge Calculator</h2>
<form action='protcharge.cgi' method='post'>
Enter the amino-acid sequence:
<textarea name="seq" rows="5" cols="40"></textarea><p>
```

```
Do you want to see the proportion of charged amino-acid?
<p><input type="radio" name="prop" value="y">Yes<br>
<input type="radio" name="prop" value="n"> No<br><p>
Job title (optional):
<input type="text" size="30" name="title"
value=""><br><p>
<input type='submit' value='Send'></form></body></html>
```

Below the code (`emphprotcharge.cgi`) that will be called when the form is used:

Listing 11.5: Web version of function to calculate the net charge of a protein and proportion of charged amino acid (py3.us/52)

```
1 #!/usr/bin/env python
2 import cgi, cgitb
3 def chargeandprop(AAseq):
4     protseq = AAseq.upper()
5     charge = -0.002
6     cp = 0
7     AACharge = {"C":-.045,"D":-.999,"E":-.998,"H":.091,
8                 "K":1,"R":1,"Y":-.001}
9     for aa in protseq:
10        charge += AACharge.get(aa,0)
11        if aa in AACharge:
12            cp += 1
13    prop = float(cp)/len(AAseq)*100
14    return (charge,prop)
15 cgitb.enable()
16 print("Content-Type: text/html\n")
17 form = cgi.FieldStorage()
18 uname = form.getvalue("username","NN")
19 seq = form.getvalue("seq","QWERTYYTREWQRTYEYTRQWE")
20 prop = form.getvalue("prop","n")
21 jobtitle = form.getvalue("title","No title")
22 charge,propvalue = chargeandprop(seq)
23 print("<html><body>Job title:"+jobtitle+"<br/>")
24 print("Your sequence is:<br/>"+seq+"<br/>")
25 print("Net charge:",charge,"<br/>")
26 if prop=="y":
27     print("Proportion of charged AA: %.2f <br/>" %propvalue)
28 print "</body></html>"
```

Code explanation: On line 17 we create an instance (`form`) of the class **cgi.FieldStorage**. This class is responsible for taking the values sent by the form and making them available in a dictionary-like fashion. From 18 to

21 we retrieve values entered by the user. In line 22, the "net charge" and "proportion of charged amino acids" are evaluated. Line 23 up to the end generates the HTML that will be sent to the browser.

11.2   mod_python

One of the problems with CGI is the execution speed of the programs. For servers with a small number of users it is not something that is easily noticed. However, if the program is executed multiple times, the difference in speed can be significant. This is because for each CGI execution, a new instance of the program is loaded in memory. When the server has a high load, it may run out of memory needed to process all the requests and as a result some of the processes will crash and others will run more slowly. This makes another option necessary. There are various CGI options to run Python applications like **mod_python**.

What is mod_python? It is an Apache Web server module that contains the Python interpreter. In this form, although there may be many simultaneous requests, only one instance of the interpreter will be executed. The difference in performance that can be obtained through this method can be up to a thousand times compared to CGI. The disadvantage is that being an Apache module, it only works with this web server, which is a minor disadvantage taking into consideration that Apache is the most installed web server and is available for all important platforms.

11.2.1   Configuring a Web Server for mod_python

Since this is an Apache module, it must be installed separately.[4] Then we have to modify the apache server configuration file adding:

```
<Directory /var/www/apache2-default>
    AddHandler mod_python .py
    PythonHandler mptest
    PythonDebug On
</Directory>
```

The directory (in this case /var/www/apache2-default) specified will be the directory where our script will be located. mod_python is activated with the line AddHandler modpython .py and the name of the script that will be

---

[4]Under Linux the package is installed with `apt-get install libapache-mod-python2.5` in Debian-based systems or with `yum install mod_python` for Red Hat. For Windows there is an installer.

used to process the request is named, in this case, mptest. At this level it is important to point out an important difference with CGI. In this case, all requests for whichever type of file *.py will be directed to a single Python script (mptest in this case). The last line (PythonDebug On) offers adequate error messages to debug the program and it is convenient to leave it while building the program.

Below there is a possible mptest.py script:

Listing 11.6: Executable program in a mod_python server

```
1 from mod_python import apache
2
3 def handler(req):
4     req.content_type = "text/plain"
5     req.write("Hello, World!")
6     return apache.OK
```

Code explanation: On the first line the Apache module is imported to the server as an interface to the web server. This line will be present each time we use mod python. On line 3 we define the handler function which will be called when we execute the script. Take note that the parameter is req, which is an object that offers us all the information about this request. On line 4 we indicate the type of content, in this case we use "text/plain" because we will test it with text. If we are going to produce HTML, we will use "text/html." On line 5 we wrote the string that the user will see. The final line indicates to Apache that the request has been processed correctly.

11.2.2 Web Program to Calculate the Net Charge of a Protein (mod_python version)

We can create the same program that we created on page 231, but using mod_python instead of CGI. This will allow us to observe the differences between both methods. As a first step we will show the HTML code for the form, that practically doesn't have any differences, except for the name of the program that will process it.

```
<html><head><title>Protein Charge Calculator</title></head>
<body><h2>Protein Charge Calculator</h2>
<form action='mptest.py' method='post'>
Enter the amino-acid sequence:
<textarea name="seq" rows="5" cols="40"></textarea><p>
Do you want to see the proportion of charged amino-acid?
<p><input type="radio" name="prop" value="y">Yes<br>
<input type="radio" name="prop" value="n"> No<br><p>
Job title (optional):
<input type="text" size="30" name="title"
```

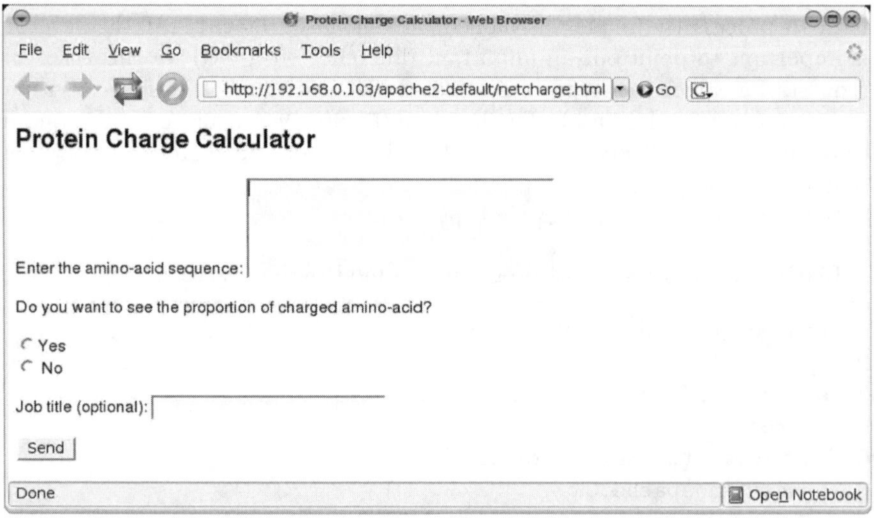

FIGURE 11.5:   HTML form to submit data to a Python script.

```
value=""><br><p>
<input type='submit' value='Send'></form></body></html>
```

This generates a form like the one in Figure 11.5.

The script that processes our form has a few significant differences:

Listing 11.7: Function to calculate the net charge of a protein proportion of charged amino acid with mod_python (py3.us/53)

```
 1 #!/usr/bin/env python
 2 from mod_python import apache
 3 from mod_python.util import FieldStorage
 4
 5 def chargeandprop(AAseq):
 6     protseq=AAseq.upper()
 7     charge = -0.002
 8     cp = 0
 9     AACharge = {"C":-.045,"D":-.999,"E":-.998,"H":.091,
10             "K":1,"R":1,"Y":-.001}
11     for aa in protseq:
12         charge += AACharge.get(aa,0)
13         if aa in AACharge:
14             cp += 1
15     prop=float(cp)/len(AAseq)*100
16     return (charge,prop)
17
```

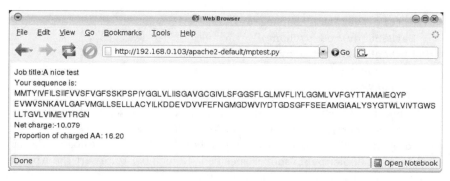

FIGURE 11.6: HTML generated by a Python script.

```
18 def handler(req):
19     req.content_type='text/html'
20     fs = dict(FieldStorage(req))
21     charge,pvalue = chargeandprop(fs['seq'])
22     req.write("<html><body>Job title:"+fs['title']+"<br/>"+
23             "Your sequence is:<br/>"+fs['seq']+"<br/>"+
24             "Net charge:"+str(charge)+"<br/>")
25     if fs['prop']=="y":
26         req.write("Proportion of charged AA: %.2f "
27                 %pvalue)
28     req.write("<br/></body></html>")
29     return apache.OK
```

Code explanation: The first two lines are those that are expected for these types of programs. On line 3 `FieldStorage` is imported in order to recover the data sent by the form. From lines 5 to 16, there is a function that calculates the net charge and the proportion of charged amino acids. The main section of the program is in reality the function `handler`, that begins from line 18. On line 19 we define the content type as "text/html." On line 20 we generate a dictionary (`fs`) with the values sent by the form. The rest of the code is similar to the CGI example.

In Figure 11.6, we see the output generated by script 11.7.

11.2.3 mod_python with Publisher

An alternative that involves **mod_python** is to use a higher level handler called **Publisher**.

Modifying Apache to Use Publisher

The Apache configuration file must be slightly modified to use Publisher:

```
<Directory /var/www/apache2-default>
    SetHandler mod_python
    PythonHandler mod_python.publisher
    PythonDebug On
</Directory>
```

The first line (`SetHandler mod python`) tells the web server that all files in this directory (texttt/var/www/apache2-default in this case) must be handled by `mod_python`. The second line (textttPythonHandler mod_python.publisher) is where the handler `publisher` is specified. Remember to restart the Apache Eeb server after modifying the configuration file.[5]

To compare both systems we will use the same example as before.

11.2.4 Web Program to Calculate the Net Charge of a Protein (mod_python.publisher version)

The web form is the same as the previous one, with the only difference in the path of the script that processes the data. Apart from pointing to the script, you have to specify the name of the function that should process the data. If the script is called `handler.py` and the function is `netc`, the associated path for this request would be: `http://yoursite/handler.py/netc`. This system creates cleaner and easier to read URLs. As an additional bonus, the site will be indexed better by search engines. Let's see the HTML form code:

```
<html>
<head><title>Protein Charge Calculator</title></head>
<body>
<h2>Protein Charge Calculator</h2>
<form action='/apache2-default/handler.py/netc' method='POST'>
Enter the amino-acid sequence:
<textarea name='seq' rows='5' cols='40'></textarea><p>
Do you want to see the proportion of charged amino-acid?<p>
<input type='radio' name='prop' value='y'>Yes<br>
<input type='radio' name='prop' value='n'>No<br><p>
Job title (optional):
<input type='text' size='30' name='title'><br><p>
<input type='submit' value='Send'>
</form>
</body>
</html>
```

The script associated with this form is different to previous cases. The first detail to take into account is that the code must be inside a function and this

---

[5]In Debian based Linux Apache is restarted with the script *apachectl*: `$ sudo /usr/sbin/apache2ctl -k restart`.

should be called like the variable `action` from the form that invokes it. In this case, the function that works as the entry point to our code is `netc`. Another difference to spot is that the variable names can be passed to the function in a straightforward manner.

Listing 11.8: Net charge of a protein with Web Publisher (py3.us/54)

```python
1  #!/usr/bin/env python
2
3  def chargeandprop(AAseq):
4      protseq = AAseq.upper()
5      charge = -0.002
6      cp = 0
7      AACharge = {"C":-.045,"D":-.999,"E":-.998,"H":.091,
8                  "K":1,"R":1,"Y":-.001}
9      for aa in protseq:
10         charge += AACharge.get(aa,0)
11         if aa in AACharge:
12             cp += 1
13     prop = float(cp)/len(AAseq)*100
14     return (charge,prop)
15
16 def netc(req,seq,title,prop):
17     req.content_type = 'text/html'
18     charge,propval = chargeandprop(seq)
19     req.write("<html><body>Job title: "+title+"<br/>"+
20               "Your sequence is:<br/>"+seq+"<br/>"+
21               "Net charge: "+str(charge)+"<br/>")
22     if prop=="y":
23         req.write("Proportion of charged AA: %.2f" %propval)
24     req.write("<br/></body></html>")
25     return None
```

The result is identical to the one in Figure 11.6, with the only difference in the URL path in the address bar.

11.3   WSGI

Before WSGI there was a lot of incompatible choices for web programming in Python. Some of them were *web frameworks*, that is, a set of programs for development of dynamic Web sites. The problem with these frameworks was that each one operated in a different way and most of them were tied to a Web server, limiting the choice of Web server/application pair.

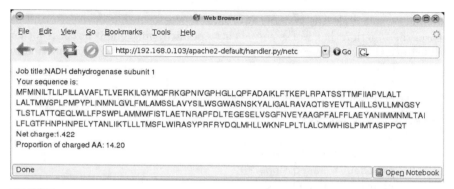

FIGURE 11.7:   HTML generated by a Python script.

WSGI was made to fill this gap, and it is defined as a "simple and universal interface between web servers and web applications or frameworks." A lot of components (or middleware) are now WSGI compatible, so the programmer doesn't need to deal directly with WSGI. Once an application works with a middleware, it can be deployed in any WSGI compliant server. WSGI is now standardized and part of the Python language (as described in PEP 333). For these reasons, WSGI is the recommended choice for Web programming.

11.3.1   Preparatory Steps

There are several ways to run a WSGI in the Apache Web server. In this book we will use **mod_wsgi**, an Apache module made to host Python application which supports the Python WSGI interface.

The module can be downloaded from the project website[6] or installed automagically by using operating system package manager.[7]

Once **mod_wsgi** is installed, you have to modify the `apache.conf` file by adding a line like this:

```
WSGIScriptAlias webpath path_in_server
```

Where `webpath` is the path seen by the user and `path_in_server` is the path to the file that will receive all the request in this directory. For example,

```
WSGIScriptAlias / /var/www/sitepath/htdocs/test.wsgi
```

That means that every request pointing to any page in the root directory of the Web server will be handled by a script localted in `/var/www/sitepath/-htdocs/test.wsgi`.

---

[6]http://code.google.com/p/modwsgi/

[7]It is called `libapache2-mod-wsgi` in Debian based systems.

11.3.2 "Hello World" in WSGI

Here is a simple "Hello World" application in WSGI:

Listing 11.9: Hello World in WSGI (py3.us/55)

```
1 def application(environ, start_response):
2     status = '200 OK'
3     output = "<html><head><title>HW WSGI</title></head>\
4              <body>Hello World!</body></html>"
5     response_headers = [('Content-type', 'text/html'),
6                         ('Content-Length', str(len(output)))]
7     start_response(status, response_headers)
8     return output
```

application is a function that the server calls for each request. It has two parameters, environ and start_response. The first one (environ) is a dictionary with the CGI defined variables and some extra information. start_response is an executable function with the information that returns the HTTP headers. As you may see, there is a lot of "low level" statements going around. This can be avoided by using *middleware*, that is, applications used as a glue between different software components (like your program and the WSGI server). Since WSGI is included with Python, there are a lot of middleware created for it.[8]

Tip: Using the Built-in Server.

You don't need a full featured web server to test your scripts. Since Python 2.5 there is a web server as part of the standard library. This server is not suitable for real life large scale applications, but it is more than enough for testing. If you don't have access to a web server, you can program your application anyway, test it in this built-in server and deploy the application in any other WSGI compatible web server when available.

In order to run the server, import make_server, initialize it with the hostname, port and application name and start the server with serve_forever(). In other words, it only takes three lines to have a working web server:

```
from wsgiref.simple_server import make_server
server = make_server('localhost', 8888, application)
server.serve_forever()
```

---

[8]See a complete list in http://wsgi.org/wsgi/Middleware_and_Utilities.

In this book we will use **yaro** and **selector**, both useful middleware developed by Luke Arno.[9] **yaro** (**Y**et **A**nother **R**equest **O**bject) hides under the hood lot of WSGI functions. In the word of the author: "is intended to be simple and useful for web developers who don't want to have to know a lot about WSGI to get the job done." `selector` is used to select which function is executed according to URL path and method (`post` or `get`). See the same code as listing 11.9 but with **yaro**:

Listing 11.10: Hello World with yaro (py3.us/56)

```
1 from yaro import Yaro
2
3 def _application(req):
4     output = "<html><head><title>HW WSGI</title></head>\
5                 <body>Hello World!</body></html>"
6     return output
7
8 application = Yaro(_application)
```

Code explanation: The main function is defined in the _application function. This function is wrapped by **Yaro** (line 8), this way **Yaro** adds functionality to _application. Note that this function takes `req` as argument. This is the *request* object and it contains useful methods, like the `environ` object.

This example shows how to use WSGI with yaro, but doesn't show what a complete application looks like. For this reason I present code 11.11 that is the version WSGI version of code 11.8:

Listing 11.11: Net charge in WSGI with yaro (py3.us/57)

```
1 #!/usr/bin/env python
2
3 from selector import Selector
4 from yaro import Yaro
5
6 rootdir = '/var/www/mywebsite/htdocs/'
7
8 def index(req):
9     return open(rootdir+'form.html').read()
10
11 def _chargeandprop(AAseq):
```

---

[9]Both programs are available at `http://lukearno.com/projects`.

```
12       protseq = AAseq.upper()
13       charge = -0.002
14       cp = 0
15       AACharge = {"C":-.045,"D":-.999,"E":-.998,"H":.091,
16               "K":1,"R":1,"Y":-.001}
17       for aa in protseq:
18           charge += AACharge.get(aa,0)
19           if aa in AACharge:
20               cp += 1
21       prop = 100.*cp/len(AAseq)
22       return (charge,prop)
23
24 def netc(req):
25       seq = req.form.get('seq','')
26       title = req.form.get('title','')
27       prop = req.form.get('prop','')
28       charge,propval = _chargeandprop(seq)
29       yield "<html><body>Job title: %s<br/>"%title+\
30               "Your sequence is:<br/>%s<br/>"%seq+\
31               "Net charge: %s<br/>"%charge
32       if prop=="y":
33           yield "Proportion of charged AA: %.2f"%propval
34       yield "<br/></body></html>"
35
36 s = Selector(wrap=Yaro)
37 s.add('/', GET=index)
38 s.add('/netc', POST=netc)
39 application = s
```

Code Explanation: This program is composed of two main functions: index (from line 8) and netc (from line 24), each function represents a page that the user receives when the function is invoked. The program begins at line 36 where a **Selector** is instanced as s. Note that wrap=Yaro part, this indicates that every function must be wrapped by the Yaro class (thus saving a step like line 8 in code 11.10). Line 37 can be read as: If the user makes a *GET* request of the root page, send him what returns function index. In the same way line 38 means: If the user makes a *POST* request of page netc, send him the result of the function netc. In fact, netc is not a function, but a generator. And it will be executed until exhausted. Note that index just returns a file. This also can be accomplished with the Static module[10] from the same author or Yaro and Selector. It even has some functions to server "template files" and give them variable content "on the fly." This is slower than letting the Web server handle a static file. You can't serve static pages

---

[10]It is available from http://lukearno.com/projects/static.

from the same directory where you are serving WSGI pages. So you should serve them from another directory or even from another subdomain. In line 39 the selected object is named `application`, that is the name of the function that is executed in the WSGI program. Note lines 25 to 27, where the method `form` of the `req` object is invoked to get the data entered by the user in the Web form.

11.4 Alternative Options for Making Python Based Dynamic Web Sites

Solutions presented up to this point are useful enough to build small and medium sized sites from the ground up. But if your website uses advanced features like multiple forms, user and session management and internationalization support, it would be better to use a framework (or Web framework) where most of these features are already covered. Since these types of application are beyond the scope of this book, I will show a table that summarizes the most important frameworks (see table 11.1). The table is sorted roughly on abstraction level. The first entries are systems with less features and requires more tweaking to achieve the same result than a higher level framwork.

No framework has received the status of "Python official web framework," so there is some usage and developer dispersion. In spite of that, **Django** is gaining momentum as the most popular web framework.

TABLE 11.1: Frameworks for Web Development (sorted by abstraction level)

Name	URL	Description
Web.py	`webpy.org`	Minimalist Web framework
Pylons	`pylonshq.com`	Rapid Web application development framework
Zope	`zope.org`	One of the first Web application servers written in Python
Twisted	`twistedmatrix.com`	An event-driven networking engine
Plone	`plone.net`	Content management system built on Zope
Django	`djangoproject.com`	High-level Python Web framework that encourages rapid development
TurboGears	`turbogears.org`	Complete Web framework that integrates: SQLObject, Cherrypy, Kid, and Mochikit
web2py	`mdp.cti.depaul.edu`	Full-stack framework for agile development of database-driven Web-based applications

There are some alternatives to Web frameworks for Python Web developing. **Pyjamas** is a Python-to-JavaScript compiler with an AJAX Framework and Widget set that can be used to write applications that run in most Web browsers. Since Web browsers can't execute Python code, this program translate Python code into JavaScript, a language fully supported in any modern Web browser. This is port of the Google Web Toolkit, a set of tools that convert Java code into JavaScript. Note that this program is made for front-end design using client-side scripting insted of server-side applications.

Another client-side alternative is **Silverlight**, a Flash-like run-time system made by Microsoft. From version 2 of Silverlight, applications can be written in any .net programming language. Since IronPython is a Python implementation that runs on .net platform, it can be used to make Silverlight applications with Python. Silverlight 2 is already available for Windows and Mac, and the Linux support is in the works via the Moonlight project from Mono.

11.5 Some Words about Script Security

If your scripts are made only for your own consumption, you can skip this section and jump to the next section on page 244.

Something to have in mind when designing web applications is that the user may (and will) enter data in an unexpected format. Even if this is not always true, when the form is publicly accessible on the Internet, this threat shouldn't be underestimated.

There will be people who will not know how to complete the online form and try whatever they think is best. There will be attackers who will test your site looking for any exploitable vulnerability.

A first barrier that can be used to avoid misuse of your scripts is to use JavaScript (JS) for form validation. It is not the purpose of this book to teach JS, so there are links in the "Additional resources" section.

JS can be used to avoid problems associated with the end user, but it is rather useless as a deterrent for anyone who is determined to attack your server. If anyone wants to interact with your script, he could do it without using the Web browser, bypassing completely your carefully created JS code.

This is why all data validation must be done "server side" instead of leave it to a random user who cannot be trusted.

A script like the one in code 11.8 does not require as much precautions as a CGI script that runs a shell session and pass user entered parameters to the Operating System.

Another critical point to watch out for is when a script accesses a database engine. There is a chance that an attacker could inject SQL commands to produce unwanted results (like listing the full contents of a table with sensitive information like usernames and passwords[11]). This kind of attack is called "SQL injection" and it will be covered in the "Python and databases" chapter.

There is no rule of thumb regarding how to sanitize every kind of input, but it depends on the particular application. Anyway there are some outlines of what to take into consideration at the moment of designing the security of our application.

1. Identify where the data can access the application. Clearly the most evident point of entry are the forms you set up for data input. But you should not overlook other points of entry like URLs, files stored on the server, other web sites if your scripts read external sources like RSS feeds.

2. Watch for escape characters used by the program your application interacts with. These should always be filtered. If your program accesses a Unix shell, filter the ";" character (semicolon) since it can be used to issue arbitrary commands. This depends on the type of shell your system is using. Some characters you should consider watching are: ;, &&, ||, \ and ".

3. Consider making a list of valid accepted characters (a "white list") to make sure that your strings have only the required characters.

4. The running privileges of the web server program must be the lowest possible. Most Unix systems use an ad-hoc user for the web server process. This is called "Principle of Least Privilege." The program is given the smallest amount of privilege required to do its job. This limits the abuse that can be done to a system if the web server process is hijacked by an attacker.

11.6 Where to Host Python Programs

If you've tested satisfactorily your scripts on your local server, it's time to put them on the Internet so that the rest of the world can enjoy them. Usually the institution that you work for has a web server where you can store your scripts, for which the first step would be to ask for support from your IT department. In the event that you don't get a satisfactory response, you would have to consider resolving the problem by yourself. It is not too

---

[11]It is not a good idea anyway to store passwords in plain text in a database. The best practice is to store a hash of the password instead, like SHA224. Python has several message digest algorithms in the `hashlib` module.

difficult. There are thousands of web hosting businesses. Look for one that explicitly supports Python.

Among the diverse plans that offer the web hosting businesses, choose the "shared" plan type if your script is very simple and does not involve installation of programs or additional modules. If your script executes programs that aren't installed on the server, as is the case with Biopython, you can ask for it to be installed. Ask before contracting the service if the install modules on demand. Another problem that can surface is with the web frameworks. Some work as a long running process, which is not permitted by the hosting agreement.

Make sure that the version of Python installed on the hosting server is compatible with your scripts. This is not a minor topic considering that the servers are used to not having the latest version of Python installed.

If the web hosting service does not allow program installation, you will have to consider the dedicated hosting type, where you have root access to a computer where there are not limits with regard to what you can install (assuming that the necessary disk space is available). These types of plans tend to be expensive as they require contracting the use of a computer for each user, but thanks to virtualization technologies, it is possible to contract a dedicated virtual hosting plan at a more than affordable price. This is because the computer is shared between various users, but it differs from the shared hosting plan type in that each user has total access to the server. For very demanding applications, this may not be the best solution and you may have to resort to the use of dedicated hosting (not virtual).

A new alternative to consider is the "Google App Engine," this system enables you to build web applications on the same scalable systems that power Google applications. Leave Google take care of Apache web server configs, startup scripts, the SQL database, server monitoring and software upgrades. You just write your Python code. Applications designed for this engine are implemented using the Python programming language. The App Engine Python runtime environment includes a specialized version of the Python interpreter. For more information on "Google App Engine," see http://code.google.com/appengine.

On the Python official site there is a listing of hosting providers with Python support. It includes both community and commercial providers: http://wiki.python.org/moin/WebProgramming.

11.7   Additional Resources

- W3Schools: "JavaScript Form Validation."
  http://www.w3schools.com/js/js_form_validation.asp

- JavaScript-Coder.com: "JavaScript Form Validation : quick and easy!"
  `http://www.javascript-coder.com`

- Python Web Server Gateway Interface v1.0.
  `http://www.python.org/dev/peps/pep-0333`

- Server Fault is a Q&A site for system administrators and IT professionals that's free.
  `http://serverfault.com`

- Armin Ronacher, "Getting Started with WSGI."
  `http://lucumr.pocoo.org/2007/5/21/getting-started-with-wsgi`

- Pesto: A library for Python Web applications.
  `http://pesto.redgecko.org`

- Microsoft Silverlight: A Flash-like web application framework.
  `http://www.silverlight.net/`

- Pyjamas: A port of Google Web Toolkit to Python.
  `http://pyjs.org/`

11.8   Self-Evaluation

1. What is CGI?

2. What is cgitb? When should you avoid its use?

3. How do you use **cgi.FieldStorage** to retrieve values sent over an HTML form?

4. What is mod_python? Name advantages of mod_python over CGI.

5. What is WSGI? Why is it the recommended choice for Web programming?

6. What is the rationale for using Yaro or any other "middleware"?

7. Python includes a limited web server. Why would you use such a Web server if there are free full featured web servers like Apache?

8. Name security considerations to take into account when running a live Web server.

9. Why is client-side data validation not useful as server-side data validation?

10. What is the difference between shared, dedicated and virtual dedicated hosting? When would you use dedicated hosting over a shared plan?

Chapter 12

XML

12.1 Introduction to XML

What Is XML?

A widespread problem in all branches of information technology is the storage and interchange of data. Each application has its own particular way of storing the generated information, which is often a problem, especially when we don't have the application that generated the data.

For example, DNA sequencers made by Applied Biosystems generate data and store it in files with the extension .ab1. If we want to access data stored in such a file, we need to know how it is structured internally. In this case, the creator of the format has released the specification of the file;[1] and it would be possible, though not necessarily easy, to write code to extract our data from these files. Usually we do not have such good luck, and it is very common to find data file formats poorly documented, or not documented at all. In many cases those who have wanted to open these files have had to resort to "reverse engineering," with mixed results. To avoid this type of problem and to make more fluid exchange of data between applications from different manufacturers, the W3C[2] developed the e**X**tensible **M**arkup **L**anguage, better known as XML.

XML is a way of representing data. What kind of data? Practically any type can be represented using XML. Configuration files, databases, web pages, spreadsheets, and even drawings can be represented and stored in XML.

For some specific applications, there have been defined subsets of XML, prepared for representing a particular type of data. So, mathematical formulas can be stored in an XML dialect called MathML,[3] vector graphics in SVG,[4] chemical formulas in CML.,[5] and page printouts can be represented with

---

[1] File format specification for ABI files are available at `www.appliedbiosystems.com/support/software_community/ABIF_File_Format.pdf`.

[2] The World Wide Web Consortium, abbreviated W3C, is an international consortium that produces standards for the World Wide Web.

[3] `http://www.w3.org/Math`

[4] `http://www.w3.org/Graphics/SVG`

[5] `http://www.xml-cml.org`

XSLFO.[6] In terms of bioinformatics, there are various formats based on XML, the most well-known being BSML[7] and INSDSeq XML.[8]

In addition to the above formats, more applications store their data in XML. This means that, by learning to read XML, we can access a multitude of files from the most diverse origins.

Before going into details on how to process this type of file, I want to share a W3C document called "XML in 10 points" that can shows the big picture:

XML in 10 Points[9]

1. XML is for structuring data: Structured data includes things like spreadsheets, address books, configuration parameters, financial transactions, and technical drawings. XML is a set of rules (you may also think of them as guidelines or conventions) for designing text formats that let you structure your data. XML is not a programming language, and you don't have to be a programmer to use it or learn it. XML makes it easy for a computer to generate data, read data, and ensure that the data structure is unambiguous. XML avoids common pitfalls in language design: it is extensible, platform-independent, and it supports internationalization and localization. XML is fully Unicode-compliant.

2. XML looks a bit like HTML: Like HTML, XML makes use of tags (words bracketed by '<' and '>') and attributes (of the form name="value"). While HTML specifies what each tag and attribute means, and often how the text between them will look in a browser, XML uses the tags only to delimit pieces of data, and leaves the interpretation of the data completely to the application that reads it. In other words, if you see "<p>" in an XML file, do not assume it is a paragraph. Depending on the context, it may be a price, a parameter, a person, a p... (and who says it has to be a word with a "p"?).

3. XML is text, but isn't meant to be read: Programs that produce spreadsheets, address books, and other structured data often store that data on disk, using either a binary or text format. One advantage of a text format is that it allows people, if necessary, to look at the data without the program that produced it; in a pinch, you can read a text format with

---

[6]http://www.w3.org/TR/xsl11

[7]Bioinformatic Sequence Markup Language[TM](http://www.bsml.org) created by the National Human Genome Research Institute.

[8]Format of the International Nucleotide Sequence Database Collaboration (http://www.insdc.org/files/documents/INSD_V1.4.dtd).

[9]Taken from http://www.w3.org/XML/1999/XML-in-10-points. Authorized by "Copyright ©[1999] World Wide Web Consortium, (Massachusetts Institute of Technology, European Research Consortium for Informatics and Mathematics, Keio University). All Rights Reserved. http://www.w3.org/Consortium/Legal/2002/copyright-documents-20021231."

your favorite text editor. Text formats also allow developers to more easily debug applications. Like HTML, XML files are text files that people shouldn't have to read, but may when the need arises. Compared to HTML, the rules for XML files allow fewer variations. A forgotten tag, or an attribute without quotes makes an XML file unusable, while in HTML such practice is often explicitly allowed. The official XML specification forbids applications from trying to second-guess the creator of a broken XML file; if the file is broken, an application has to stop right there and report an error.

4. XML is verbose by design: Since XML is a text format and it uses tags to delimit the data, XML files are nearly always larger than comparable binary formats. That was a conscious decision by the designers of XML. The advantages of a text format are evident (see point 3), and the disadvantages can usually be compensated at a different level. Disk space is less expensive than it used to be, and compression programs like zip and gzip can compress files very well and very fast. In addition, communication protocols such as modem protocols and HTTP/1.1, the core protocol of the Web, can compress data on the fly, saving bandwidth as effectively as a binary format.

5. XML is a family of technologies: XML 1.0 is the specification that defines what "tags" and "attributes" are. Beyond XML 1.0, "the XML family" is a growing set of modules that offer useful services to accomplish important and frequently demanded tasks. XLink describes a standard way to add hyperlinks to an XML file. XPointer is a syntax in development for pointing to parts of an XML document. An XPointer is a bit like a URL, but instead of pointing to documents on the Web, it points to pieces of data inside an XML file. CSS, the style sheet language, is applicable to XML as it is to HTML. XSL is the advanced language for expressing style sheets. It is based on XSLT, a transformation language used for rearranging, adding and deleting tags and attributes. The DOM is a standard set of function calls for manipulating XML (and HTML) files from a programming language. XML Schemas 1 and 2 help developers to precisely define the structures of their own XML-based formats. There are several more modules and tools available or under development. Keep an eye on W3C's technical reports page.

6. XML is new, but not that new: Development of XML started in 1996 and it has been a W3C Recommendation since February 1998, which may make you suspect that this is rather immature technology. In fact, the technology isn't very new. Before XML there was SGML, developed in the early '80s, an ISO standard since 1986, and widely used for large documentation projects. The development of HTML started in 1990. The designers of XML simply took the best parts of SGML, guided

by the experience with HTML, and produced something that is no less powerful than SGML, and vastly more regular and simple to use. Some evolutions, however, are hard to distinguish from revolutions... And it must be said that while SGML is mostly used for technical documentation and much less for other kinds of data, with XML it is exactly the opposite.

7. XML leads HTML to XHTML: There is an important XML application that is a document format: W3C's XHTML, the successor to HTML. XHTML has many of the same elements as HTML. The syntax has been changed slightly to conform to the rules of XML. A format that is "XML-based" inherits the syntax from XML and restricts it in certain ways (e.g, XHTML allows "<p>", but not "<r>"); it also adds meaning to that syntax (XHTML says that "<p>" stands for "paragraph", and not for "price", "person", or anything else).

8. XML is modular: XML allows you to define a new document format by combining and reusing other formats. Since two formats developed independently may have elements or attributes with the same name, care must be taken when combining those formats (does "<p>" mean "paragraph" from this format or "person" from that one?). To eliminate name confusion when combining formats, XML provides a namespace mechanism. XSL and RDF are good examples of XML-based formats that use namespaces. XML Schema is designed to mirror this support for modularity at the level of defining XML document structures, by making it easy to combine two schemas to produce a third which covers a merged document structure.

9. XML is the basis for RDF and the Semantic Web: W3C's Resource Description Framework (RDF) is an XML text format that supports resource description and metadata applications, such as music play-lists, photo collections, and bibliographies. For example, RDF might let you identify people in a Web photo album using information from a personal contact list; then your mail client could automatically start a message to those people stating that their photos are on the Web. Just as HTML integrated documents, images, menu systems, and forms applications to launch the original Web, RDF provides tools to integrate even more, to make the Web a little bit more into a Semantic Web. Just like people need to have agreement on the meanings of the words they employ in their communication, computers need mechanisms for agreeing on the meanings of terms in order to communicate effectively. Formal descriptions of terms in a certain area (shopping or manufacturing, for example) are called ontologies and are a necessary part of the Semantic Web. RDF, ontologies, and the representation of meaning so that computers can help people do work are all topics of the Semantic Web Activity.

10. XML is license-free, platform-independent and well-supported: By choosing XML as the basis for a project, you gain access to a large and growing community of tools (one of which may already do what you need!) and engineers experienced in the technology. Opting for XML is a bit like choosing SQL for databases: you still have to build your own database and your own programs and procedures that manipulate it, but there are many tools available and many people who can help you. And since XML is license-free, you can build your own software around it without paying anybody anything. The large and growing support means that you are also not tied to a single vendor. XML isn't always the best solution, but it is always worth considering.

12.2 Structure of an XML Document

We do not need to know the details of the internal structure of an XML document. This is because Python has its own tools for accessing this type of file. The developers of Python had to deal with the internals of XML in order to construct these tools; however I think that is necessary is to have a minimal notion of the structure of XML files, in order to make better use of the facilities Python provides us for handling this type of document.

Let's see a sample XML document, in this case a Uniprot record:[10]

---

[10]This record was altered sligthly to fit the page. Information left out was not relevant for explanation in this book.

Listing 12.1: UNIPROT record in XML

```
<?xml version="1.0" encoding="UTF-8"?>
<uniprot xmlns="http://uniprot.org/uniprot"
 xmlns:xsi="http://www.w3.org/2001/XMLSchema-instance"
 xsi:schemaLocation="http://uniprot.org/uniprot
 http://www.uniprot.org/support/docs/uniprot.xsd">
<entry dataset="TrEMBL" created="2000-10-01" version="35">
  <accession>Q9JJE1</accession>
  <organism key="1">
    <name type="scientific">Mus musculus</name>
    <lineage>
      <taxon>Eukaryota</taxon>
      <taxon>Metazoa</taxon>
      <taxon>Chordata</taxon>
      <taxon>Craniata</taxon>
      <taxon>Vertebrata</taxon>
      <taxon>Euteleostomi</taxon>
      <taxon>Mammalia</taxon>
      <taxon>Eutheria</taxon>
      <taxon>Euarchontoglires</taxon>
      <taxon>Glires</taxon>
      <taxon>Rodentia</taxon>
      <taxon>Sciurognathi</taxon>
      <taxon>Muroidea</taxon>
      <taxon>Muridae</taxon>
      <taxon>Murinae</taxon>
      <taxon>Mus</taxon>
    </lineage>
  </organism>
  <dbReference type="UniGene" id="Mm.248907" key="5"/>
  <sequence length="393" checksum="E0C0CC2E1F189B8A">
MPKKKPTPIQLNPAPDGSAVNGTSSAETNLEALQKKLEELELDEQQRKRL
EAFLTQKQKVGELKDDDFEKISELGAGNGGVVFKVSHKPSGLVMARKLIH
LEIKPAIRNQIIRELQVLHECNSPYIVGFYGAFYSDGEISICMEHMDGGS
LDQVLKKAGRIPEQILGKVSIAVIKGLTYLREKHKIMHRDVKPSNILVNS
RGEIKLCDFGVSGQLIDSMANSFVGTRSYMSPERLQGTHYSVQSDIWSMG
LSLVEMAVGRYPIPPPDAKELELLFGCHVEGDAAETPPRPRTPGGPLSSY
GMDSRPPMAIFELLDYIVNEPPPKLPSGVFSLEFQDFVNKCLIKNPAERA
DLKQLMVHAFIKRSDAEEVDFAGWLCSTIGLNQPSTPTHAASI
  </sequence>
</entry>
</uniprot>
```

In broad outlines, the structure of an XML document is very simple. It generally consists of a prologue, a body, and an epilogue.[11]

Prologue

The prologue is an optional section that marks the beginning of the XML data and gives important information to the parser. A prologue might have only one line, like this one,

```
<?xml version="1.0" encoding="UTF-8"?>
```

Or several lines:

```
<?xml version="1.0"?>
<!DOCTYPE BlastOutput PUBLIC "-//NCBI//NCBI BlastOutput/EN"
"http://www.ncbi.nlm.nih.gov/dtd/NCBI_BlastOutput.dtd">
<!-- edited with XMLSPY (http://www.xmlspy.com) by Andy -->
```

The first line is the **XML declaration** where the XML version and character code are specified. Character code information is optional only if the document is encoded in UTF-8 or UTF-16.

The second line is the **DOCTYPE declaration**, whose purpose is to relate the XML document with a document type definition (DTD). This DTD file contains information about the particular structure of the XML file: it says which tags and attributes are permitted, as well as where they can be found. In some cases, in place of a DTD reference, there are references to an alternate method to DTD called XML Schema which serves the same function but with better performance, and with a syntax based on XML. The structure of a DTD or XML Schema file is beyond the scope of this book; however, there are several, quite complete, references on the Internet (see **Additional Resources** at the end of this chapter).

The third line, in this case, is a comment. It is equivalent to # in Python. It begins with "<!--" and ends with "-->", and can be in the prologue as well as in the body of an XML document. It is the same type of comment that is used in HTML and it can span multiple lines.

Body

The body is where reside the **elements**, the true protagonists of XML files. An element is the information from the beginning of the start tag to the end of the end tag, including all that lies in between.

An example of an element that can be found in the body of an XML document:

---

[11]This feature is seldom used, so the prologue and the body are the most important parts of an XML file.

```
<taxon>Eukaryota</taxon>
```

Where <taxon> is the start tag, </taxon> is the end tag, and the contents (Eukaryota), is that which is between the two tags.

Elements may show up empty. It is valid to write, for example:

```
<accession></accession>
```

While in this case it doesn't make much sense to have nothing contained in the "accession" element (a Uniprot base should always have a number of accessions), it is possible that in some circumstances the contents of an element will be optional.

There is an abbreviated way to represent empty elements, called an "empty element tag," and consists of the name of the element followed by a forward slash (/), all enclosed by angle brackets, for example:

```
<accession/>
```

The elements can be "nested" inside one another. In listing 12.1 we can see how the element "taxon" is nested within "lineage". This gives an idea of a hierarchical structure: there are elements which are subordinate to others. We see that "taxon" is an element of "lineage", which is an element of "organism". Normally this type of structure is compared to a tree. The first element is called the "Document Element" (in this case, "uniprot"), from which hangs all the rest, which are its "children." To obtain a graphical representation of this tree, one can use a program like XML Viewer.[12], which shows something similar to figure 12.1

Some elements have "attributes," that is, additional information about the element. The general syntax of an element with an attribute is,

```
<element attributeName="value">
```

Continuing with the example of listing 12.1, we come across other elements with attributes as, for example,

```
<name type="scientific">
```

In this case the element called "name" is an attribute of "type", which has a value of "scientific." Additionally it can have more than one attribute, as in the element "sequence":

```
<sequence length="393" checksum="E0C0CC2E1F189B8A">
```

---

[12]XML Viewer is available at http://sourceforge.net/projects/ulmxmlview.

FIGURE 12.1:   Screenshot of XML viewer: Tree viewer shows the structure of the document in listing 12.1.

Here the attributes are "length" and "checksum", whose values are "393" and "E0C0CC2E1F189B8A", respectively.

At this level already we have elements which give us an idea of the data contained in an XML file. Of the record of listing 12.1 we can say that the element "sequence" contains a nucleotide sequence which has a length of 393bp, a known signature, and has as an ID "Q9JJE1" of the UniProt base. All this without prior knowledge of the data structure and without the use of a special program. Try to open an **.ab1** file to see if you can find any recognizable element.

Despite having a general overview of the structure of XML files, you will find the format has other particularities that go beyond the scope of this book. If you are interested in knowing more about XML, see the list of resources at the end of the chapter.

The following section shows how to access the contents of XML documents using Python.

12.3   Methods to Access Data inside an XML Document

Regardless of the programming language you use, there are two strategies that you can use to gain access to the information contained in an XML file.

On one hand, you can read the file in its entirety, analyze the relationships between the elements, and build a tree-type structure, by which the application can navigate the data. This is called the Document Object Model (DOM) and is the manner recommended by the W3C in parsing XML documents. In this chapter we will see two parsers of this type: **Minidom** and **ElementTree**.

Another possibility is that the application detects and reports events such as the start and the end of an element, without the necessity of constructing a tree-type representation. In the case that a tree representation is needed, this task is left to the programmer. This is the method used by the **S**imple **API** for **XML** (SAX). Generally these types of parsers are called "event driven parsers." In this chapter we will see, as an example of an event-based parser, **Iterparse** of cElementTree.

In some cases it is convenient to use DOM, while in other cases SAX is the preferred option. DOM usually implies saving the whole tree in memory for later traversal. This can present a problem at the time of parsing large documents, especially when what you want to do is simply detect the presence of a single element's value. In these cases a SAX is the most efficient parser. Nevertheless, many applications require operating on all the elements within the tree, for which we must turn to DOM. From the perspective of the programmer, the DOM interface is easier to use than SAX as it doesn't require event-driven programming.

12.3.1 DOM: Minidom

One of the DOM parsers in Python is "Minidom." This provides a method called "parse," which is what does the dirty work of reading all the archive, analyzing the structure contained in the XML, and reproduces in a representation accessible from our Python program.

Let's see an XML parse in action:

```
>>>from xml.dom.minidom import parse
>>>midom=parse("smallUniprot.xml")
```

This creates an object called "midom" that contains a representation, in tree form, of the data contained in `smallUniprot.xml`. From now on we will refer to this object to extract the information in the XML file. In this case, I am interested in the information of the protein sequence and its associated data, although the methods described here serve to extract whatever information is contained in any XML file.

This tree can be navigated with the following methods:

childNodes: A method which returns a list of all the nodes contained in the node to which this method is applied. If we apply this to our object derived from the parser:

```
>>> midom.childNodes
[<DOM Element: uniprot at 0xb65efccc>]
```

The result is a list with only one item: the reference to an element called `uniprot`. This was expected, since each XML file has one node from which hang all the others, and this is the element found. A way of directly accessing this particular element is by using **documentElement**:

```
>>> midom.documentElement
<DOM Element: uniprot at 0xb65efccc>
```

To see inside `uniprot` element, we have to apply **childNodes** again. The only difference is that this time it should be done on the first item returned by the first **childNodes** function:

```
>>> a = midom.childNodes[0].childNodes
>>> print a
[<DOM Text node "
">, <DOM Element: entry at 0xb65ef50c>, <DOM Text node "
">]
```

The result in this case is a list with three items, two text-type elements and one of type "element". The text objects correspond to what is found between the tags. If we peruse the original document, we see that the only thing between the "uniprot" tag and "entry" is a linefeed ('\n'). Use the method **data** to get that linefeed from Python:

```
>>>a[0].data
"\n"
```

In the case of the element names, the method to use is **nodeName**:

```
>>>a[1].nodeName
"entry"
```

To know what type of node is involved, there is **nodeType** method, which returns an integer (from 1 to 12) representing a node type. The node class enumerates all the constants symbolically. The list of node types follows:
ELEMENT_NODE, ATTRIBUTE_NODE, TEXT_NODE,
CDATA_SECTION_NODE, ENTITY_NODE, PROCESSING_INSTRUCTION_NODE,
COMMENT_NODE, DOCUMENT_NODE, DOCUMENT_TYPE_NODE, NOTATION_NODE
To know what class of element is "entry":

```
>>> a[1].nodeType
1
```

Or check its type using these constants:

```
>>> a[1].nodeType==x[1].TEXT_NODE
False
>>> a[1].nodeType==a[1].ELEMENT_NODE
True
```

To resume traversing the tree, apply `childNodes` as ascend and descend the tree structure (see what happens with a[1].childNodes), or search for a particular element using **getElementsByTagName(TAGNAME)**:

```
>>> a[1].getElementsByTagName("sequence")
[<DOM Element: sequence at 0xb65fff6c>]
```

Combining all we've seen so far, we can get the sequence:

```
>>> seq=a[1].getElementsByTagName("sequence")[0]
>>> seq.childNodes[0].data
'\nMPKKKPTPIQLNPAPDGSAVNGTSSAETNLEALQKKLEELELDEQQRKRL\nEAFLTQK...'
```

To remove line-feeds (\n), use the **replace()** method:

```
>>> seq.childNodes[0].data.replace("\n","")
'MPKKKPTPIQLNPAPDGSAVNGTSSAETNLEALQKKLEELELDEQQRKRLEAFLTQK...'
```

It doesn't end here; the element tagged "sequence" has some attributes that can be recovered. As a first step, check if "sequence" has attributes:

```
>>> seq.hasAttributes()
True
```

The list of attributes is provided by attributes.keys() and the contents of these can be obtained with `attributes.get(ATTRIB).value`:

```
>>> seq.attributes.keys()
[u'checksum', u'length']
>>> seq.attributes.get("checksum").value
u'E0C0CC2E1F189B8A'
>>> seq.attributes.get("length").value
u'393'
```

12.3.2 ElementTree

ElementTree is an alternative to **minidom**. It's not a module designed especially for XML, but stores in memory anything in a hierarchical structure. Because of this it can be applied to XML.

To compare different methods, the same sample file (12.1, `uniprot.xml`) is used. Following the same pattern as before, we will use the parser provided by ElementTree (parse), which since version 2.5 of Python resides in `xml.etree.ElementTree`:

```
>>> import xml.etree.ElementTree as ET
>>> tree=ET.parse("/home/sb/bioinfo/smallUniprot.xml")
```

Unlike Minidom, it doesn't return an element, but a tree instance:

```
>>> tree
<xml.etree.ElementTree.ElementTree instance at 0xb64ac94c>
```

To obtain the first element, use the **getroot()** method:

```
>>> tree.getroot()
<Element {http://uniprot.org/uniprot}uniprot at b64b0ccc>
```

The result is very similar to what is obtained by **childNodes** in minidom, with the difference that it does not returns a list with only one element, but the root element. Another difference is the presence of the **namespaceURI** in the name of the element.

Each element has one required property and others that are optional. The required property is the tag which identifies the element, the equivalent of minidom's **nodeName**.

```
>>> tree.getroot().tag
'{http://uniprot.org/uniprot}uniprot'
```

Among the optional element properties are:

attrib: Works like a dictionary where the names and values of the attributes are stored. Remembering the element "entry" that has three attributes:

```
<entry dataset="TrEMBL" created="2000-10-01" version="35">
```

Applying **attrib** on this element, it returns:

```
{'created': '2000-10-01', 'version': '35', 'dataset': 'TrEMBL'}
```

text: Returns the text of an element, for example, from,

```
<taxon>Vertebrata</taxon>
```

It returns:

```
'Vertebrata'
```

Child elements: The elements which hang from a higher-level element, and can be obtained by any of these methods:

getiterator(*tag*): It returns a list[13] of sub-elements matching the given tag, searching in all levels beneath the node in which it is invoked. The elements are returned in the order they appear in the document:

```
>>> root=tree.getroot()
>>> root.getiterator("{http://uniprot.org/uniprot}name")
[<Element {http://uniprot.org/uniprot}name at b64b434c>]
```

To see the contents of an element of the list, use the property "text":

```
>>> root.getiterator("{http://uniprot.org/uniprot}name")[0].text
'Mus musculus'
```

getiterator() can be invoked without arguments, and in that case returns a list[14] with all the elements beneath the node from which it is invoked.

```
>>> root.getiterator()
[<Element {http://uniprot.org/uniprot}uniprot at b64b0ccc>,<=
<Element {http://uniprot.org/uniprot}entry at b64b420c>,<=
<Element {http://uniprot.org/uniprot}accession at b64b428c>,<=
... cut ...
<Element {http://uniprot.org/uniprot}taxon at b64b4a8c>,<=
<Element {http://uniprot.org/uniprot}dbReference at b64b4b4c>,<=
<Element {http://uniprot.org/uniprot}sequence at b64b4c0c>]
```

---

[13]In ElementTree it returns a list, but in cElementTree it returns a *generator object*.

[14]As commented in previous footnote, ElementTree returns a list, but cElementTree returns a *generator object*.

Using this method the `namespaceURI` information can ve removed:

```
>>> for elem in root.getiterator():
        elem.tag=elem.tag.split("}",1)[1]
>>> root.getiterator()
[<Element uniprot at b6eac9cc>, <Element entry at b65b872c>,
<Element accession at b65b898c>, <Element organism at b65b8a8c>,
<Element name at b65b8b0c>, <Element lineage at b65b8b6c>,
<Element taxon at b65b8bac>, <Element taxon at b65b8bec>,
... cut ...
<Element taxon at b65be10c>, <Element dbReference at b65be1cc>,
<Element sequence at b65be28c>]
```

With a shorter name, it's easier to refer to them:

```
>>> [x.text for x in root.getiterator("taxon")]
['Eukaryota', 'Metazoa', 'Chordata', 'Craniata',<=
'Vertebrata', 'Euteleostomi', 'Mammalia', 'Eutheria',<=
'Euarchontoglires','Glires', 'Rodentia', 'Sciurognathi',<=
'Muroidea', 'Muridae', 'Murinae', 'Mus']
```

find(*pattern*): Returns the first sub-element which matches the given pattern, or "None" in case there are no matching elements. Note that it searches only in the level just below the element upon which it is called:

```
>>> root.find("entry")
<Element entry at b65c69ec>
>>> root.find("entry").tag
'entry'
>>> root.find("entry").find("accession")
<Element accession at b65c648c>
```

Use **text** to retrieve the value of the text attribute:

```
>>> root.find("entry").find("accession").text
'Q9JJE1'
```

Or using instead **findtext(*pattern*)**:

```
>>> root.find("entry").findtext("accession")
'Q9JJE1'
```

Both **find** and **findtext** find the first matching element:

```
>>> root.find("entry/organism/lineage").find("taxon")
<Element taxon at b65c61cc>
```

To have a list of all matching elements, use **findall(*pattern*)**:

```
>>> lineage=root.find("entry/organism/lineage")
>>> [x.text for x in lineage.findall("taxon")]
['Eukaryota', 'Metazoa', 'Chordata', 'Craniata', 'Vertebrata', <=
'Euteleostomi', 'Mammalia', 'Eutheria', 'Euarchontoglires', <=
'Glires', 'Rodentia', 'Sciurognathi', 'Muroidea', 'Muridae', <=
'Murinae', 'Mus']
```

As in the case of **Minidom**, we now look for the sequence and its attributes:

```
>>> root.getiterator("sequence")[0].text
'\nMPKKKPTPIQLNPAPDGSAVNGTSSAETNLEALQKKLEELELDEQQRKRL\nEAFLT...'
```

For the attributes, **attrib** works as a dictionary:

```
>>> root.getiterator("sequence")[0].attrib["checksum"]
'E0C0CC2E1F189B8A'
>>> root.getiterator("sequence")[0].attrib["length"]
'393'
```

cElementTree

Since Python 2.5, there is a C version of ElementTree (cElementTree) that is optimized to parse quickly and with less use of memory. cElementTree is used the same way as ElementTree. If your installation has this module available, it is advisable to use it. Another advantage of cElementTree is a function (not found in ElementTree) called *Iterparse*. This function provides us the use of an event based parser, which will be explained in the next section.

12.3.3   SAX: cElementTree Iterparse

cElementTree Iterparse isn't SAX, but it is included here because, unlike the other parsers, it is based on events.

Iterparse returns a flow iterable by tuples in the form (event, element). It is used to iterate over the elements and processing them on the fly. Once we've learned to use **cElementTree**, the path to understanding how to use *Iterparse* is less steep.

Let's first look for the protein sequence and its attributes:

```
>>> import xml.etree.cElementTree as cET
>>> for event, elem in cET.iterparse("smallUniprot.xml",
    events=("start", "end")):
        if event=="end" and "sequence" in elem.tag:
            print elem.text
            print elem.attrib["checksum"]
            print elem.attrib["length"]
            elem.clear()
```

```
MPKKKPTPIQLNPAPDGSAVNGTSSAETNLEALQKKLEELELDEQQRKRL
EAFLTQKQKVGELKDDDFEKISELGAGNGGVVFKVSHKPSGLVMARKLIH
LEIKPAIRNQIIRELQVLHECNSPYIVGFYGAFYSDGEISICMEHMDGGS
LDQVLKKAGRIPEQILGKVSIAVIKGLTYLREKHKIMHRDVKPSNILVNS
RGEIKLCDFGVSGQLIDSMANSFVGTRSYMSPERLQGTHYSVQSDIWSMG
LSLVEMAVGRYPIPPPDAKELELLFGCHVEGDAAETPPRPRTPGGPLSSY
GMDSRPPMAIFELLDYIVNEPPPKLPSGVFSLEFQDFVNKCLIKNPAERA
DLKQLMVHAFIKRSDAEEVDFAGWLCSTIGLNQPSTPTHAASI
```

E0C0CC2E1F189B8A
393

Looks like text and attributes are accessed the same way as with **Element-Tree**. The difference is that it's necessary to iterate it by the tuples iterparse returns, and that we use the **clear** method. The first element of the tuple is the "event" and can be one of two values: "start" or "end". If the event we received is "start", it means that we can access the name of the element and its attributes, but not necessarily its text. When we receive "end", we can be assured that we've processed all the components of that element. For this reason the previous code checked not only that we had reached the chosen element, but that we had also found the "end" event.[15] If the parser were to return only "end", there would be no need for this check:

```
>>> for event, elem in cET.iterparse("smallUniprot.xml"):
        if "sequence" in elem.tag:
            print elem.text
            print elem.attrib["checksum"]
            print elem.attrib["length"]
            elem.clear()
```

As for the **clean** method, it is used to "clean up" the node after it's used, because unlike a classic SAX parser like ElementTree, iterparse constructs a complete tree. The problem with this code is that the primary element remains with all its (now empty) children, and that uses memory. In this simple example, this behavior is not problematic, but it could very well be when processing large files. The ideal would be to access the parent node in order to clean it up.

A way to do this is to save a reference to the first variable; for this we create an iterator and obtain from it the first element, calling it "root":

---

[15]In the current implementation, the parser goes along reading 16Kb chunks, so in this case the whole sequence could be read from the "start" element. To make sure that you pick up all the elements you should read it after an "end" element.

```
>>> allelements = iterparse(source, events=("start", "end"))
>>> allelements = iter(allelements)
>>> event, root = allelements.next()
```

Now we process it the same as before, only this time we can delete the parent element specifically:

```
>>> for event, elem in allelements:
        if event=="end" and "sequence" in elem.tag:
            print elem.text
            root.clear()
```

To see the cElementTree parser in action, please turn to page 342, where there is a script that parses the product of a BLAST run.

12.4 Summary

XML means e**X**tensible **M**arkup **L**anguage and was created to enable a standard way of storing and exchanging data. One of the advantages of XML is that it is supported by various programming languages, among which is Python. XML documents consist of a prologue, a body, and an epilogue. The prologue contains information on the version, the encoding, and the structure of that document. The body contains all the information of the document, divided into hierarchically ordered elements. Each element consists of a tag with its text. Optionally, an element can have attributes. There also exist elements without text at all, called "empty elements."

Without regard of the programming language used, there are two major strategies used when accessing these types of files. On one hand, it can analyze the relationships between all the elements, and construct the corresponding tree. This implies having the whole file structure in memory, and is called the **D**ocument **O**bject **M**odel (DOM). The other option is to recurse over the file and generate events by which we can then travel, recursing on each distinct element. At each event we can process our data. These are called "event driven parsers" and the most well known is **S**imple **A**PI for **X**ML (SAX).

In this chapter we presented as examples of DOM, Minidom and Element-Tree. As an example of a parser based on events, we saw the use of Iterparse, provided by cElementTree. DOM is often easier to use because it does not involve event handling; however, on some occasions it's more convenient to use a parser based on events, especially for large files. There are more parsers than the ones presented here, like BeautifulSoup,[16] a parser that can be used both for XML and HTML.

---

[16]Available at http://www.crummy.com/software/BeautifulSoup.

12.5   Additional Resources

- Extensible Markup Language (XML). Links to W3C Recommendations, Proposed Recommendations and Working Drafts.
  `http://www.w3.org/XML`

- Software Carpentry course, by Greg Wilson. "XML."
  `http://swc.scipy.org/lec/xml.html`

- O'Reilly Media Bioinformatics XML reference page.
  `http://www.xml.com/pub/rg/Bioinformatics`

- Mark Pilgrim. Dive into Python. Chapter 9. XML Processing.
  `http://diveintopython.org/xml_processing/`

- Python and XML: An Introduction.
  `http://www.boddie.org.uk/python/XML_intro.html`

- Resources on DTD:
  `http://www.w3schools.com/dtd/`, `http://www.xmlfiles.com/dtd`, and `http://www.w3.org/TR/REC-xml/#dt-doctype`.

- Resources on XML Schema:
  `http://www.w3schools.com/schema/schema_intro.asp`.

12.6   Self-Evaluation

1. What does the OpenOffice format have in common with RSS feeds and GoogleEarth's geographic coordinates?

2. What are the benefits of using XML for data storage and information interchange?

3. When you will not use XML?

4. Why an XML parser should not read a malformed XML document?

5. Distinguish between the terms: tag, element, attribute, value, DTD, and Schema.

6. In the example XML file (listing 12.1) there is one empty-element tag. Which one is it?

7. What is the difference between the SAX and DOM models of XML file processing?

8. If you have to parse an XML file that has a size approaching or exceeding available RAM, what is the recommended parser?

9. In **cElementTree.iterparse** there are both **start** and **end** event types. By default it returns only **end** event. When would you use the information in a **start** event?

10. Make two programs to parses all hit names in an XML formated BLAST output. One program should use Python XML tools and the other should read the input file as text.

Chapter 13

Python and Databases

13.1 Introduction to Databases

The amount of data that is handled in a typical bioinformatics project forces us to use something more versatile than the data structures bundled with Python. Lists, tuples, and dictionaries are very flexible, but they are not suitable to model all the complexity associated with real world data. Sometimes it is necessary to have a permanent data repository (in computer terms this is called **data persistence**), since data structures are available only while the program is running. While it is possible to write out the data to a file using **pickle**, this is not as efficient as using a database engine designed for that purpose.

What Is a Database?

A database is an ordered collection of related data. Generally, they are constructed to model real-world situations: a person's video collection, the students of a university, a firm's inventory, etc. The database stores relevant data for the users of our program. In modeling the students of a university, we have to take into account the first and last names, the year of entry, and the subjects studied; we wouldn't care about the hair color or height of the student. Designing a database is like modeling a natural process. The first step is to determinate what are the relevant variables.

One advantage of databases is that, in addition to data storage, they provide search tools. Some of these searches have immediate replies, such as "how many students are there?" Others are trickier, involving combining different information sources to enable a response, as for example "How many different subjects, on average, did each 2007 freshman take?" In a biological database a typical question might be "What are the proteins with a weight of less than 134 kDa that have been crystallized?" It's interesting to note that one need not anticipate all the questions that could be asked; but having an idea of the most common questions will help the design process.

In any case, the advantage of having a database is that we can ask these questions and receive these answers without having to program the search mechanism. That is the job of the **database engine**, which is optimized to quickly handle large amounts of data. Using Python, we communicate with the database engine and process its responses, without having to worry about the internal processes. This doesn't mean we have to totally disengage from the functioning of the database, as the more we understand the internals, the better results we can achieve.

Database Types

Not all databases are the same. There are different theoretical models for describing both the structure of the database and the interrelationships of the data. Some of the most popular models are: Hierarchical, Network, Relational, and Entity-relationship. Choosing between the different models is more a job for IT professionals than for bioinformatics researchers. In this chapter we will spend our time only with the **Relational model** due to the flexibility it offers, the many implementations available, and (why not?) its popularity.

A relational database is a database that groups data using common attributes. The resulting sets of organized data can be handled in a logical way.

For example, a data set containing all the real estate transactions in a town can be grouped by the year the transaction occurred; or it can be grouped by the sale price of the transaction; or it can be grouped by the buyer's last name; and so on.

Such a grouping uses the relational model. Hence such a database is called a "relational database." To manage a relational database, you must use a specific computer language called SQL (Structured Query Language). It allows a programmer to create, query, update and delete data from a database. Although SQL is an ANSI standard, due to several reasons there are multiple non compatible implementations. Even if they are different, since all versions are based in the same published standard, it is not hard to transfer your knowledge from one SQL dialect to another.

Among the different implementations of relational databases and query languages, this book focus on two of them: MySQL and SQLite. MySQL (it is pronounced "My Ess Cue Ell") is the most popular database used in web applications, with more than 10 million installations. The great majority of small and medium websites use MySQL. While many system administrators would not consider using MySQL for very demanding applications, there are many high-traffic sites successfully using it. One example of this is YouTube.com, which needs no introduction. Other popular MySQL-based sites are Wikipedia.org, Flickr.com, and Slashdot.org.[1] SQLite's target is much more narrowly defined: it is made for small embedded systems, both

---

[1]Granted, they are not default installations running on commodity hardware, but highly optimized installations running on branded hardware.

hardware and software. The Firefox browser uses SQLite internally, as do Symbian cell phones; even the operating system like OS X and Solaris 10. This versatility is due to its small size (about 250KB), its lack of external dependencies, and its storage of a database in a single file. These advantages of small size and simplicity are offset by a lack of features, but for its unique niche this is not a problem.

In both cases, the fundamentals are similar; and the concepts explained in this chapter are applicable to all relational databases. When a characteristic is exclusive to a database in particular, this will be pointed out.

13.1.1 Database Management: RDBMS

RDBMS stands for **R**elational **D**ata**B**ase **M**anagement **S**ystem. It is software designed to act as an interface between the database engine, the user, and the applications. The just mentioned MySQL and SQLite are examples of RDBMS.[2]

In the case of MySQL, the RDBMS is separated into two components: A server and a client. The server is the program that accomplishes the hard work associated with the database engine; it can work on our own computer or on a remotely accessible server. The client is the program that gives us an interface to the server. MySQL provides its own client (mysql), that is a command line program, but nothing prevents us from using any other compatible client. A popular client is PhpMyAdmin,[3] that requires a Web server to run, but provides to the final user a nice Web-based front-end to the MySQL server (see Figure 13.1). There are also desktop clients with the same function, like MySQL GUI,[4] SQLyog,[5] and Navicat[6] among others. There is a screenshot of MySQL GUI in Figure 13.2.

SQLite, on the other hand, is available as a library to include into your programs or as a stand-alone executable. Since version 2.5 Python has a built-in module (**sqlite3**) to interface with SQLite and it works "out of the box" if Python was compiled with SQLite present. It also can be linked to an external executable file with the module **pysqlite2.dbapi2**.

13.1.2 Components of a Relational Database

The first concept of databases we need to understand is that of **entities**. Formally, an entity is defined as *every significant element that should be stored*. We should distinguish between an **entity type** and **occurrence** of

---

[2]Other well known RDBMS are Oracle, DB2, and PostgreSQL.

[3]http://www.phpmyadmin.net

[4]http://dev.mysql.com/downloads/gui-tools/5.0.html

[5]http://webyog.com/en/

[6]http://www.navicat.com

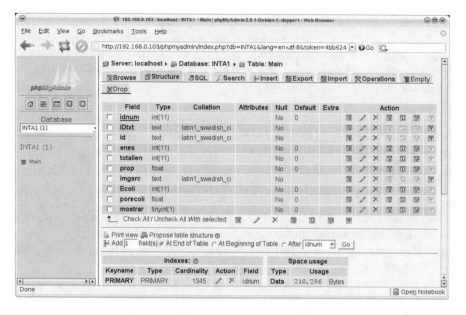

FIGURE 13.1: Screenshot of PhpMyAdmin: Easy to use HTML front end to administrate a MySQL database.

an entity. In an administration database, `Students` is an entity type, while each student in particular is an occurrence of this entity.

Each **entity** has its own **attributes**. The **attributes** are the data associated to an **entity**. Let's go back to the college administration database we have just schemed. *name*, *lastname*, *DateJoined*, and *OutstandingBalance* are attributes of the **entity** *Students*.

In turn, each entity has its own attributes. The attributes are the data associated with an entity. Let's create a college administration database, where `Name`, `Lastname`, `DateJoined`, and `OutstandingBalance` are attributes of the entity Students.

The data in a relational database are not isolated, but, as the name implies, they are represented by **relations**. A relation maps a key, or a grouping of keys, with a grouping of rows. Each key corresponds to an occurrence of one entity, which relates to the group of attributes associated with that occurrence. These relationships are displayed as tables, independently of how they are stored physically. A database can have multiple tables. Continuing the example of the university administration database, we might have a table with information on the students and another on the professors, as each entity has its own attributes.

In Table 13.1 we can see an example of the `students` relation.

FIGURE 13.2: Screenshot of MySQL Query Browser.

TABLE 13.1:   Students in Python University

Name	LastName	DateJoined	OutstandingBalance
Joe	Campbell	2006-02-10	No
Joe	Doe	2004-02-16	No
Rick	Hunter	2005-03-20	No
Laura	Ingalls	2001-03-15	Yes
Virginia	Gonzalez	2003-04-02	No

A Key Concept: Primary Key

Every table has to have a means of identifying a row of data; it must
have an attribute, or group of attributes, that serve as a unique identifier.
This attribute is called a **primary key**. In the case that no single attribute
can be used as a primary key, several can be taken simultaneously to make a
composite key. Returning to Table 13.1, we can see that the attribute Name
cannot be used as a primary key, as there is more than one occurrence of an
entity with the same attribute (Joe Campbell and Joe Doe share the same
first name). One solution to this problem would be to use `Name` and `LastName`
as a composite key; but this would not be the best solution, because it's still
possible to have more than one occurrence of an entity sharing this particular
composite key, such as another Joe Doe. For this reason, normally we add to
the table an ID field–a unique identifier–instead of depending on the data to
have a primary key. In most databases there are mechanisms for automatically
generating such a primary key when we insert data. Let us look at a version
of Table 13.1 with a new attribute that can be used as the primary key:

TABLE 13.2:   Table with Primary Key

ID	Name	LastName	DateJoined	OutstandingBalance
1	Joe	Campbell	2006-02-10	No
2	Joe	Doe	2004-02-16	No
3	Rick	Hunter	2005-03-20	No
4	Laura	Ingalls	2001-03-15	Yes
5	Virginia	Gonzalez	2003-04-02	No

13.1.3   Database Data Types

As in programming languages, databases have their own data types. For
example, in Python we have *int*, *float* and *string* (among others); databases
have their own data types such as *tinyint*, *smallint*, *mediumint*, *int*, *bigint*,
float, *char*, *varchar*, *text*, and others. You may be wondering why there are
so many data types (such as five different data types for integers). The main

reason is that with so many options it is possible to the make best use of available resources. If we need a field where we wish to store the age of the students, we can achieve that with a field of type tinyint, as it supports a range of values between -128 and 127 (which can be stored in one byte). Of course, we can just as well store it in a field of type int, which supports a range between -2147483648 to 2147483647 (that is, 4 bytes); but that would be a waste of memory, as the system must unnecessarily reserve space. Because of the difference in the number of bytes, a number stored as int occupies 4 times as much RAM and disk space as one stored as tinyint. The difference between one and four bytes may seem insignificant and not worth mentioning, but then multiply it by the number of data entries you have; when the dataset is large enough, disk space and access time could be an issue. That is why you should be aware of the data type storage requirements.[7]

Table 13.3 summarizes the characteristics of the main data types in MySQL. Note that some of the minor characteristics may vary depending on the version of MySQL used, which is why it is advisable to consult the documentation for your particular version.[8] In the case of SQLite, there are only 5 data types: INTEGER, REAL, TEXT, BLOB and NULL. However, one must realize that SQLite is typeless, and that any data can be inserted into any column. For this reason, SQLite has the idea of "type affinity": it treats the data-types as a recommendation, not a requirement.[9]

13.2 Connecting to a Database

To connect to the MySQL database server, you need a valid user; and to set up a user, you need to connect to the database. This catch-22 is solved by accessing the server with the default user and password (user: "root", password: ""). From the command line, if the server is in the same computer, it is possible to access with this command:

```
$ mysql -u root -p
```

---

[7]Estimating what data types are adequate for the situation is no minor issue. In the online multi-player game World of Witchcraft, some players found they could not receive more gold when they had reached the limit of the variable in which money was stored, a signed 32-bit integer. Much more serious was the case of the software in the Ariane 5 rocket when a 64-bit real was converted to a 16-bit signed integer. This led to a cascade of problems culminating in destruction of the entire flight, costing 370 million US dollars.

[8]MySQL has a complete online reference manual. Data Type documentation for MySQL 5.1 is available at http://dev.mysql.com/doc/refman/5.1/en/data-types.html.

[9]For more information about the idea of "type affinity" I recommend the section "Datatypes In SQLite Version 3" (utlhttp://www.sqlite.org/datatype3.html) of the SQLite online documentation.

TABLE 13.3:   Most Used MySQL Data Types

Data type	Comment
TINYINT	±127 (0-255 UNSIG.)
SMALLINT	±32767 (0-65535 UNSIG.)
MEDIUMINT	±8388607 (0-16777215 UNSIG.)
INT	±2147483647 (0-4294967295 UNSIG.)
BIGINT	±9223372036854775807 (0-18446744073709551615 UNSIG.)
FLOAT	A small number with a floating decimal point.
DOUBLE	A large number with a floating decimal point.
DATETIME	From '1000-01-01 00:00:00' to '9999-12-31 23:59:59'
DATE	From '1000-01-01' to '9999-12-31'
CHAR(n)	A fixed section with n characters long (up to 255).
VARCHAR(n)	A variable section with n characters long (up to 255).
TEXT	A string with a maximum length of 65535 characters.
BLOB	A binary string version of TEXT.
MEDIUMTEXT	A string with a maximum length of 16777215 characters.
MEDIUMBLOB	Binary string equivalent to MEDIUMTEXT.
LONGTEXT	A string with a maximum length of 4294967295 characters.
LONGBLOB	Binary string equivalent to LONGTEXT.
ENUM	String value taken from a list of allowed values.

```
Enter password:
Welcome to the MySQL monitor.  Commands end with ; or \g.
Your MySQL connection id is 8
Server version: 5.0.45-Debian_1ubuntu3 Debian etch distribution

Type 'help;' or '\h' for help. Type '\c' to clear the buffer.

mysql>
```

From now on, interaction with MySQL server will be shown by using the phpMyAdmin front-end, as shown in Figure 13.3.

13.3   Creating a MySQL Database

Before working with a database, we should create one. You could skip this step if you plan to access a database previously created. But it is likely that sooner or later you will need to create your own database. Creating a database is a simple task and will help you to understand the data you are going to handle, and create more effective queries.

Since database creation is something that is done only once for each database,

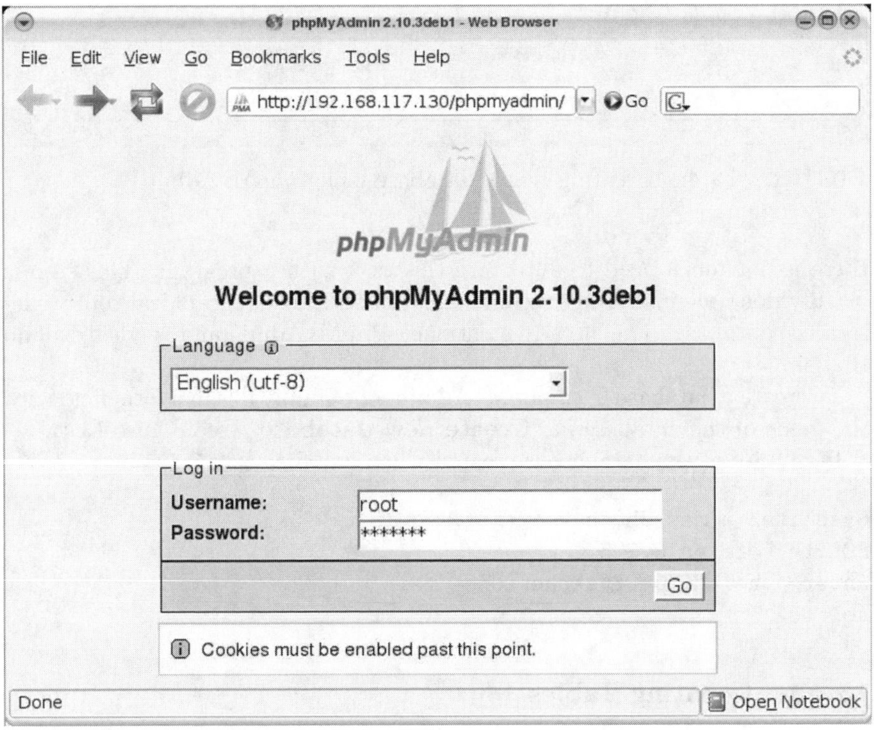

FIGURE 13.3: Login screen to the MySQL database using phpMyAdmin.

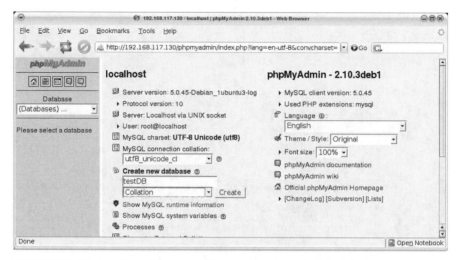

FIGURE 13.4: Creating a new database using phpMyAdmin.

there is not much need to automate this task with a program. This step is usually done manually. My recommendation is to use a graphical tool to design the database. The already mentioned PhpMyAdmin and Navicat will do the job.

To create a database from phpMyAdmin, one simply fills in a form field with the name of the database in "**Create new database**" (see Figure 13.4).

In SQLite a new database is created when one first calls it, as in:

```
$ sqlite3 a_new_db.db
SQLite version 3.3.5
Enter ".help" for instructions
sqlite>
```

13.3.1 Creating Tables

Once we have a newly created database, the next step is to create the tables where the data will be stored. Creating the tables using this kind of software doesn't seem a problem worth mentioning in this book, so we will focus more on the table structure rather than on the procedure of dealing with a GUI tool.

We must keep in mind that a table represents a relationship between the data; it makes no sense to create a table for one entity and then populate it with data of another entity. Continuing with the example of our "Python University," we can think about what information related to students we need to store in the Students table.

As we saw earlier, in the table Students we assigned the following fields: ID, Name, LastName, DateJoined and OutstandingBalance.

The assignment of fields is somewhat arbitrary. There are no special criteria for deciding which fields should be in which table; nor even is the definition of each field set in stone. But that does not mean there are no rules of thumb, or at least "good practices," for database design. You could even write a book on the subject; in fact, there are a number of such books. It is certainly not easy to convey in this space the necessary knowledge to achieve an efficient design for every situation; in any case, good database design is something that one learns with practice.

Let's see how we define each field in this case:

ID: Is a unique id for each registrant. Since Python University is expected to have several students, an unsigned INT data type is used (up to 4294967295). There is no need to use negative numbers in an ID, so this field should be set as unsigned.

Name: Since the size of a name is variable with less than 255 characters, VARCHAR is used. The maximum size for names in characters, according to my arbitrary criteria, is 150.

LastName: This field was set with the same criteria as the former field. The only difference is in the maximum size for a last name; that is set to 200 characters.

DateJoined: There is not much choice here. A simple DATE field would do it best.

OutstandingBalance: This field represents whether the student has paid the tuition in full or not. Since there are only two possible values (paid or not paid), a BINARY data type is chosen. This data type stores a 0 or a 1. It is up to the programmer to assign a meaning to this values, but in mathematical notation 0 stands for FALSE and 1 for TRUE, so this convention is generally used.

The last choice is the table type (**InnoDB** or **MyISAM**). In this case it is OK to leave the default option (MyISAM), which will be appropriate for most uses. Please see Advanced Tip: MyISAM vs InnoDB on page 281 for a brief discussion on both table types.

If you want to manually create the table, here are the commands you should type into the MySQL prompt (also available at **py3.us/58**):

```
CREATE TABLE 'Students' (
'ID' INT UNSIGNED NOT NULL AUTO_INCREMENT PRIMARY KEY ,
'Name' VARCHAR( 150 ) NOT NULL ,
'LastName' VARCHAR( 200 ) NOT NULL ,
'DateJoined' DATE NOT NULL ,
'OutstandingBalance' BINARY NOT NULL
) ENGINE = MYISAM COMMENT = 'Principal table.';
```

No wonder I recommended the use of a GUI to design the table!

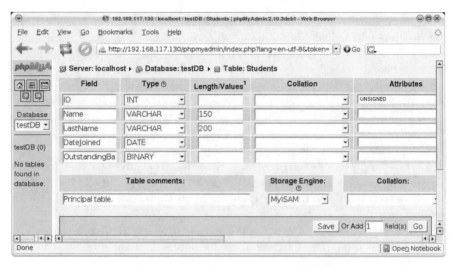

FIGURE 13.5:   Creating a new table using phpMyAdmin.

Tip: Creating a Database Using Another Database as a Template.

Instead of manually defining each field on each table, you could import a "MySQL dump" from another database and create a database in one step. There are two different kinds of dump files: "Structure only" and "structure and data" dump files. Both files are imported the same way into a newly created database:

```
$ mysql -p database_name < dbname.sql
```

You may be wondering where do you get the dump file in the first place. You can get a dump file from the backup of another database or from the installation files of a program that requires a database.

13.3.2   Loading a Table

Once we have the table created, it is time to load the data into it. This operation can be done from any MySQL front-end, either row by row or in batch. Since there are several data to load at the beginning, and the manual data load is intuitive, let's see how to load data in batch mode.

The most common way to upload data is by using csv files. This kind of file was reviewed in section 5.3 (page 92). To upload the data that is seen in table 13.2, we can prepare a csv file (**data.txt**) with the following format:

```
1,Joe,Campbell,2006-02-10,No
2,Joe,Doe,2004-02-16,No
```

ID	Name	LastName	DateJoined	OutstandingBalance
1	Joe	Campbell	2006-02-10	N
2	Joe	Doe	2004-02-16	N
3	Rick	Hunter	2005-03-20	N
4	Laura	Ingalls	2001-03-15	Y
5	Virginia	Gonzalez	2003-04-02	N

FIGURE 13.6: View of the Student table.

```
3,Rick,Hunter,2005-03-20,No
4,Laura,Ingalls,2001-03-15,Yes
5,Virginia,Gonzalez,2003-04-02,No
```

To upload the csv file into the MySQL database, use the LOAD DATA INFILE command[10] at the MySQL prompt:

```
LOAD DATA INFILE 'data.txt' INTO TABLE Students
FIELDS TERMINATED BY ',';
```

An alternative way, using **INSERT** statements:

```
INSERT INTO 'Students'
VALUES ('1', 'Joe', 'Campbell', '2006-02-10', 'No');
INSERT INTO 'Students'
VALUES ('2', 'Joe', 'Doe', '2004-02-16', 'No');
INSERT INTO 'Students'
VALUES ('3', 'Rick', 'Hunter', '2005-03-20', 'No');
INSERT INTO 'Students'
VALUES ('4', 'Laura', 'Ingalls', '2001-03-15', 'Yes');
INSERT INTO 'Students'
VALUES ('5', 'Virginia', 'Gonzalez', '2003-04-02', 'No');
```

Once the data is loaded into the database, the table looks like the one in figure 13.6.

Advanced Tip: MyISAM versus InnoDB.

There are several formats for the internal data structures in MySQL tables. The most commonly used formats are InnoDB and MyISAM. MyISAM is used by default and is characterized by its higher reading speed (using SELECT operations), and uses less disk space. It is slower than InnoDB at data writing, since when a datum is being recorded, the table is momentarily blocked

---

[10]For a complete reference of this command, see the MySQL online manual at http://dev. mysql.com/doc/refman/5.1/en/load-data.html.

StudentID	Python_101	Python_101-TERM	Mathematics_for_CS	Mathematics_for_CS-TERM
1	7	2002-1	8	2003-1
2	6	2002-2	9	2002-1
3	5	2003-1	7	2003-1

FIGURE 13.7: An intentionally faulty "Score" table.

until finished, so all other operations must wait to complete. This limitation doesn't exist in a InnoDB table. The main advantage of this format is that it allows secure transactions and has a better crash recovery. InnoDB is thus recommended for intensively updated tables and for storing sensitive information. To sum up, since you can use different table types in the same database, choose the appropriate table type according to the operation that will be performed most on each table.

13.4 Planning Ahead

Making a database requires some planning. This is specially true when there is a large amount of data and we want to optimize the time it takes to answer our queries. A bad design can make a database unusable. In this section I present a sample database to show the basis of database design. To have a better idea of designing relational databases, you should read elsewhere about "Database normalization," but this section should give you a brief glimpse of this field.

13.4.1 PythonU: Sample Database

Let's keep the example of a student database from the fictional Python University (whose database is called PythonU), to store student data and subjects taken. To store student data, we already have the table Students. We have to make a table to store the scores associated with each subject (a "Score" table). As in many other aspects of programming, there is more than one way to accomplish this. We will start by showing some non optimal ways, to better understand why there is a recommended way.

Score Table

In the table Score we want to store for each student: subjects studied, score, and date each course was taken.

A proposed Score table for two courses is depicted in Figure 13.7.

StudentID	Course	Score	Term
1	Python_101	7	2002-1
1	Mathematics_for_CS	8	2003-1
2	Python_101	6	2002-2
2	Mathematics_for_CS	9	2002-1
3	Python_101	5	2003-1
3	Mathematics_for_CS	7	2003-1

FIGURE 13.8: A better "Score" table.

This design (schema) has several flaws. The first design error in the table is its inflexibility. If we want to add a new course, we have to modify the table. This is not considered good programming practice; a change in the structure of an already populated table is an expensive operation which must be avoided whenever possible. The other problem that arises from this design, as seen in the diagram, is that there is no place to store the score of a student who has taken a course more than once. How do we solve this? With a more intelligent design. An almost optimal solution can be seen in Figure 13.8.

The first problem, the need to redesign the table for entering new subjects, we solve by entering the course name as a new field: `Course`. This field can be of type TEXT or VARCHAR. And the problem of being able to keep track of when a student took a course more than once, we solved with the field Term. While this is a decidedly better design than the previous one, it is far from being optimal. It is evident that storing the name of each subject for each student is an unnecessary waste of resources. A way to save this space is to use the datatype ENUM in the field `Course`; in this way we can save a substantial amount of space, because MySQL internally uses one or two bytes for each entry of this type. The table remains the same as seen before (Figure 13.8), and only changes the way the field Course is defined, saving disk space as mentioned.

Is this now the best way? The problem with using ENUM with the field `Course` is that when we wish to add a new subject, we still have to alter the table structure. This modification, to add a new option to the ENUM, is not as costly as adding a new column, but conceptually it is not a good idea to modify the definition of a new table in order to accommodate a new type of data. In cases like these, we resort to "lookup tables."

Courses Table

A lookup table is a reference table that is used to store values that are used as a content of a column located in another table. Continuing with the example of Python University, we can make a lookup table for the subjects (see Figure 13.9).

This table `Courses` contains a field for storing the ID of the course (`CourseID`) and another for the name of the course (`Course_Name`). For this scheme to

CourseID	Course_Name
1	Python 101
2	Mathematics for CS

FIGURE 13.9:   Courses table: A lookup table.

StudentID	Course	Score	Term
1	1	7	2002-1
1	2	8	2003-1
2	1	6	2002-2
2	2	9	2002-1
3	1	5	2003-1
3	2	7	2003-1

FIGURE 13.10:   Modified "Score" table.

work, we must change the field `Course` of the table `Score`; in place of an ENUM field, we now use an INT field (see Figure 13.10).

The data in `CourseID` now correspond to that of the field `Course` in the table `Score`. Using a single lookup, we can then link the ID with the corresponding course name. This way we save the same amount of space in the `Students` table as when we used an ENUM for the `Course` field, with the additional advantage that we can expand the list of subjects simply by adding one element to the `Courses` table.

Tip: ENUM field type versus Lookup Table.

We have seen how convenient it is to use a lookup table in place of an ENUM field. You are probably wondering how to decide when to use one strategy or the other when designing your database. ENUM is better than TEXT or VARCHAR in the cases where the number of possibilities is limited and not expected to vary: for example, a list of colors, the months of the year, and other options that by their very nature have a set range. One disadvantage that should be taken into account with regard to ENUM, is that it is a datatype specific to MySQL which may not be available on other DB engines; this limits the potential portability of the database.

Now we have the `PythonU` database with 3 tables: `Students`, `Score`, and `Courses`. It's time to learn how to construct queries.

13.5 SELECT: Querying a Database

The most useful operation in a database, once it is created and populated, is querying its contents. We can extract information from one table or many tables simultaneously. For example, to have a list of students, the table Students must be queried. On the other hand, if we want to know a student's average scores, we need to query the Students and Scores tables. In addition, there are cases where one must query 3 tables simultaneously, as when finding out a student's score in one particular subject.

Let's look at each case:

Simple Query

To obtain a listing of students (first and last names) from the Students table, we would use the following command at the MySQL prompt:

```
mysql> SELECT Name, LastName FROM Students;
```

While this command is quite self-explanatory, we will see later what options there are for constructing a query.

Combining Two Queries

To obtain the average scores of a particular student, we need to extract all the scores corresponding to that student. As the scores are in the Score table and the names in the Students table, we need to query both tables in order to receive a reply to our question. First we need to query Students for the ID of the student; then with this ID we must search for all corresponding records.

Supposing that the student in question is Joe Campbell:

```
SELECT AVG(Score) FROM Scores
WHERE StudentID = (SELECT ID FROM Students
WHERE Name='Joe' AND LastName='Campbell' );
```

We can also accomplish it with a single query, without using the nested SELECT:

```
SELECT AVG(Score) FROM Scores, Students
WHERE Scores.StudentID=Students.ID
AND Students.Name='Joe' AND Students.LastName='Campbell';
```

There are two new things to understand in this example: when we use fields from more than one table, we should prepend the table name to avoid ambiguities in the field names. Thus, StudentID becomes Scores.StudentID. The following statement is equivalent to the above:

```
SELECT AVG(Score) FROM Scores, Students
WHERE StudentID=ID AND Name='Joe' AND LastName='Campbell';
```

If the field name is present only in one table, there is no need to add the table name, but it makes the query easer to parse for the programmer.

The other feature worth pointing out on this example is that instead of looking only the student ID, there is a condition that matches the IDs of both tables (`Scores.StudentID = Students.ID`).

In either case, the result is the same: `7.5`

Querying Several Tables

To retrieve the average score of one student (Rick Hunter) in one particular course (Python 101), there is a need to build a query using more than one table:

```
SELECT Scores.Score FROM Scores, Courses, Students
WHERE Courses.CourseID = Scores.Course
AND Courses.Course_Name = 'Python 101'
AND Students.ID = Scores.StudentID
AND Students.Name = 'Rick' AND Students.LastName = 'Hunter';
```

13.5.1   Building a Query

The general syntax of `SELET` statements is,

```
SELECT field(s)_to_retrieve FROM table(s)_where_to_look_for
WHERE condition(s)_to_met] [ORDER BY ordering_criteria]
[LIMIT limit_the_records_returned];
```

To use grouping functions, include at the end of your query:

```
GROUP BY variable_to_be_grouped HAVING condition(s)_to_met
```

The aggregating functions are: **AVG()**, **COUNT()**, **MAX()**, **MIN()** and **SUM()**.

Note that `HAVING` works like `WHERE`. The difference is that `HAVING` is used only with `GROUP BY` since it restricts the records after they have been grouped.

These constructs can be understood better with actual examples. The following cases shows how to execute the queries from the MySQL command line.

To get all the elements of a table, use wildcards,

```
mysql> select * from Students;
+----+----------+----------+------------+-------------------+
| ID | Name     | LastName | DateJoined | OutstandingBalance |
```

```
+----+----------+----------+------------+--------------------+
|  1 | Joe      | Campbell | 2006-02-10 | N                  |
|  2 | Joe      | Doe      | 2004-02-16 | N                  |
|  3 | Rick     | Hunter   | 2005-03-20 | N                  |
|  4 | Laura    | Ingalls  | 2001-03-15 | Y                  |
|  5 | Virginia | Gonzalez | 2003-04-02 | N                  |
+----+----------+----------+------------+--------------------+
5 rows in set (0.07 sec)
```

To obtain a count of all elements in a table,

```
mysql> select COUNT(*) from Students;
+----------+
| COUNT(*) |
+----------+
|        5 |
+----------+
1 row in set (0.00 sec)
```

To see the average score of all students,

```
mysql> select avg(Score) from Scores GROUP BY StudentID;
+------------+
| avg(Score) |
+------------+
|     7.5000 |
|     7.5000 |
|     6.0000 |
+------------+
3 rows in set (0.21 sec)
```

To retrieve the best score of one particular student (Joe Campbell):

```
mysql> select max(Scores.Score) from Scores,Students
WHERE studentID=ID AND Students.Name = 'Joe'
AND Students.Lastname='Campbell';
+-------------------+
| max(Scores.Score) |
+-------------------+
|                 8 |
+-------------------+
1 row in set (0.00 sec)
```

Which courses have the string "101" in their names?

```
mysql> SELECT Course_Name FROM Courses
```

```
WHERE Course_Name LIKE '%101%';
+-------------+
| Course_Name |
+-------------+
| Python 101  |
+-------------+
1 row in set (0.00 sec)
```

Note that % is used as a wildcard character when working with strings.

How many students have flunked a class? Supposing that the passing score is 7, this query is equivalent to asking how many scores are below 7.

```
mysql> SELECT Name,LastName,Score FROM Students,Scores
WHERE Scores.Score<7 and Scores.StudentID=Students.id;
+------+----------+-------+
| Name | LastName | Score |
+------+----------+-------+
| Joe  | Doe      |     6 |
| Rick | Hunter   |     5 |
+------+----------+-------+
2 rows in set (0.00 sec)
```

The above was simply an example of the possibilities of the **SELECT** command. For more complex queries I recommend the resources indicated in "Additional Resources."

13.5.2 Updating a Database

While values can be changed using any of the aforementioned GUI tools, it's good to know the syntax for updating data, to enable implementing it from Python when necessary.

The general syntax is:[11]

```
UPDATE table_name(s) SET variable1=expr1 [,variable2=expr2 ...]
[WHERE condition(s)];
```

Suppose you want the database to reflect the fact that Joe Campbell didn't pay his tuition, therefore we must make sure the **OutstandingBalance** field in the **Students** table is set to **Y**. Here is the SQL command with the server's response:

---

[11] For more information on this command see the MySQL manual at: http://dev.mysql.com/doc/refman/5.1/en/update.html.

```
mysql> UPDATE Students SET OutstandingBalance='Y'
WHERE Name='Joe' and LastName='Campbell';
Query OK, 1 row affected (0.67 sec)
Rows matched: 1  Changed: 1  Warnings: 0
```

It is also possible, instead of changing a specific value, to apply a function [12] to all values in a column. For example, to subtract one point from all scores:

```
mysql> UPDATE Scores SET Score = Score-1;
Query OK, 6 rows affected (0.00 sec)
Rows matched: 6  Changed: 6  Warnings: 0
```

13.5.3 Deleting a Record from a Database

To delete a record use the *DELETE* command:

```
mysql> DELETE from Students WHERE ID = "5";
```

As in SELECT, the WHERE clause specifies the conditions that identify which rows to delete. Without the WHERE clause, all rows are deleted. But this is not the best way to delete a whole table. Instead of deleting all records row by row, you can use the TRUNCATE command that drops and recreates the table. This is faster for large tables.

To limit the number of records to delete, there is the LIMIT clause:

```
mysql> DELETE from Students WHERE ID = "5" LIMIT=1;
```

In this case there is no difference since there is only one record that matches the WHERE clause.

13.6 Accessing a Database from Python

Now that we know how to access our data using SQL, we can take advantage of Python's tools for interfacing with databases.

13.6.1 MySQLdb Module

This module allows accessing MySQL databases from Python. It's not installed by default; there are even webservers lacking the module (in a shared webhosting environment you may have to request the installation of the

---

[12] Any valid MySQL function can be used. To see a list with available functions, check the MySQL manual at: http://dev.mysql.com/doc/refman/5.1/en/functions.html.

MySQLdb Python module). To know if the module is installed, try importing it. If you get an import error, it's not installed:

```
>>> import MySQLdb
Traceback (most recent call last):
  File "<stdin>", line 1, in ?
ImportError: No module named MySQLdb
>>>
```

To install it, download it from its web site[13], or from your software repository if using Linux.[14]

13.6.2   Establishing the Connection

There is the `connect` method in the MySQLdb module. This method returns a connection object which we'll need to act upon later, so we should give it a name (in the same way we would give a name to the object resulting from opening a file):

```
>>> import MySQLdb
>>> db = MySQLdb.connect(host="localhost", user="root",
... passwd="mypassword", db="PythonU")
```

13.6.3   Executing the Query from Python

Once the connection to the database is established, we have to create a **cursor**. A cursor is a structure used to walk through the records of the result set.

The method used to create the cursor has a clever name, **cursor()** :

```
>>> cursor = db.cursor()
```

The connection is established, and the cursor has been created. It is time to execute some SQL commands:

```
>>> cursor.execute("SELECT * FROM Students")
5L
```

The `execute` method is used to execute SQL commands. Note that there is no need to add a semicolon (;) at the end of the command. Now the question is how to retrieve data from the cursor object. To get one element, use **fetchone()**:

---

[13]http://sourceforge.net/projects/mysql-python
[14]In this case, the package name is **python-mysqldb**.

```
>>> cursor.fetchone()
(1L, 'Joe', 'Campbell', datetime.date(2006, 2, 10), 'N')
```

fetchone() returns a row with the elements of the first record of the table. Remaining records can be extracted one by one in the same way:

```
>>> cursor.fetchone()
(2L, 'Joe', 'Doe', datetime.date(2004, 2, 16), 'N')
>>> cursor.fetchone()
(3L, 'Rick', 'Hunter', datetime.date(2005, 3, 20), 'N')
```

In contrast, `fetchall()` extracts all the elements at once:

```
>>> cursor.fetchall()
((1L, 'Joe', 'Campbell', datetime.date(2006, 2, 10), 'N'),
(2L, 'Joe', 'Doe', datetime.date(2004, 2, 16), 'N'),
(3L, 'Rick', 'Hunter', datetime.date(2005, 3, 20), 'N'),
(4L, 'Laura', 'Ingalls', datetime.date(2001, 3, 15), 'Y'),
(5L, 'Virginia', 'Gonzalez', datetime.date(2003, 4, 2), 'N'))
```

Which method to use depends on the amount of data returned, the available memory in the PC, and above all, what we're trying to accomplish. When working with limited datasets, there's no problem using fetchall(); but if the database is too large to fit in memory, one must implement a strategy like that found in listing 13.1.

Listing 13.1: Reading results once at a time (py3.us/59)

```
1 import MySQLdb
2 db = MySQLdb.connect(host="localhost",
3     user="root",passwd="secret", db="PythonU")
4 cursor = db.cursor()
5 recs = cursor.execute("SELECT * FROM Students")
6 for x in range(recs):
7     print(cursor.fetchone())
```

While indeed the code in listing 13.1 works flawlessly, in fact it was shown as an example of using fetchone(). Conceptually it is easier to iterate directly over the cursor object.[15] (as seen in code 13.2):

Listing 13.2: Iterating directly over the DB cursor (py3.us/60)

---

[15]If we pay strict attention to the language rules, we should iterate using `iter(cursor.fetchone, None)`. In this case it isn't necessary because both MySQLdb and sqlite3 support direct iteration over the cursor.

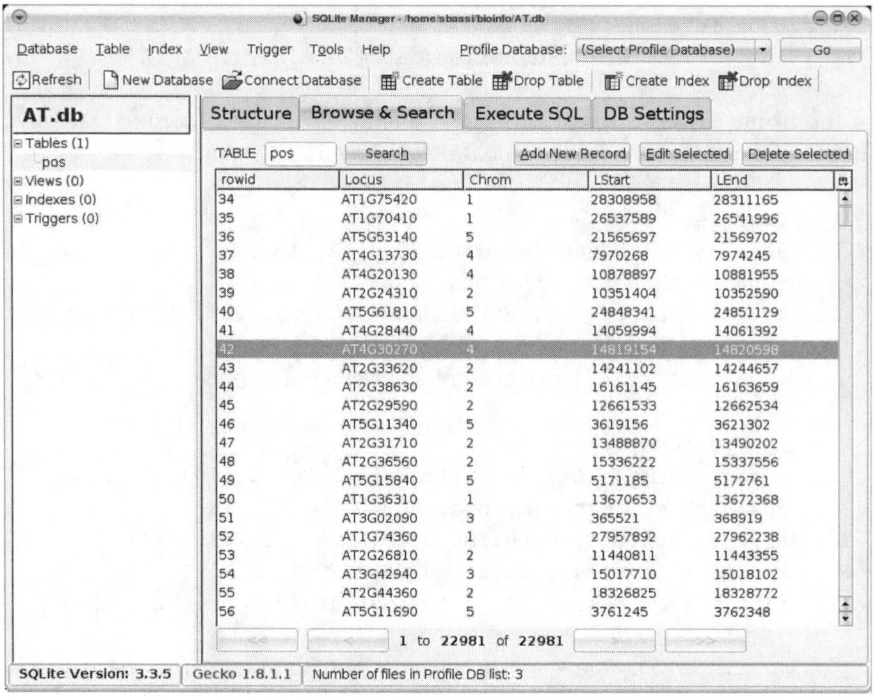

FIGURE 13.11: Screenshot of SQLite manager: A SQLite GUI as a Firefox add-on.

```
1 import MySQLdb
2 db = MySQLdb.connect(host="localhost",
3     user="root",passwd="secret", db="PythonU")
4 cursor = db.cursor()
5 cursor.execute("SELECT * FROM Students")
6 for row in cursor:
7     print row
```

13.7   SQLite

The following example shows that, practically speaking, there is no difference in working with one database type or another:

Listing 13.3: Same script as 13.2, but with SQLite (py3.us/61)

```
1 import sqlite3
2 db = sqlite3.connect('PythonU.db')
3 cursor = db.cursor()
4 cursor.execute("Select * from Students")
5 for row in cursor:
6     print(row)
```

The only thing that changed in listing 13.3 with respect to 13.2 was the first two lines. In line 1, module sqlite3 was imported instead of MySQLdb. Meanwhile, in line 2 the connection code is far simpler, as it does not require a password nor a username to connect to an SQLite database.[16]

There is an externally maintained version of sqlite (pysqlite2.dbapi2) that can be downloaded from `http://pysqlite.org`. This version is always adding new features, does not depend on Python release schedules, and is also available for older Python releases such as 2.3 and higher.

Tip: Creating a Table in SQLite.

It is possible, and even recommendable, to create a table in SQLite using one of the GUIs already mentioned. But if you wish to do it from the command line, here is an example:

```
$ sqlite3 PythonU.db #Creating a DB
SQLite version 3.4.2
Enter ".help" for instructions
sqlite> create table Students(ID int, Name text, LastName char,
DateJoined datetext, OutstandingBalance Boolean);
sqlite> .separator ,
sqlite> .import mybackup.csv Students
```

The first step is creating an empty database file (`$ sqlite3 PythonU.db`). Then create tables with the `create table` command. The next step is to set the data separator (a comma in this case) with `.separator`. The last step is to populate the table by importing it from `mybackup.csv` file with the `.import` command.

As with MySQL, there are some GUI for SQLite. SQLite Administrator[17] is a Windows application[18] that allows the user to create new databases or modify existing ones. SQLite Manager[19] has similar capacities but is available

---

[16]The author argues that access permissions can be applied by using the normal file access permissions of the underlying operating system.

[17]Available at `http://sqliteadmin.orbmu2k.de`.

[18]It also works on Linux with Wine.

[19]Available from `http://www.sqlabs.net/sqlitemanager.php`.

both for Windows and Mac OSX. A multi-platform SQLite front-end is the SQLite Manager Firefox add-on,[20] it should work on any platform the Firefox browser runs. See Figure 13.11 for a screen-shot of SQLite Manager.

13.8 Additional Resources

- Marc-Andre Lemburg. "Python Databases API Specification."
  http://www.python.org/dev/peps/pep-0249/

- "Database Interfaces in Python."
  http://wiki.python.org/moin/DatabaseInterfaces

- Robin Schumacher and Arjen Lentz. "Dispelling the Myths."
  http://dev.mysql.com/tech-resources/articles/dispelling-the--myths.html

- MySQL queries examples.
  http://www.pantz.org/software/mysql/mysqlcommands.html

- Richard Hipp. "SQLite Lecture."
  http://video.google.com/videoplay?docid=-5160435487953918649

- SQLite FAQ.
  http://www.sqlite.org/cvstrac/wiki/wiki?p=SqliteWikiFaq

- SQLite Applications Comparison, comparison of GUIs, mostly for Mac OSX.
  http://www.tandb.com.au/sqlite/compare

- Software:
    - MySQL homepage
      http://www.mysql.com
    - SQuirreL SQL Client - JDBC SQL GUI Client
      http://www.squirrelsql.org/
    - SQLite homepage
      http://www.sqlite.org/
    - SQLite Administrator
      http://sqliteadmin.orbmu2k.de/
    - PostgreSQL home page
      http://www.postgresql.org

---

[20] Available at http://code.google.com/p/sqlite-manager.

- PL/Python - Python Procedural Language (This language allows PostgreSQL functions to be written in Python)
  http://www.postgresql.org/docs/8.2/interactive/plpython.html

- Alternative Solutions:

  - Choosing a non-relational database; why we migrated from MySQL to MongoDB.
    http://bit.ly/1N97s

  - The CouchDB Project
    http://couchdb.apache.org

  - HyperTable: Performance and scalability.
    http://www.hypertable.org

  - libcloud: a unified interface to the cloud
    http://libcloud.org

13.9   Self-Evaluation

1. What is a database?

2. Give some examples of databases.

3. What is a relational database?

4. Define the following terms: entity, attributes, and relationships.

5. What is SQL?

6. What is a query?

7. Translate this query into English:
   `SELECT LastName,Score FROM Student,Scores WHERE Scores.Score>3;`

8. What is the difference between MySQL and SQLite?

9. When is it appropriate to use SQLite?

10. What are the limitations of SQLite with regard to MySQL?

Chapter 14

Collaborative Development: Version Control

14.1 Introduction to Version Control

While programs usually start as a single file handled by a single developer, sometimes these grow to include tens or hundreds of files shared by many people working in different places with different timetables. It is possible that more than one programmer might work on the same portion of code or that one many work on the basis of an outdated version. Without a proper order, it is a recipe for disaster. There is also the case where there is a single programmer who may want to work from different locations (like home and work) and keep track of different versions, without moving files from one side to the other. This kind of problem could be applied to any text file, not only computer code. So solutions found on this chapter could be applied in any document.[1]

VCS, CVS, SCM, Version Control, Revision Control. What Is in a Name?

There are many different names for the same concept: "The management of multiple revisions of the same unit of information." Sometimes it is referred to as CVS (**C**oncurrent **V**ersion **S**ystem) because this is the name of one of the first programs of its class and it became a kind of "generic term." This is why in some texts you will find terms like *CSV server* when it is referring to a specific version control server program that is not CSV. A more generic term is VCS, for **V**ersion **C**ontrol **S**ystem.

---

[1]There are specific web services for sharing documents like Zoho Writer (`http://writer.zoho.com`) and Google Docs (`http://docs.google.com`).

14.1.1 Little History

In the mid 1980s, Dick Grune, a professor of Vrije University (Amsterdam), created *cmt*, the antecessor of one of the most used programs for version control: CVS.

Dick was working in a C compiler with two students. They faced the type of problem described above, because they were not working together since they all had different schedules. The first system consisted of a series of shell scripts that contained the core functions of the program.

When the project was completed, *cmt* kept evolving in an independent way to generate the C written CVS program that is known today. One of the reasons for its popularity, apart from being the first program of its kind, is that it was adopted by high profile projects such as the development of the Linux kernel. Another reason for its success is that the developer site http://sourceforge.net/ provides CVS hosting for open source software. In regards to bioinformatics, http://www.bioinformatics.org also provides free access to a CSV server.

14.2 Version Control Terminology

Whatever the version control program you decide to use, there is a set of terminology that is shared among them. Some version control program documentation uses the terminology in its own way, so this guide should not be taken literally. Here I introduce the most used terms, sorted in a coherent way, where each term can be understood taking into account the former term definition.

Repository: The place where all the shared files and complete revision history are stored. Some version control systems follow a client-server approach and have a central repository. Some other version control systems have a peer-to-peer approach where each peer has a full working copy of the codebase.

Trunk: By convention, the unique line of development that is not a branch (sometimes also called *Baseline* or *Mainline*).

Branch: A set of files under version control which may be branched or forked at a point in time so that, from that time forward, two copies of those files may be developed at different speeds or in different ways independently of the other.

Commit (also known as **check-in**): When a change made by a programmer is written into the repository (either personal or shared).

Revision: A snapshot of the files you're working with. A revision also has some metadata associated with it, such as who committed it, when it was committed and a commit message.

Change: Represents a specific modification to a document under version control.

Check-out: Creates a copy of the code from the repository. Usually the latest version is requested, but also a specific version can be retrieved if needed.

Merge: A merge brings together two sets of changes from a set of files into a unified version of these files.

Conflict: When two changes are made by different programmers to the same document, and the system is unable to reconcile the changes, a user must resolve the conflict by combining the changes, or by selecting one change in favor of the other.

Export: Similar to a check-out except that it creates a clean directory tree without the version control metadata. Often used prior to publishing the contents. This is generally output in a compressed file, like tarballs (.tar.gz, tar.bz2).

14.3   Centralized versus Distributed

One of the most important features of version control systems is the repository localization. As mentioned in the glossary, a version control system can be distributed or centralized.

Centralized version control systems have only one repository from which all programmers check-out code. To submit code (commit), they have to connect to the repository, so they are compelled to work online. Since there is one central repository, only a small group of "core developers" can write to it.

On the other hand, in distributed version control systems (DVCS), each programmer has his or her own repository. There is no need to connect to an external repository to make a commit, permitting everybody to take advantage of revision control. Network access is only required when publishing changes or when accessing changes from somebody else's repository. Since each programmer has his or her own repository, he or she can commit locally without having access to the main repository (trunk). This speeds up the development process.

Open source developers are increasingly adopting DVCS for their projects. The Linux Kernel and the Gnome Desktop migrated from centralized control systems to Git. Python has chosen Mercurial, MySQL uses Bazaar and Biopython will move to Git[2].

14.4 Bazaar: Distributed Revision Control System

Choosing a version control program is not easy. There are several alternatives,[3] most with their own unique advantages. Sometimes the choice is limited by the options given by the leader of the project you are working on. But when you work in your own project, you have to make the choice. For this book, **Bazaar** was chosen because of the following features:

- Coded purely in Python (it has some optional C optimizations)

- Decentralized

- Provides multi-platform support

- Ease of use

- Provides complete documentation

- Easy to integrate into Launchpad.net

- Open Source

Bazaar has other nice features such as plug-in architecture, commercial training and support. These features were not taken into account for this book since they are not used for entry level projects. I mention them since they might be good selling points for larger projects.

14.4.1 Installing Bazaar

Since Bazaar can work without a central server, the installation is straightforward. An often highlighted feature of Bazaar is that *If you can run Python 2.4, then you can run Bazaar*. While this is true, there are some optional dependencies. It is better to have *cElementTree* installed, as well as other libraries like Pyrex. With Python 2.5 or later *cElementTree* is not needed since it is already included. *Paramiko* must be installed together with *pyCrypto* if you want to use sftp to upload your code.

---

[2]Please see http://biopython.org/wiki/GitMigration for more information.

[3]See *Additional Resources* on page 311 for other programs.

Under Windows there is a standard installer.[4] In Linux the software is in the distribution repositories[5] and can be installed (in any Debian based system) as:

```
$ sudo apt-get install bzr
```

As an alternative, you can download the latest version from the Bazaar website and install it like any other Python program:

```
$ sudo python setup.py install
```

To test that everything went OK, try the command *bzr version* and you should get an output similar to this one:

```
$ bzr version
Bazaar (bzr) 1.1.0
  Python interpreter: /usr/local/bin/python2.5 2.5.0.final.0
  Python standard library: /usr/local/lib/python2.5
  bzrlib: /mnt/hda2/bzr-1.1/bzrlib
  Bazaar configuration: /home/sbassi/.bazaar
  Bazaar log file: /home/sbassi/.bzr.log

Copyright 2005, 2006, 2007, 2008 Canonical Ltd.
http://bazaar-vcs.org/

bzr comes with ABSOLUTELY NO WARRANTY.  bzr is free software, and
you may use, modify and redistribute it under the terms of the GNU
General Public License version 2 or later.
```

14.5   Using Bazaar for the First Time

The first step is to identify yourself to the program. This way your contributions will be properly credited in the revision logs.

The usual way to do it is by using your full name and email address, like:

```
$ bzr whoami "Sebastian Bassi sbassi@gmail.com"
```

---

[4]The latest version of the Windows installer can be downloaded from http://bazaar-vcs. org/releases/win32/bzr-setup-latest.exe.

[5]Since the program develops at a fast pace, in general the version available in the repository is not the last version. To make sure you have the most updated version, you have to include in the repositories list file, the repositories included at this page: https://launchpad.net/ ~bzr/+archive.

To check that everything went OK:

```
$ bzr whoami
Sebastian Bassi sbassi@gmail.com
```

As a general rule, the syntax of the **bzr** command is `bzr command [options]` `[arguments]`.

14.6   Different Ways to Use a VCS

There are several ways to use a VCS program. Here are some common scenarios:

- A single user

- Two users without a central server

- Multiple users with a central server

Even if you plan to follow only one of the proposed settings, I recommend reading all of them in the order presented here since most concepts exposed in one case are used in the following:

14.6.1   Workflow: Single User

A programer working alone can benefit by using a VCS a software tool. While he is not going to incorporate versions of code from other parties or publish their work for others to make new derivated code, he can use the VCS capability of tracking revisions to return to a previous state if necessary.

The samples below are based on a programmer (let's call him *Lone Ranger*) using Bazaar:

The fist step is to create the working directory (dna_calculator):

```
loneranger@hp:~$ mkdir dna_calculator
loneranger@hp:~$ cd dna_calculator
```

Bazaar has to be initialized in this directory:

```
loneranger@hp:~/dna_calculator$ bzr init
```

If there is no message, then there was no error (that is a UNIX legacy of "no news, good news"). The next step is editing the files (**data.txt** and **main.py**).

```
loneranger@hp:~/dna_calculator$ pico data.txt
loneranger@hp:~/dna_calculator$ pico main.py
loneranger@hp:~/dna_calculator$ ls
data.txt  main.py
```

Once the programmer has his files ready, he **adds** them to the project:

```
loneranger@hp:~/dna_calculator$ bzr add
added data.txt
added main.py
```

Note that since no additional arguments where passed to **add**, Bazaar added all the versioned files available in the directory recursively.

The second step is to **commit** the files. This will be revision 1 of the Lone Ranger program:

```
$ bzr commit -m "My first commit"
Committing to: /home/loneranger/dna_calculator/
added data.txt
added main.py
Committed revision 1.
```

If the programmer edits one of the files (like **main.py**), he can see the difference between the edited version and the version already included in the repository.

```
loneranger@hp:~/dna_calculator$ pico main.py
loneranger@hp:~/dna_calculator$ bzr diff
=== modified file 'main.py'
--- main.py     2008-03-01 17:39:05 +0000
+++ main.py     2008-03-01 17:44:13 +0000
@@ -1,3 +1,5 @@
 #!/usr/local/bin/python2.5

 import __hello__
+
+print "This works in Python 2.5"
```

When the programmer is satisfied with the changes, he can **commit** it again:

```
$ bzr commit -m "A print statement added"
Committing to: /home/loneranger/dna_calculator/
modified main.py
Committed revision 2.
```

This procedure (modify and commit) can be repeated anytime that it is needed. To see the change log, use the **log** command:

```
loneranger@hp:~/dna_calculator$ bzr log
------------------------------------------------------------------
revno: 6
committer: Lone Ranger <loneranger@hp>
branch nick: dna_calculator
timestamp: Sat 2008-03-01 15:07:56 -0300
message:
  comment
------------------------------------------------------------------
revno: 5
committer: Lone Ranger <loneranger@hp>
branch nick: dna_calculator
timestamp: Sat 2008-03-01 15:06:38 -0300
message:
  Corrected data
------------------------------------------------------------------
revno: 4
committer: Lone Ranger <loneranger@hp>
branch nick: dna_calculator
timestamp: Sat 2008-03-01 15:06:04 -0300
message:
  New data
------------------------------------------------------------------
revno: 3
committer: Lone Ranger <loneranger@hp>
branch nick: dna_calculator
timestamp: Sat 2008-03-01 14:45:51 -0300
message:
  Print for Python 3000
------------------------------------------------------------------
revno: 2
committer: Lone Ranger <loneranger@hp>
branch nick: dna_calculator
timestamp: Sat 2008-03-01 14:44:52 -0300
message:
  A print statement added
------------------------------------------------------------------
revno: 1
committer: Lone Ranger <loneranger@hp>
branch nick: dna_calculator
timestamp: Sat 2008-03-01 14:39:05 -0300
message:
  My first commit
```

This **log** command can be applied to a specific file:

```
loneranger@hp:~/dna_calculator$ bzr log data.txt
------------------------------------------------------------
revno: 5
committer: Lone Ranger <loneranger@hp>
branch nick: dna_calculator
timestamp: Sat 2008-03-01 15:06:38 -0300
message:
  Corrected data
------------------------------------------------------------
revno: 4
committer: Lone Ranger <loneranger@hp>
branch nick: dna_calculator
timestamp: Sat 2008-03-01 15:06:04 -0300
message:
  New data
------------------------------------------------------------
revno: 1
committer: Lone Ranger <loneranger@hp>
branch nick: dna_calculator
timestamp: Sat 2008-03-01 14:39:05 -0300
message:
  My first commit
```

If Lone Ranger wants to go back to the state in revision number 3, he can accomplish this by using the **revert** command:

```
$ bzr revert -r 3
 M  data.txt
 M* main.py
```

If he only wants to go back to the previous state, he can use:

```
$ bzr revert filename
```

After making the changes, he has to *commit* again:

```
$ bzr commit -m "Going back to version 3"
Committing to: /home/loneranger/dna_calculator/
modified data.txt
modified main.py
Committed revision 7.
```

To publish his code,[6] he can use any protocol he desires (i.e., FTP, sFTP, SSH, etc.).[7] The fact that the programmer doesn't need a special server is a

---

[6]Lone Ranger programs alone, but he shares his code when it is finished.

[7]Remember that for using sFTP you need the *paramiko* module.

big selling point for Bazaar. In this case Lone Ranger has a (fictional) server called **site.com** to which he uploads files using FTP,[8]:

```
$ bzr push <=
ftp://loneranger%40site.com@site.com/dna_calc
FTP loneranger@site.com@site.com password:
Created new branch.
```

If unlike Lone Ranger you don't have a web-server, there are some free project hosting servers like **Launchpad.net**:[9]

```
$ bzr push bzr+ssh://lone.ranger@bazaar.launchpad.net/~lone.<=
ranger/py4bio/newbranch
Enter passphrase for key '/home/loneranger/.ssh/id_dsa':
Created new branch.
```

From now on, anybody can get a copy of your code with the command:

```
$ bzr branch http://bazaar.launchpad.net/~lone.ranger/py4bio/<=
newbranch
Branched 7 revision(s).
```

A feature of Launchpad is that any user can see the code from a web browser, without using Bazaar. This is made possible thanks to Loggerhead (`http://www.lag.net/loggerhead`), a program to view, annotate, search and syndicate projects made with Bazaar. This example can be seen online.[10]

Another way to publish your code is using the *export* function:

```
$ bzr export ../last_release/dnacalc-0.9.tar.gz
```

14.6.2   Workflow: Two Users Sharing Code without a Central Server

In this scenario we have two programmers (Harry and Sally) who want to sync a shared branch. There are several steps in common with the previously described process. Each user has his or her own directory to save the project files:

---

[8]Note that some commercial hosting providers will give usernames in the form of "user@domain". In this case you should write your user as **user%40domain** so the full URL would be: **ftp://user%40domain@domain**.

[9]Launchpad.net describes itself as a "free software hosting and development website." It is a service from Canonical Ltd, the commercial sponsor of Ubuntu Linux. There are other similar services like `sourceforge.net`, `bioinformatics.org` and `http://code.google.com/hosting`. Launchpad is featured here because it is open source and is written in Python.

[10]At this URL: `http://bazaar.launchpad.net/~lone.ranger/py4bio/newbranch/changes`.

```
harry@hp:~$ mkdir projectY
harry@hp:~$ cd projectY
harry@hp:~/projectY$
```

The same with Sally,

```
sally@ibm:~$ mkdir projectY
sally@ibm:~$ cd projectY
sally@ibm:~/projectY$
```

Harry works on "project Y" and generates two files: **main.py** and **data.txt**. The first step is to initialize Bazaar in this directory:

```
harry@hp:~/projectY$ bzr init
```

The second step is to include his files:

```
harry@hp:~/projectY$ bzr add
added data.txt
added main.py
```

Now, it is time for the first commit:

```
harry@hp:~/projectY$ bzr commit -m "First commit"
Committing to: /home/harry/projectY/
added data.txt
added main.py
Committed revision 1.
```

At this time, Sally wants to work with Harry's code. As the first step, she has to make a *branch* of Harry's code. This *branch* will be called by Sally **projectY-fromH**:

```
sally@ibm:~/projectY$ bzr branch
sftp://harry@192.168.0.1/home/sally/projectY/ projectY-fromH
Branched 1 revision(s).
```

Work on this branch is just a matter of changing the working directory and editing those files:

```
sally@ibm:~/projectY$ cd projectY-fromH
sally@ibm:~/projectY/projectY-fromH$ pico main.py
```

Once the changes are done, she can see the differences between the files by using the *diff*:

```
sally@ibm:~/projectY/projectY-fromH$ bzr diff main.py
=== modified file 'main.py'
--- main.py      2008-02-21 01:16:21 +0000
+++ main.py      2008-02-21 01:27:50 +0000
@@ -1,8 +1,8 @@
 #This program is made for the Python Book
 protseq=raw_input("Enter your protein sequence: ")
-protweight={"A":89,"V":117,"L":131,"I":131,"P":115,"F":165,\
-            "W":204,"M":149,"G":75,"S":105,"C":121,"T":119,\
-            "Y":181,"N":132,"Q":146,"D":133,"E":147,\
+protweight={"A":89,"V":117,"L":131,"I":131,"P":115,"F":165,
+            "W":204,"M":149,"G":75,"S":105,"C":121,"T":119,
+            "Y":181,"N":132,"Q":146,"D":133,"E":147,
             "K":146,"R":174,"H":155}
 totalW=0
 for aa in protseq:
```

The *diff* output tells Harry that Sally has removed the symbols "\" from the lines starting with the sign "-". The backslash is used to break a line into multiple lines, but it is not needed here since there are brackets ({}).

Since Sally is happy with this change, she commits it into her branch:

```
sally@ibm:~/projectY/projectY-fromH$ bzr commit -m <=
"Some slash removed"
Committing to: /home/user2/projectY/projectY-fromH/
modified main.py
Committed revision 2.
```

Now it is time to contribute her modification to Harry's code. To this end, Sally prepares a patch:

```
sally@ibm:~/projectY/projectY-fromH$ bzr send -o sally.patch
sftp://harry@192.168.0.1/home/sally/projectY/
```

This file (`sally.patch`) is created based on Sally's code and Harry's code base. Now Sally can email this file to Harry. He will check if he likes the change; if so, this file will be merged into Harry's code:

```
harry@hp:~/projectY$ bzr merge sally.patch
 M  main.py
All changes applied successfully.
```

When Harry merges Sally's patch into his project, he is warned that the file main.py was modified. Now Harry runs a *diff* as if he were the one who modified the main.py file:

```
harry@hp:~/projectY$ bzr diff
=== modified file 'main.py'
--- main.py     2008-02-21 01:16:21 +0000
+++ main.py     2008-02-22 17:59:35 +0000
@@ -1,8 +1,8 @@
 #This program is made for the Python Book
 protseq=raw_input("Enter your protein sequence: ")
-protweight={"A":89,"V":117,"L":131,"I":131,"P":115,"F":165,\
-          "W":204,"M":149,"G":75,"S":105,"C":121,"T":119,\
-          "Y":181,"N":132,"Q":146,"D":133,"E":147,\
+protweight={"A":89,"V":117,"L":131,"I":131,"P":115,"F":165,
+          "W":204,"M":149,"G":75,"S":105,"C":121,"T":119,
+          "Y":181,"N":132,"Q":146,"D":133,"E":147,
           "K":146,"R":174,"H":155}
 totalW=0
 for aa in protseq:
```

It is time to *commit* the change into Harry's personal repository:

```
harry@hp:~/projectY$ bzr commit -m "Sally patch"
Committing to: /home/harry/projectY/
modified main.py
Committed revision 2.
```

14.6.3 Workflow: Multiple Users Sharing Code with a Central Server

In this case there is a group of people who want to work together to develop a program.

The first step is to put a version of the code in a place available for all parties. This is done by creating a central shared repository:

```
$ bzr init-repo sftp://user1@myserver.com/projectX1
sFTP user1@myserver.com password:
```

A common way to populate a central branch is by making a local branch and pushing it into the newly created central branch:

```
$ bzr init-repo projectX1
$ bzr init projectX1/trunkv2
$ cd projectX1/trunkv2
```

After editing your files:

```
$ bzr add
$ bzr commit -m "Initial import"
$ bzr push sftp://user1@myserver.com/projectX1/trunkv2
```

Now we use *bind* to keep the local *commits* to the server in synch:

```
$ bzr bind sftp://user1@myserver.com/projectX1/trunkv2
```

From now on, each *commit* in our local branch well be replicated into the central branch:

```
$ bzr commit -m "new code with range"
FTP user1@myserver.com password:
Committing to: ftp://user1@myserver.com/projectX1/MyTrunk/
modified main.py
Committed revision 2.
[====================] Running post_commit hooks - Stage 6/6
```

To go back to apply *commits* in local mode, use *unbind*

```
$ bzr unbind
```

From now on, the *commit* are local again. This is equivalent to use `bzr commit --local` each time you use **commit**. To interact again with the server, use *bind*. Another option is to use *update* to sync the changes with the server. In this case we have to use **commit** to send the changes to the central server.

When other users want to modify the code, they will have to **checkout** to have a copy, and then each time they make a *commit*, it will be updated in the server:

```
$ bzr checkout ftp://user2@myserver.com/projectX1/MyTrunk/
FTP ftp://user2@myserver.com password:
$ cd MyTrunk/
$ ls
main.py
$ pico main.py
$ bzr commit -m"print x"
FTP ftp://user2@myserver.com password:
Committing to: ftp://user2@myserver.com/projectX1/MyTrunk/
modified main.py
Committed revision 3.
```

Use *update* to make sure you are working with the last available version:

```
$ bzr update
FTP ftp://user2@myserver.com password:
Tree is up to date at revision 3.
```

14.7 VCS Conclusion

There are several ways to use a VCS program. This chapter barely touches the surface of this subject. It is up to the reader to keep researching on this matter (please see *Additional Resources* for more information).

14.8 Additional Resources

- Jennifer Vesperman. "Introduction to CVS."
  `http://www.linuxdevcenter.com/pub/a/linux/2002/01/03/cvs_intro.html`

- Dave O Connor. "Getting Started with CVS."
  `http://www.linux.ie/articles/tutorials/cvs.php`

- "Git for Computer Scientists" by Tommi Virtanen (aka Tv).
  `http://eagain.net/articles/git-for-computer-scientists`

- "Easy Git – git for mere mortals."
  `http://www.gnome.org/~newren/eg/`

- "An Introduction to (Easy) Git," by Elijah Newren.
  http://www.gnome.org/ newren/eg/presentations/git-introduction.pdf

- "Bazaar Tutorial"
  `http://doc.bazaar-vcs.org/bzr-0.15/tutorial.htm`

- "Bazaar Tutorial Screencasts And Videos."
  `http://showmedo.com/videos/bazaar`

- "PEP-374: Migrating from svn to a distributed VCS."
  `http://www.python.org/dev/peps/pep-0374/`

- Noble WS, 2009 "A Quick Guide to Organizing Computational Biology Projects." PLoS Comput Biol 5(7): e1000424.
  `http://dx.doi.org/10.1371/journal.pcbi.1000424`

- Software:

  - Bazaar
    `http://bazaar-vcs.org`
  - Git
    `http://git.or.cz`

- Easy Git
  http://gitorious.org/projects/eg
- Mercurial
  http://www.selenic.com/mercurial/wiki

14.9   Self-Evaluation

1. What is version control software?

2. Name advantages of using version control.

3. Why would a single programmer use such a program?

4. What is the difference between centralized and distributed version control?

5. Define (in the context of version control): repository, branch, commit, merge and check-out.

6. What is the difference between local and remote commit?

7. What kind of server is needed to publish a branch using Bazaar?

8. Name advantages of Bazaar over other version control software.

9. What is a patch file and how do you submit one?

10. What is Launchpad.net and how is it related to Bazaar?

Part IV

Python Recipes with
Commented Source Code

Chapter 15

Sequence Manipulation in Batch

15.1 Problem Description

15.2 Problem One: Create a FASTA File with Random Sequences

There are some statistical tests where random sequences are useful. Random sequences can also be used to test programs when you don't have real data in the required amount.

In code 15.1 we assume that we need to generate 5000 sequences, each one with a length between 4000 and 15000 nucleotides.

15.2.1 Commented Source Code

Listing 15.1: Generate random sequences (py3.us/62)

```
1 import random
2
3 from Bio.SeqRecord import SeqRecord
4 from Bio.Seq import Seq
5 from Bio import SeqIO
6
7 def new_rnd_seq(sl):
8     """ Generate a random DNA sequence with a sequence length
9         of "sl" (int).
10    """
11    s = ''
12    for x in range(sl):
13        s += random.choice('ATCG')
14        # s += random.sample('ATCG',1)[0] is not so fast.
15    return s
16
17 newfh = open('randomseqs.txt','w')
```

```
18 for i in range(1,501):
19     # Creates a random number in the range of 4000-15000
20     rsl = random.randint(4000,15000)
21     # Generate the random sequence
22     rawseq = new_rnd_seq(rsl)
23     # Generate a correlative name
24     seqname = 'Sequence_number_' + str(i)
25     rec = SeqRecord(Seq(rawseq),id=seqname,description='')
26     SeqIO.write([rec],newfh,'fasta')
27 newfh.close()
```

Code explanation: Generation of the random sequence is done in the *new_rnd_seq* function (fron lines 7 to 15). This function is called inside the *for loop* and it is stored as *rawseq*. In line 25 a *SeqRecord* object is created. This object is passed to *SeqIO.write* in line 26.

15.3   Problem Two: Filter Not Empty Sequences from a FASTA File

Sometimes you need to get rid of malformed sequences from a FASTA file. Some programs choke when they receive in the input file an empty sequence. Formatdb, the program used to format BLAST databases, is known behave like this. The code in listing 15.2 asumes that you have a FASTA file like this:

```
>SSR86 [ssr] : Tomato-EXPEN 2000 map, chr 3
AGGCCAGCCCCCTTTTCCCTTAAGAACTCTTTGTGAGCTTCCCGCGGTGGCGGCCGCTCTAG
>SSR91 [ssr]
>SSR252 [ssr]
TGGGCAGAGGAGCTCGTANGCATACCGCGAATTGGGTACACTTACCTGGTACCCCACCCGGG
TGGAAAATCGATGGGCCCGCGGCCGCTCTAGAAGTACTCTCTCTCT
>SSR257 [ssr]
TGAGAATGAGCACATCGATACGGCAATTGGTACACTTACCTGCGACCCCACCCGGGTGGAAA
ATCGATGGGCCCGCGGCC
>SSR92 [ssr]   : Tomato-EXPEN 2000 map, chr 1
```

And it should produce a version of the file without the empty records:

```
>SSR86 [ssr] : Tomato-EXPEN 2000 map, chr 3
AGGCCAGCCCCCTTTTCCCTTAAGAACTCTTTGTGAGCTTCCCGCGGTGGCGGCCGCTCTAG
>SSR252 [ssr]
TGGGCAGAGGAGCTCGTANGCATACCGCGAATTGGGTACACTTACCTGGTACCCCACCCGGG
TGGAAAATCGATGGGCCCGCGGCCGCTCTAGAAGTACTCTCTCTCT
>SSR257 [ssr]
```

TGAGAATGAGCACATCGATACGGCAATTGGTACACTTACCTGCGACCCCACCCGGGTGGAAA
ATCGATGGGCCCGCGGCC

15.3.1 Commented Source Code

Listing 15.2: Filter a FASTA file (py3.us/63)

```
1 from Bio import SeqIO
2 # Name of the input file
3 fh = open('out22.fas')
4 # Name of the output file
5 newfh = open('out22-GOOD.fas','w')
6
7 def retseq(seqfh):
8     """ Parse a fasta file and store non empty records
9         into the fullseqs list.
10    """
11    # Empty list to store good sequences
12    fullseqs = []
13    for record in SeqIO.parse(seqfh,'fasta'):
14        if len(record.seq)!=0:
15            fullseqs.append(record)
16    seqfh.close()
17    return fullseqs
18
19 SeqIO.write(retseq(fh),newfh,'fasta')
20 newfh.close()
```

Although this program does its job, it is not an example of efficient use of computer resources. The list **fullseqs** ends up with the information on every non-empty sequence in the file. For short sequence files this is not noticieable. In a realistic scenario, an input file of 500 Mb long can easily bring a server to its knees.

The same program can be adapted for low memory usage. This is accomplished in code 15.3 by the use of a *generator*. A *generator* is a special kind of function. Syntactically, a generator and a function are very alike, both has a header with the *def* keyword, a name and parameters (if any). The most visible difference is that instead of having the word *return* as an exit point, generators has *yield*. The conceptual diference between a function and a generator is that the generator keeps its internal state after being called. The next time the generator is called, it resumes its execution from the point it was before. This property is used to yield several values, one at a time.

Listing 15.3: Filter a FASTA file with a generator (py3.us/64)

```
1 from Bio import SeqIO
2 # Name of the input file
3 fh = open('out22.fas')
4 # Name of the output file
5 newfh = open('out22-GOOD.fas','w')
6
7 def retseq(seqfh):
8     for record in SeqIO.parse(seqfh,'fasta'):
9         if len(record.seq)!=0:
10             yield record
11
12 SeqIO.write(retseq(fh),newfh,'fasta')
13 newfh.close()
14 fh.close()
```

Code explanation: This code is very similar to 15.2. The first difference
that is apparent when they are compared line by line is that in this code there
is no empty list to store the sequences (it was called *fullseqs* in listing 15.2).
Another difference is that in the first code, *retseq* is a function while in the last
version it is a generator. Both differences are tightly related: Since generators
return elements one by one, there is no need to use a list. The generator yields
one record to *SeqIO.write*, which keeps on calling the generator until it gets
a StopIteration exception.

There is another way to do the same task without using generators and
functions, while still consuming an optimal amount of RAM:

Listing 15.4: Yet another way to filter a FASTA file (py3.us/65)

```
1 from Bio import SeqIO
2 fh = open('out22.fas')
3 newfh = open('out22-GOOD.fas','w')
4 for record in SeqIO.parse(fh,'fasta'):
5     if len(record.seq)!=0:
6         SeqIO.write([record],newfh,'fasta')
7         # Try (record,) as an alternative to [record]
8 newfh.close()
9 fh.close()
```

Note that there is no list creation (hence no RAM abuse) and there is no
generator because the *SeqIO.write* function saves the records as soon as it
receives them.[1] If code 15.4 was shown in the first place, I wouldn't have the
chance to show the difference between using a function and a generator.

---

[1]There is some internal small RAM caching but it is not relevant in terms of how this
function works.

15.4 Problem Three: Modify Every Record of a FASTA File

In this problem we have a FASTA file that looks the one in listing 15.5:

Listing 15.5: Input file

```
>Protein-X
NYLNNLTVDPDHNKCDNTTGRKGNAPGPCVQRTYVACH
>Protein-Y
MEEPQSDPSVEPPLSQETFSDLWKLLPENNVLSPLPSQAMDDLMLSPDDIEQWFTEDPGPDA
>Protein-Z
MKAAVLAVALVFLTGCQAWEFWQQDEPQSQWDRVKDFATVYVDAVKDSGRDYVSQFESST
```

The goal of the exercise is to modify all the sequences by adding the species tag in each sequence name. This kind of file modification may be required for sequence submission for a genetic data bank. A modified FASTA file would look like this:

Listing 15.6: Input file

```
>Protein-X [Rattus norvegicus]
NYLNNLTVDPDHNKCDNTTGRKGNAPGPCVQRTYVACH
>Protein-Y [Rattus norvegicus]
MEEPQSDPSVEPPLSQETFSDLWKLLPENNVLSPLPSQAMDDLMLSPDDIEQWFTEDPGPDA
>Protein-Z [Rattus norvegicus]
MKAAVLAVALVFLTGCQAWEFWQQDEPQSQWDRVKDFATVYVDAVKDSGRDYVSQFESST
```

Note that in listing 15.6 there is the tag [Rattus norvegicus] in the name of each record. There are several ways to accomplish this task. Here is a version with Biopython Bio.SeqIO module (listing 15.7) and another that uses just the Standard Python Library (listing 15.8)

15.4.1 Commented Source Code

Listing 15.7: Add a tag in a FASTA sequence with Biopython (py3.us/66)

```
1 from Bio import SeqIO
2
3 # Name of the input file
4 fh = open('out22.fas')
5 # Name of the output file
6 newfh = open('out3.fas','w')
7 for record in SeqIO.parse(fh,'fasta'):
```

```
8      # Modify description
9      record.description += '[Rattus norvegicus]'
10     SeqIO.write([record],newfh,'fasta')
11 newfh.close()
12 fh.close()
```

Even if you can use Biopython to modify a FASTA sequence, sometimes this is overkill. The following code shows how to accomplish the same task without Biopython:

Listing 15.8: Add a tag in a FASTA sequence (py3.us/67)

```
1 # Name of the input file
2 fh = open('out22.fas')
3 # Name of the output file
4 newfh = open('out3.fas','w')
5 for line in fh:
6     if line.startswith('>'):
7         line = line.replace('\n','')+' [Rattus norvegicus]\n'
8     newfh.write(line)
9 newfh.close(); fh.close()
```

Chapter 16

Web Application for Filtering Vector Contamination

16.1 Problem Description

DNA sequences are usually inserted into a cloning vector for manipulation. When sequencing, these constructs frequently produce raw sequences that include segments derived from a vector. If the vector part of the raw sequence is not removed, the finished sequenced will be contaminated, spoiling further analysis. There are multiple sources of DNA contamination, like transposons, insertion sequences, organisms infecting our samples and other organisms used in the same laboratory (e.g., cross contamination from dirty equipment).

Sequence contamination is not a minor issue, since it can lead to several problems like:[1]

- Time and effort wasted on meaningless analyses

- Misassembly of sequence contigs and false clustering

- Erroneous conclusions drawn about the biological significance of the sequence

- Pollution of public databases

- Delay in the release of the sequence in a public database

In order to identify the vector part of a sequence, a BLAST can be done against a vector sequence database (or against any other database that you think your sequence could be contaminated by). To help in removing those sequences, this program take a sequence or a group of sequences in FASTA format and makes the BLAST against a user selected database. It identifies the match and the contamination is masked by using "N" character in the sequence input by the user.

This program works as a web application, so there is an HTML form for the user to enter the data and a Python file to process it.

---

[1]For information regarding each item please see NCBI VecScreen program at http://www. ncbi.nlm.nih.gov/VecScreen/contam.html.

16.1.1   Commented Source Code

HTML form

Listing 16.1: Primer design out of one sequence (py3.us/68)

```
1 <html><head><title>Vector filter</title></head>
2 <body>Paste sequences in FASTA format:
3 <form action="../cgi-bin/filtro.py"
4 enctype="multipart/form-data" method="post">
5 <textarea name="seqs" rows="25" cols="80"></textarea>
6 <br /><p>Or enter the sequence(s) in a file:<br>
7 <input type="file" name="seqdatafile" size="40">
8 <p>Filter by:
9 <select name="blastdb">
10 <option value="customdb">Custom vectors</option>
11 <option value="ncbivector">NCBI Vector DB</option>
12 </select><br /><input type="submit" value="  Filter  " />
13 </form></body></html>
```

This code produces, when rendered by a web browser, a page like the one shown in figure 16.1.

In line 4 there is a call to the Python script that processes the form. This script is shown in listing 16.2:

Listing 16.2: Web script to filter a DNA sequence (py3.us/69)

```
1 #!/usr/bin/python
2
3 import cgi, cgitb
4 from Bio import SeqIO
5 from Bio.SeqRecord import SeqRecord
6 from Bio.Blast import NCBIXML, NCBIStandalone
7 from tempfile import NamedTemporaryFile
8
9 blast_exe = '/var/www/blast-2.2.18/bin/blastall'
10
11 cgitb.enable()
12 mask = "N" # Mask character
13 form = cgi.FieldStorage()
14 # Get sequence data from text area
15 seqs = form.getvalue("seqs")
16 # Check if the textarea is empty
17 if not seqs:
18     # Since the textarea is empty, check the uploaded file
19     seqs = form.getvalue("seqdatafile")
```

```
20
21 blast_db = form.getvalue("blastdb",'customdb')
22 if blast_db == 'customdb':
23     db = '/var/www/blast/db/lauravect'
24 elif blast_db == 'ncbivector':
25     db = '/var/www/blast/db/vector'
26 elif blast_db == 'plantmito':
27     db = '/var/www/blast/db/plantmitogenomes'
28 else:
29     # In case someone sends an unexpected string to
30     # the script, it defaults to the custom vector DB
31     db = '/var/www/blast/db/lauravect'
32
33 def create_rel(XMLin):
34     """ Create a dictionary that relate the sequence name
35     with the region to mask """
36     bat1 = {}
37     b_records = NCBIXML.parse(XMLin)
38     for b_record in b_records:
39         for alin in b_record.alignments:
40             for hsp in alin.hsps:
41                 qs = hsp.query_start
42                 qe = hsp.query_end
43                 if qs>qe:
44                     qe,qs=qs,qe
45                 if b_record.query not in bat1:
46                     bat1[b_record.query] = [(qs,qe)]
47                 else:
48                     bat1[b_record.query].append((qs,qe))
49     return bat1
50
51 def maskseqs(ffh,bat1):
52     """ Take a FASTA file and apply the mask using the
53         positions in the dictionary"""
54     outseqs = []
55     for record in SeqIO.parse(ffh, "fasta"):
56         if record.id in bat1:
57             # Generate a mutable sequence object to store
58             # the sequence with the "mask".
59             mutable_seq = record.seq.tomutable()
60             coords = bat1[record.id]
61             for x in coords:
62                 mutable_seq[x[0]:x[1]] = mask*(x[1]-x[0])
63             seq_rec = SeqRecord(mutable_seq,record.id,'','')
64             outseqs.append(seq_rec)
```

```
65          else:
66              # Leave the sequence as found when its name is
67              # not in the dictionary.
68              outseqs.append(record)
69      return outseqs
70
71 # Create a temporary file
72 fasta_in_fh = NamedTemporaryFile()
73 # Write the user entered sequence into this temporary file
74 fasta_in_fh.write(seqs)
75 # Flush the data to disk without closing and deleting the file,
76 # since that closing a temporary file also deletes it
77 fasta_in_fh.flush()
78 # Get the name of the temporary file
79 file_in = fasta_in_fh.name
80 # Run the BLAST query
81 rh, eh = NCBIStandalone.blastall(blast_exe, "blastn", db,
82                                  file_in, expectation='1e-6')
83 # Create contamination position and store it in a dictionary
84 bat1 = create_rel(rh)
85 # Reset the pointer position to the begining of the file
86 fasta_in_fh.seek(0)
87 # Get the sequences masked
88 newseqs = maskseqs(fasta_in_fh,bat1)
89 # Close and delete the temporary file
90 fasta_in_fh.close()
91 # Creates a new temporary file
92 fasta_out_fh = NamedTemporaryFile()
93 # Write the masked sequence into this temporary file
94 SeqIO.write(newseqs,fasta_out_fh,'fasta')
95 # Reset the pointer position to the begining of the file
96 fasta_out_fh.seek(0)
97 # Read the file
98 finalout = fasta_out_fh.read()
99 # Close and delete the temporary file
100 fasta_out_fh.close()
101
102 print 'Content-type: text/html\n'
103 print """<html><head><title>Vector Filter Output</title></head>
104 <body>Filtered sequences:<br/><p></p><pre>%s</pre>
105 </body></html>"""%(finalout)
```

Note that this code assumes that there is a BLAST formated database (from line 19 to 27). To create such a base you should run the *formatdb* utility that is included in the NCBI BLAST package.[2]

If your sequence file is called `mito.nt`, a *formatdb* command may look like this:

```
$ ./formatdb -i mito.nt -p F -o T
```

Where `-i` means "input file", `-p F` stands for *nucleotide* (`-p T` is for protein) and `-o T` is used only when you want to index the database to create links to the NCBI Web site (use `-o T` when the file is downloaded from the NCBI ftp server and `-o F` if the fasta file is made in situ). The *formatdb* manual is available from the command line with the option `--help` or posted at `ftp://ftp.ncbi.nih.gov/blast/documents/formatdb.html`.

16.2 Additional Resources

- K. W. Liao, Y. W. Chang, S. R. Roffler, "Presence of cloning vector sequences in the untranslated region of some genes in Genbank," J. Biomed. Sci., 7(6):529-30, 2000.

- C. Miller, J. Gurd, A. Brass, "A RAPID algorithm for sequence database comparisons: application to the identification of vector contamination in the EMBL databases," Bioinformatics, 15(2):111-21, 1999.

- G. A. Seluja, A. Farmer, M. McLeod, C. Harger, P. A. Schad, "Establishing a method of vector contamination identification in database sequences," Bioinformatics, 15(2):106-10, 1999.

- C. Savakis, R. Doelz, "Contamination of cDNA sequences in databases," Science, 259(5102):1677-8, 1993.

---

[2]Look for the appropriate package for your system on this FTP site: `ftp://ftp.ncbi.nih.gov/blast/executables/LATEST`.

FIGURE 16.1:   HTML form for sequence filtering.

Chapter 17

Searching for PCR Primers Using Primer3

17.1 Problem Description

Primers are small DNA strands (from 15 to 30 base pairs long) that are complementary to a specific spot in a DNA molecule. They are needed for DNA replication to take place. In molecular biology, primers are used for a DNA amplification chain reaction called PCR (**P**olymerase **C**hain **R**eaction). PCR primers have their own characteristics like: specific melting temperature, primer length, need to avoid self-complementarity and other parameters.[1]

PCR primer design is one of the most ubiquitous tasks done in a molecular biology laboratory. There are several programs that help researchers to pick good primers. Some programs are Web based, some of them are stand-alone GUI applications like VectorNTI Suite[2] and Oligo.net. These programs are suitable for case by case study of a few sequences, but they are not the chosen option for automatic batch generation of hundreds of primers, a task that is routinely done in sequencing and fingerprinting projects. One of the most used programs is *primer3*.[3] This is due to the high quality of proposed primers and because it can be run in batch and generate multiple primers at once.

Primer3 takes care of primer design, what is left for us is to prepare the input file for *primer3*. This input file holds the sequence for which the primer should be picked and other required and optional parameters like desired primer name, primer size, product size, regions to exclude and others.

A *primer3* input file looks like this:

---

[1]For more information on primer design please see "Additional Resources."

[2]VectorNTI Web site: http://www.informax.com.

[3]This software is available at http://sourceforge.net/projects/primer3, please see the included documentation for how to cite this software if used in a publication.

```
PRIMER_SEQUENCE_ID=<Name>
SEQUENCE=<DNA Sequence in one line>
TARGET=<start>,<length>
PRIMER_OPT_SIZE=<size>
PRIMER_MIN_SIZE=<size>
PRIMER_MAX_SIZE=<size>
PRIMER_NUM_NS_ACCEPTED=<int>
PRIMER_EXPLAIN_FLAG=<int>
PRIMER_PRODUCT_SIZE_RANGE=<start>-<end>
=
```

Each parameter is detailed in the *primer3* README.txt file, although most of them are self-explanatory. The = character is used to terminate the record. Several records can be included in one *primer3* input file.

This recipe chapter is divided in two tasks. The first task will be to generate an input file for *primer3* based in a FASTA file with one sequence inside and one restriction. The second task involves analysis of several sequences for the generation of a multiple sequence primer3 input file.

17.2 Primer Design Flanking a Variable Length Region

This script should read a FASTA formated file with one sequence inside. This sequence has a microsatellite[4] of variable length where we should avoid doing primer design over it. In fact, we need to assure that the chosen primer flanks this region. We don't know either the length neither the position of the microsatellite, but we know it is a repeat of the "AAT" sequence and it can be present between 5 to 15 times.

This task could be divided in the following steps:

1. Read the sequence from the FASTA file: Biopython provides the **SeqIO** module that will be used to read the sequence.

2. Detect the region with the microsatellite and store its position: A sliding windows approach will be used. On each possible 45 base pair long window[5] we will count how many times our repeated sequence is present. The chosen window will be the one with the highest number of repetitions inside. We need to store the position of this window.

---

[4]A **microsatellite** is a region in the chromosome that is characterized by having a small DNA sequence repeated a variable number of times. Since they are inheritable, they can be used to trace relationship between individuals.

[5]This size is estimated by calculating a 3 letter repeat (AAT) and this sequence can be repeated up to 15 times.

3. Generate the primer3 input file: This is trivially accomplished by using the retrieved sequence and the previously stored position as a target.

17.2.1   Commented Source Code

Listing 17.1: Primer design out of one sequence (py3.us/70)

```
1 from Bio import SeqIO
2
3 sfile = open('/home/sb/bioinfo/seqwrep.fasta')
4 # mysel stores a SeqRecord object generated from the
5 # first (and only) record in the fasta file.
6 myseq = SeqIO.read(sfile, "fasta")
7 # title stores the "id" attribute of the SeqRecord object.
8 title = myseq.id
9 # seq stores the sequence converted into string and
10 # uppercased.
11 seq = str(myseq.seq).upper()
12 win_size = 45
13 i = 0
14 number_l = []
15 # This while is used to walk over the sequence.
16 while i<=(len(seq)-win_size):
17     # Each position of number_l stores the amount of 'AAT'
18     # found on each window.
19     number_l.append(seq[i:i+win_size].count('AAT'))
20     i += 1 # This is the same as i = i+1
21 # pos stores the position of the window with the highest
22 # amount of 'AAT'
23 pos = number_l.index(max(number_l))
24 fout = open('/home/sb/bioinfo/swforprimer3.txt','w')
25 fout.write(
26 '''PRIMER_SEQUENCE_ID=%s
27 SEQUENCE=%s
28 TARGET=%s,%s
29 PRIMER_OPT_SIZE=18
30 PRIMER_MIN_SIZE=15
31 PRIMER_MAX_SIZE=20
32 PRIMER_NUM_NS_ACCEPTED=0
33 PRIMER_EXPLAIN_FLAG=1
34 PRIMER_PRODUCT_SIZE_RANGE=%s-%s
35 =''' % (title,seq,pos,win_size,win_size,len(seq)))
36 fout.close()
37 sfile.close()
38 # Saves the data formated as the input file needed by
```

```
39 # primer3.
```

This program could process a file like this one:[6]

```
>Upstream region of mitochondrial ATP
AATGAAGAAAGCATCTCAATTGGAGAAAAGTTTGTTTTCCCGGGGAATTTGCTTGTCAAC
GAAATTCCACAATAATAATAATAATACTGGCGATAAGCGGATATTTCATAAGTAGGTTCA
CATCGTGATCTAAGTTCCATTTCCCATCGAGAGGTTATGATACTGGTAAAGAGTCCTATT
CTAATAGCTCCGGGC
```

The result of code 17.1 run with the former sequence as an input, produces a file like this one:

```
PRIMER_SEQUENCE_ID=Upstream
SEQUENCE=AATGAAGAAAGCATCTCAATTGGAGAAAAGTTTGTTTTCCCGGGGAA<=
TTTGCTTGTCAACGAAATTCCACAATAATAATAATAATACTGGCGATAAGCGGATA<=
TTTCATAAGTAGGTTCACATCGTGATCTAAGTTCCATTTCCCATCGAGAGGTTATG<=
ATACTGGTAAAGAGTCCTATTCTAATAGCTCCGGGC
TARGET=40,45
PRIMER_OPT_SIZE=18
PRIMER_MIN_SIZE=15
PRIMER_MAX_SIZE=20
PRIMER_NUM_NS_ACCEPTED=0
PRIMER_EXPLAIN_FLAG=1
PRIMER_PRODUCT_SIZE_RANGE=45-195
=
```

This file is used as input into **primer3** in this way:

```
$ ./primer3_core < swforprimer3.txt > primer3out.txt
```

17.3   Batch Primer Design from Multiple Sequences

For this recipe we have 40 files, each file with a sequence pair (that is, a total of 80 sequences) that we are going to search primers for. Each sequence is from an end of a clone and we want to find out the region between each end. To this end we need to amplify that sequence using a primer from a known region. The idea is to search for the left primer over the first sequence and for the right primer over the second one.

This task could be divided in the following steps:

---

[6]To change the name of the input file you have to change line three.

1. Read all the ".fasta" file names and store them into a list.

2. Read the sequences from each fasta file (by walking over the previously created list), using the same **SeqIO** module just used in the 17.1 program. The first FASTA record should be read "as is," while the second one should be converted into its inverse complementary sequence.[7] This is done in order to leave both sequences in the same polarity.

3. Both sequences should be concatenated into one. This "new sequence" is a dummy sequence since it doesn't exist as is. The region between them is missing. This is why we are picking the primers in the first place, to sequence this unknown region. Remember that the idea behind this task is to search for a primer on each end. This way we will flank the missing region.

4. With these sequences, we are ready to write the multirecord primer3 input file. To command the program to search for primers into the desired region, a TARGET directive should be specified in the input file. This target must be in the region where both ends are joined.

17.3.1 Commented Source Code

Listing 17.2: Primer design out of several sequences (py3.us/71)

```
1  import glob
2
3  from Bio import SeqIO
4
5  fout = open('/home/sb/bioinfo/mfdir/forprimer3.txt','w')
6  for x in glob.glob('/home/sb/bioinfo/mfdir/*.fasta'):
7      # Read both records in each fasta file.
8      seq1 = SeqIO.parse(open(x), "fasta").next()
9      seq2 = SeqIO.parse(open(x), "fasta").next()
10     # Get the title of each fasta record.
11     seq1title = seq1.description
12     seq2title = seq2.description
13     # Get the sequence. In seq2, get the reverse complement.
14     seq1 = str(seq1.seq)
15     seq2 = str(seq2.seq.reverse_complement())
16     # Generate a dummy sequence title in the form:
17     # "title1--title2".
```

---

[7]Complementary inverse sequence is obtained by inverting the DNA sequence and obtaining the complentary sequence from this reverted sequence. The inverse complementary sequence of AAAGTCC is GGACTTT.

```
18      totaltitle = '%s--%s' % (seq1title,seq2title)
19      # Generate the dummy sequence.
20      totalseq = seq1+seq2
21      fout.write(
22      '''PRIMER_SEQUENCE_ID=%s
23 SEQUENCE=%s
24 TARGET=%s,2
25 PRIMER_OPT_SIZE=18
26 PRIMER_MIN_SIZE=15
27 PRIMER_MAX_SIZE=20
28 PRIMER_NUM_NS_ACCEPTED=0
29 PRIMER_EXPLAIN_FLAG=1
30 PRIMER_PRODUCT_SIZE_RANGE=30-5000
31 =
32 ''' % (totaltitle,totalseq,len(seq1)))
33 # Write the primer3 input file.
34 fout.close()
```

17.4   Additional Resources

- Steve Rozen and Helen J. Skaletsky. 2000. Primer3 on the WWW for general users and for biologist programmers. (Bioinformatics Methods and Protocols: Methods in Molecular Biology. Totowa, NJ: Human Press, 365-386). Source code available at http://sourceforge.net/projects/primer3. The paper above is available at http://jura.wi.mit.edu/rozen/papers/rozen-and-skaletsky-2000-primer3.pdf.

- Integrating PCR theory and bioinformatics into a research-oriented primer design exercise. (CBE Life Sci Educ. 2008 Spring;7(1):89-95)

- Enhancements and modifications of primer design program Primer3. (Bioinformatics. 2007 May 15;23(10):1289-91. Epub 2007 Mar 22)

Chapter 18

Calculating Melting Temperature from a Set of Primers

18.1 Problem Description

In this case we have a text file full of PCR primers. These primers were obtained from different sources, so their Tm was calculated under different programs and conditions. A researcher may want to uniform the criteria of Tm of his set of primers.

The first version of the program will output the file formated as a csv (comma separated value). This kind of file could be opened with a spreadsheet or with a custom made program. The second version will output the file as an Excel spreadsheet (xls).

Proposed steps to get the melting temperatures of a set of primers:

1. Read the input file line by line.

2. For each line, calculate the melting temperature (Tm) by using the MeltingTemp module from Biopython Bio.SeqUtils.

3. Print the primer sequence, a comma and its Tm value.

4. In the xls case, print the primer sequence in a cell and its Tm value in the next cell in the same row, using pyExcelerator.

18.1.1 Commented Source Code

Listing 18.1:Primer Tm calculation (py3.us/72)

```
1 from Bio.SeqUtils import MeltingTemp as MT
2 primerfile = 'primerlist.txt'
3 for line in open(primerfile,'rU'):
4     # prm stores the primer, without EOL character.
5     prm = line.replace('\n','')
6     # %2.2f is used to print up to two integers, the
7     # decimal separator and two decimal numbers.
8     print '%s,%2.2f' % (prm, MT.Tm_staluc(prm))
```

Version of the same code with Excel output:

Listing 18.2: Primer Tm calculation, excel output (py3.us/73)

```
1 from Bio.SeqUtils import MeltingTemp as MT
2 import pyExcelerator
3 primerfile = 'primerlist.txt'
4 # w is the name of a newly created workbook.
5 w = pyExcelerator.Workbook()
6 # ws is the name of a new sheet in this workbook.
7 ws = w.add_sheet('Result')
8 # These two lines writes the titles of the columns.
9 ws.write(0,0,'Primer Sequence')
10 ws.write(0,1,'Tm')
11 i = 1
12 for line in open(primerfile):
13     # For each line in the input file, write the primer
14     # sequence and the Tm
15     j = 0
16     primer = line.replace('\n','')
17     ws.write(i,j,primer)
18     ws.write(i,j+1,'%2.2f' %(MT.Tm_staluc(primer)))
19     i += 1
20 # Save the spreadsheel into a file.
21 w.save('/home/sb/bioinfo/primerout.xls')
```

18.2   Additional Resources

- PCR Primer Design Guidelines.
  http://www.premierbiosoft.com/tech_notes/PCR_Primer_Design.html

- Molecular Biology Techniques Manual, Vernon E Coyne, M Diane James, Sharon J Reid, and Edward P Rybicki eds.
  http://www.mcb.uct.ac.za/pcroptim.htm

- 10 Tips for Designing PCR Primers That Work.
  http://smartnote.miraibio.com/blog/?p=12

- Nicolas Le Novère. MELTING, computing the melting temperature of nucleic acid duplex.(Bioinformatics 2001 17: 1226-1227).

Chapter 19

Filtering Out Specific Fields from a Genbank File

Genomes for whole organisms are available at Genbank, the most complete genetic sequence database. The National Center for Biotechnology Information (NCBI) at the National Library of Medicine (NLM), National Institutes of Health (NIH) is responsible for producing and distributing the GenBank Sequence Database. Genbank is also the name of the format in which Genbank records are stored (GenBank Flat File Format). Biopython has reading support for this kind of file (with the **Bio.SeqIO** module).

19.1 Extracting Selected Protein Sequences

A researcher wants to extract the protein sequences of each NADH found in the *Nicotiana tabacum* mitochondria.

19.1.1 Commented Source Code

Listing 19.1: Extract sequences from a Genbank file (py3.us/34)

```
1 from Bio import SeqIO, SeqRecord, Seq
2 from Bio.Alphabet import IUPAC
3
4 gbfile = open("MTtabacum.gbk") # file at: py3.us/mt.html
5 # mr stores the genbank record.
6 mr = SeqIO.read(gbfile, "genbank")
7 seqsforfasta = []
8 for x in mr.features:
9     # Each Genbank record is full of features, the program
10    # will walk over all the features.
11    qf = x.qualifiers
12    # Each feature has several parameters
13    # Pick selected parameters.
14    if 'NADH' in qf.get('product',[''])[0] and \
```

```
15      'product' in qf and 'translation' in qf:
16          id = qf['db_xref'][0][3:]
17          desc = qf['product'][0]
18          s = Seq.Seq(qf['translation'][0],IUPAC.protein)
19          # 's' is a NADH protein sequence
20          srec = SeqRecord.SeqRecord(s,id=id,description=desc)
21          # 'srec' is a SeqRecord object from s sequence.
22          seqsforfasta.append(srec)
23          # Add this SeqRecord object into seqsforfasta list.
24 outf = open('/home/sb/t4.txt','w')
25 SeqIO.write(seqsforfasta,outf,'fasta')
26 # Write all the sequences as a FASTA file.
27 outf.close()
```

19.2   Extracting the Upstream Region of Selected Proteins

Regulatory elements are found mostly upstream of the beginning of the genes. They include polyadenylation signals, TATA box, enhancers and more.

For this program we have a Genbank file and list of genes (cox2, atp6, atp9, cob) whose sequences we want to extract their sequence plus the upstream region, up to 1000 base pairs.

19.2.1   Commented Source Code

Listing 19.2: Extract upstream regions (py3.us/75)

```
1 from Bio import SeqIO
2 from Bio.SeqRecord import SeqRecord
3
4 gbfile = open("MTtabacum.gbk") # file avail. at: py3.us/mt.html
5 # The first genbank record is named mr
6 mr = SeqIO.read(gbfile, "genbank")
7 gbfile.close()
8 seqsforfasta = []
9 tg = (['cox2'],['atp6'],['atp9'],['cob'])
10 for x in mr.features:
11     if x.qualifiers.get('gene') in tg and x.type=='gene':
12         # Get the name of the gene
13         genename = x.qualifiers.get('gene')
14         # Get the start position
```

```
15          startpos = x.location.start.position
16          # Get the required slice
17          newfrag = mr.seq[startpos-1000:startpos]
18          # Build a SeqRecord object
19          newrec = SeqRecord(newfrag, genename[0]+
20                              ' 1000bp upstream','','')
21          seqsforfasta.append(newrec)
22 outf = open('t4.txt','w')
23 # Write all the sequences as a FASTA file.
24 SeqIO.write(seqsforfasta,outf,'fasta')
25 outf.close()
```

19.3 Additional Resources

- Benson DA, Karsch-Mizrachi I, Lipman DJ, Ostell J, Wheeler DL. "Gen-Bank." Nucleic Acids Res. 36(Database issue), D25-30 (2008).
  http://www.ncbi.nlm.nih.gov/pubmed/18073190

- GenBank Flat File Format. Sample record with detailed specifications.
  http://www.ncbi.nlm.nih.gov/Sitemap/samplerecord.html

Chapter 20

Converting XML BLAST File into HTML

20.1 Problem Description

The command line standalone version of NCBI BLAST[1] can generate its output in several formats, being HTML, Text and XML being the most popular. HTML and Text are most used since they are the best format to display data for human consumption. XML is also popular because it is a structured format that can be easily parsed by most moderm programming languages (Python included).

Sometimes there is a need to generate both XML and HTML output, the XML version for your program needs and the HTML for publishing in a web site. The current NCBI BLAST[2] outputs only to one format, XML or HTML. If you run multiple BLAST you know how time consuming it can be. It is not unusual to wait two weeks to get the result of a 7000 sequence set against the NCBI non-redundant (NR) database. Running a BLAST algorigthm twice just for a cosmetic reason is not my idea of fun.

Since a Python program can read any XML file, it could convert an XML BLAST output into an HTML page.

There are several ways to accomplish this. By using Python capabilities to parse XML (as shown in chapter 12) and by using Biopython NCBI XML parser (as shown in Chapter 10). Once parser choice is made, there are two ways to present the result, a single HTML file with all the BLAST results together or several HTML files, each one with the result of one BLAST run.

A version with and without Biopython, with output to one single file is presented for the reader to have as a reference for the use of cElementTree XML parser and as a Biopython reference. A version with multiple file output by using Biopython is also shown.

Note that this code introduces the **optparse** module. This module is a powerful command line option parser. It allows to set options that a program

---

[1]This program was commented on page 191.

[2]Up to version 2.2.18, which is the last release at time of writing.

will accept from a command line and generates usage and help messages in
an automatic way.

20.1.1 XML to HTML without Biopython Commented Source Code

Listing 20.1: Convert from XML to HTML (py3.us/76)

```python
1 #!/usr/bin/env python
2
3 # From BLAST XML to HTML. By Sebastian Bassi.
4 # Tested with BLASTN xml files from 2.2.16 to 2.2.18.
5 # BLASTN xml files < 2.2.16 are not properly formatted.
6 # Converts a single BLAST XML to one HTML file.
7
8 import xml.etree.cElementTree as cET
9 from optparse import OptionParser
10
11 helpstr = '''XML2HTML converts a BLAST XML file into one,
12 or multiple HTML files. This version requires Biopython 1.45
13 with CSV fixes or higher.
14
15 Author: Sebastian Bassi (sbassi@genesdigitales.com)
16 Thanks to Yoan Jacquemin for help in testing.
17 License: GPL 3.0 (http://www.gnu.org/licenses/gpl-3.0.txt)'''
18 usage = helpstr + '\n\nusage: %prog input_file [options]'
19 parser = OptionParser(usage=usage)
20 parser.add_option("-o", "--output", dest="o_file", default=None,
21                     help="name of the output file")
22 parser.add_option("-v", '--descriptions', dest="desc_n",
23                     default=None, type="int",
24                     help="descriptions keep in output file")
25 parser.add_option("-b", '--alignments', dest="align_n",
26                     default=None, type="int",
27                     help="alignments keep in output file")
28
29 def htmlhead(f_in,outf):
30     tree = cET.parse(f_in)
31     root = tree.getroot()
32     version_date = root.find('BlastOutput_version').text
33     application = root.find('BlastOutput_program').text
34     reference = root.find('BlastOutput_reference').text[12:]
35     fo = open(outf,'w')
36     fo.write('''<HTML>
```

```
37 <TITLE>BLAST Search Results</TITLE>
38 <BODY BGCOLOR="#FFFFFF" LINK="#0000FF" \
39 VLINK="#660099" ALINK="#660099">
40 <!-- Generated from %s by XML2HTML (Sebastian Bassi) -->
41 <PRE>''' %(f_in))
42     fo.write('<b>%s %s</b>' %(application,version_date))
43     fo.write('\n<b><a href="http://www.ncbi.nlm.nih.gov/entrez/\
44 query.fcgi?db=PubMed&cmd=Retrieve&list_uids=9254694&dopt=\
45 Citation">Reference</a>:</b>'+reference.replace('~',' ')
46 +'\n')
47     return (fo,root)
48
49 def htmlfoot(fo,f_in,root):
50     b_version = root.findtext('BlastOutput_db')
51     num_letter_db = root.findtext(
52     'BlastOutput_iterations/Iteration/Iteration_stat/Statistics/\
53 Statistics_db-num')
54     num_seqs_db = root.findtext(
55     'BlastOutput_iterations/Iteration/Iteration_stat/Statistics/\
56 Statistics_db-len')
57     lambd = root.findtext(
58     'BlastOutput_iterations/Iteration/Iteration_stat/Statistics/\
59 Statistics_lambda')
60     kappa = root.findtext(
61     'BlastOutput_iterations/Iteration/Iteration_stat/Statistics/\
62 Statistics_kappa')
63     entrop = root.findtext(
64     'BlastOutput_iterations/Iteration/Iteration_stat/Statistics/\
65 Statistics_entropy')
66     b_prg = root.findtext('BlastOutput_program')
67     p_sc_match = root.findtext('BlastOutput_param/Parameters/\
68 Parameters_sc-match')
69     p_sc_mismatch = root.findtext('BlastOutput_param/Parameters/\
70 Parameters_sc-mismatch')
71     p_gap_open = root.findtext('BlastOutput_param/Parameters/\
72 Parameters_gap-open')
73     p_gap_extend = root.findtext('BlastOutput_param/Parameters/\
74 Parameters_gap-extend')
75     fo.write('''<PRE>
76  Database: %s
77  Number of letters in database: %s
78  Number of sequences in database:  %s
79
80 Lambda      K       H
81    %.2f    %.3f     %.2f
```

```
82
83 Matrix: %s matrix:%s %s
84 Gap Penalties: Existence: %s, Extension: %s
85 Number of Sequences: %s
86 Length of database: %s
87 </PRE>
88 </BODY>
89 </HTML>''' %(b_version,num_letter_db,num_seqs_db,
90                 float(lambd),float(kappa),
91                 float(entrop),b_prg,p_sc_match,
92                 p_sc_mismatch,p_gap_open,p_gap_extend,
93                 num_seqs_db,num_letter_db))
94     fo.close()
95     return None
96
97 def prettyalign(fo,q,qs,qe,m,s,ss,se):
98     """ Format the alignment in slices of 60 characters
99     """
100     #fo=file handler
101     #q query sequence
102     #qs query_start (or query_from)
103     #qe query_end (or query_to)
104     #m match sequence
105     #s, ss and se are the equivalent for subject/hit
106     pos = 0
107     qr=range(qs,qe-1,-1) if qs>qe else range(int(qs),int(qe)+61)
108     qini = qs
109     qend = qe
110     sr = range(ss,se-1,-1) if ss>se else range(ss,ss+len(s))
111     mx = max(len(str(qr[-1])),len(str(sr[-1])))
112     q_desp = 0
113     s_desp = 0
114     if max(len(q),len(s))>=60:
115         finant_u = (pos+1 if ss>se else pos-1)
116         finant_d = (pos+1 if ss>se else pos-1)
117         while pos<max(len(q)-(len(q)%60),len(s)-(len(s)%60)):
118             q_desp += (q[pos:pos+60].count('-')
119                     if '-' in q[pos:pos+60] else 0)
120             s_desp += (s[pos:pos+60].count('-')
121                     if '-' in s[pos:pos+60] else 0)
122             fo.write('Query: %-*s %s %s\n'%(mx,
123                     qr[finant_u-1 if ss>se else finant_u+1],
124                     q[pos:pos+60],qr[pos+59-q_desp]))
125             fo.write('       '+' '*mx+' '+m[pos:pos+60]+'\n')
126             fo.write('Sbjct: %-*s %s %s\n\n'%
```

```
127                          (mx,sr[finant_d-1 if ss>se else finant_d+1],
128                          s[pos:pos+60],sr[pos+59-s_desp]))
129                  finant_u = pos+59-q_desp
130                  finant_d = pos+59-s_desp
131                  pos += 60
132          if len(q)%60!=0:
133              q_desp += (q[pos:pos+60].count('-')
134                          if '-' in q[pos:pos+60] else 0)
135              s_desp += (s[pos:pos+60].count('-')
136                          if '-' in s[pos:pos+60] else 0)
137              fo.write('Query: %-*s %s %s\n'%(mx,qr[pos-q_desp],
138                          q[pos:pos+60],qend))
139              fo.write('      '+' '*mx+' '+m[pos:pos+60]+'\n')
140              fo.write('Sbjct: %-*s %s %s\n\n'%(mx,sr[pos-s_desp],
141                          s[pos:pos+60],sr[-1]))
142      else:
143          fo.write('Query: %-*s %s %s\n'%(mx,qini,
144                      q[pos:pos+60],qend))
145          fo.write('      '+' '*mx+' '+m[pos:pos+60]+'\n')
146          fo.write('Sbjct: %-*s %s %s\n\n'%(mx,sr[pos],
147                      s[pos:pos+60],sr[-1]))
148      return None
149
150 def blastconv(f_in,fo,de,al):
151      i_hits = {}
152      hits = {}
153      hsps = {}
154      for ev,x in cET.iterparse(f_in):
155          if 'BlastOutput_query-def' in x.tag:
156              b_query_def = x.text
157          elif 'BlastOutput_query-len' in x.tag:
158              b_query_len = x.text
159          elif 'BlastOutput_db' in x.tag:
160              b_db = x.text
161          elif 'Statistics_db-num' in x.tag:
162              s_db_num=x.text
163          elif 'Statistics_db-len' in x.tag:
164              s_db_len=x.text
165          elif 'Parameters_expect' in x.tag:
166              p_expect = x.text
167          elif 'Parameters_filter' in x.tag:
168              p_filter = x.text
169          elif 'Iteration_query-def' in x.tag:
170              i_query_def = x.text
171          elif 'Iteration_iter-num' in x.tag:
```

```
172                i_iter_num = x.text
173            elif 'Iteration_query-ID' in x.tag:
174                i_query_id = x.text
175            elif 'Iteration_query-len' in x.tag:
176                i_query_len = x.text
177            elif 'Iteration'==x.tag:
178                i_hits[int(i_iter_num)] = (i_query_id, i_query_def,
179                                           i_query_len, hits)
180                hits = {}
181            elif 'Hit_num' in x.tag:
182                h_num = x.text
183            elif 'Hit_id' in x.tag:
184                h_id = x.text
185            elif 'Hit_def' in x.tag:
186                h_def = x.text
187            elif 'Hit_accession' in x.tag:
188                h_accession = x.text
189            elif 'Hit_len' in x.tag:
190                h_len = x.text
191            elif 'Hit'==x.tag:
192                hits[int(h_num)] = (h_id,h_def,h_accession,h_len,
193                                    hsps)
194                hsps = {}
195            elif 'Hsp_num' in x.tag:
196                hsp_num = x.text
197            elif 'Hsp_bit-score' in x.tag:
198                hsp_bit_score = x.text
199            elif 'Hsp_score' in x.tag:
200                hsp_score = x.text
201            elif 'Hsp_evalue' in x.tag:
202                hsp_evalue = x.text
203            elif 'Hsp_query-from' in x.tag:
204                hsp_query_from = x.text
205            elif 'Hsp_query-to' in x.tag:
206                hsp_query_to = x.text
207            elif 'Hsp_hit-from' in x.tag:
208                hsp_hit_from = x.text
209            elif 'Hsp_hit-to' in x.tag:
210                hsp_hit_to = x.text
211            elif 'Hsp_query-frame' in x.tag:
212                hsp_query_frame = x.text
213            elif 'Hsp_hit-frame' in x.tag:
214                hsp_hit_frame = x.text
215            elif 'Hsp_identity' in x.tag:
216                hsp_identity = x.text
```

```
217        elif 'Hsp_positive' in x.tag:
218            hsp_positive = x.text
219        elif 'Hsp_align-len' in x.tag:
220            hsp_align_len = x.text
221        elif 'Hsp_qseq' in x.tag:
222            hsp_qseq = x.text
223        elif 'Hsp_hseq' in x.tag:
224            hsp_hseq = x.text
225        elif 'Hsp_midline' in x.tag:
226            hsp_mid = x.text
227        elif 'Hsp'==x.tag:
228            try:
229                hn = (hsp_bit_score,hsp_score,hsp_evalue,
230                      hsp_query_from,hsp_query_to,
231                      hsp_hit_from,hsp_hit_to,
232                      hsp_query_frame,hsp_hit_frame,
233                      hsp_identity,hsp_positive,
234                      hsp_align_len,hsp_qseq,hsp_hseq,hsp_mid)
235            except UnboundLocalError:
236                hn = (hsp_bit_score,hsp_score,hsp_evalue,
237                      hsp_query_from,hsp_query_to,
238                      hsp_hit_from,hsp_hit_to,
239                      hsp_query_frame,hsp_query_frame,
240                      hsp_identity,hsp_positive,
241                      hsp_align_len,hsp_qseq,hsp_hseq,hsp_mid)
242            hsps[int(hsp_num)]= hn
243        elif 'Statistics_hsp-len' in x.tag:
244            s_hsp_len = x.text
245    ihits = i_hits.keys()
246    ihits.sort()
247    # Iterations with no BLAST result are missing, so the script
248    # can't iterate over a range from one to the end.
249    for x in ihits:
250        fo.write('<b>Query=</b> %s\n' %(i_hits[x][1]))
251        fo.write('          (%s letters)\n' %(i_hits[x][2]))
252        fo.write('<b>Database:</b> %s\n' %(b_db))
253        fo.write('          %s sequences; %s total letters\n'
254                %(s_db_num,s_db_len))
255        fo.write('''Searching.................................\
256 ...done
257 <PRE>
258
259
260                                                          \
261 Score     E
```

```
262 Sequences producing significant alignments:                    \
263 (bits) Value
264
265 ''')
266         for y in range(1,len(i_hits[x][3])+1)[:de]:
267             k = i_hits[x][3][y][0]
268             desc = i_hits[x][3][y][1]
269             bs = i_hits[x][3][y][4][1][0]
270             sc = i_hits[x][3][y][4][1][2]
271             if 'gi|' in k:
272                 m = k.index('gi|')+3
273                 gi = k[m:k[m:].index('|')+3]
274                 fo.write('<a href="http://www.ncbi.nlm.nih.gov/\
275 entrez/query.fcgi?cmd=Retrieve&db=Nucleotide&list_uids=%s&dopt=\
276 GenBank" >%s</a> %s <a href = #%s> %s</a>    %s\n' %(gi,
277                     k.replace('gi|'+gi+'|',''),desc[:36]+'...',gi,
278                             bs,sc))
279             else:
280                 fo.write('><a name=%s></a>%s\n'%(k,desc))
281         fo.write('\n</PRE>\n')
282         for y in range(1,len(i_hits[x][3])+1)[:al]:
283             fo.write('<PRE>\n')
284             k = i_hits[x][3][y][0]
285             desc = i_hits[x][3][y][1]
286             if 'gi|' in k:
287                 m = k.index('gi|')+3
288                 gi = k[m:k[m:].index('|')+3]
289                 fo.write('><a href="http://www.ncbi.nlm.nih.gov/\
290 entrez/query.fcgi?cmd=Retrieve&db=Nucleotide&list_uids=%s&dopt=\
291 GenBank" >%s</a> %s \n' %(gi,k.replace('gi|'+gi+'|',''),desc))
292             else:
293                 fo.write('>%s %s\n' %(k,desc))
294             fo.write(' Length = '+i_hits[x][3][y][3]+'\n')
295             # Walk over all the hsps
296             for z in xrange(1,len(i_hits[x][3][y][4])+1):
297                 bs = i_hits[x][3][y][4][z][0]
298                 hsc = i_hits[x][3][y][4][z][1]
299                 sc = i_hits[x][3][y][4][z][2]
300                 h_id = i_hits[x][3][y][4][z][10]
301                 h_pos = i_hits[x][3][y][4][z][11]
302                 q_frame = i_hits[x][3][y][4][z][7]
303                 h_frame = i_hits[x][3][y][4][z][8]
304                 q_from = int(i_hits[x][3][y][4][z][3])
305                 q_to = int(i_hits[x][3][y][4][z][4])
306                 h_from = int(i_hits[x][3][y][4][z][5])
```

```
307                    h_to = int(i_hits[x][3][y][4][z][6])
308                    qseq = i_hits[x][3][y][4][z][12]
309                    hseq = i_hits[x][3][y][4][z][13]
310                    mid = i_hits[x][3][y][4][z][14]
311                    qf = 'Plus' if int(q_frame)>0 else 'Minus'
312                    hf = 'Plus' if int(h_frame)>0 else 'Minus'
313                    fo.write('Score = %s bits (%s), Expect = %s\n'
314                            %(bs,hsc,sc))
315                    fo.write('Identities = %s/%s (%.0f%%)\n'
316                            %(h_id,h_pos,
317                               float(int(h_id))/int(h_pos)*100))
318                    fo.write('Strand = %s/%s\n\n\n' %(qf,hf))
319                    prettyalign(fo,qseq,q_from,q_to,mid,hseq,
320                                h_from,h_to)
321                    fo.write('</PRE>\n')
322     return fo
323
324 def doconvert(f_in,outfile,desc,align):
325     fo,root = htmlhead(f_in,outfile,)
326     fo = blastconv(f_in,fo,desc,align)
327     htmlfoot(fo,f_in,root)
328     return None
329
330 (opts, args) = parser.parse_args()
331 if len(args)<1:
332     errmsg = "Bad or missing option in input."
333     errmsg += " This program requires an input file"
334     errmsg += " Please see the help with -h or --help"
335     parser.error(errmsg)
336 elif len(args)==1:
337     f = args[0]
338     if opts.o_file is None:
339         opts.o_file = f[:-3]+'html'
340     doconvert(f, opts.o_file, opts.desc_n, opts.align_n)
341 elif len(args)>1:
342     for f in args:
343         outfile = f[:-3]+'html'
344         doconvert(f, outfile, opts.desc_n, opts.align_n)
```

20.1.2   Biopython Version Commented Source Code

Listing 20.2: From XML to HTML using Biopython (py3.us/77)

```
1 #!/usr/bin/env python
```

```
 2
 3 # From BLAST XML to HTML. By Sebastian Bassi.
 4 # Tested with BLASTN xml files from 2.2.16 to 2.2.18.
 5 # BLASTN xml files < 2.2.16 are not properly formatted.
 6
 7 from optparse import OptionParser
 8
 9 from Bio.Blast import NCBIXML
10
11 helpstr='''XML2HTML converts a BLAST XML file into an HTML file.
12 This version requires Biopython 1.45 with CSV fixes or higher.
13
14 Author: Sebastian Bassi (sbassi@genesdigitales.com)
15 Thanks to Yoan Jacquemin for help in testing.
16 License: GPL 3.0 (http://www.gnu.org/licenses/gpl-3.0.txt)'''
17 usage = helpstr + '\n\nusage: %prog input_file [options]'
18 parser = OptionParser(usage=usage)
19 parser.add_option("-o", "--output", dest="o_file", default=None,
20                     help="name of the output file")
21 parser.add_option("-v", '--descriptions', dest="desc_n",
22                     default=None, type="int",
23                     help="descriptions keep in output file")
24 parser.add_option("-b", '--alignments', dest="align_n",
25                     default=None, type="int",
26                     help="alignments keep in output file")
27
28 def htmlhead(fr,oid,f_in):
29     fo = open(oid,'w')
30     fo.write('''<HTML>
31 <TITLE>BLAST Search Results</TITLE>
32 <BODY BGCOLOR="#FFFFFF" LINK="#0000FF" \
33 VLINK="#660099" ALINK="#660099">
34 <!-- Generated from %s by XML2HTML (Sebastian Bassi) -->
35 <PRE>''' %(f_in))
36     fo.write('<b>%s</b>'
37     %(fr.application+' '+fr.version+' '+fr.date))
38     fo.write('\n<b><a href="http://www.ncbi.nlm.nih.gov/entrez/\
39 query.fcgi?db=PubMed&cmd=Retrieve&list_uids=9254694&dopt=\
40 Citation">Reference</a>:</b>'+fr.reference.replace('~',' ')
41 +'\n')
42     return fo
43
44 def htmlfoot(fo,fr):
45     try:
46         nldb = fr.num_letters_in_database
```

```
47      except:
48          nldb = fr._num_letters_in_database
49      fo.write('''<PRE>
50   Database: %s
51   Number of letters in database: %s
52   Number of sequences in database:  %s
53
54 Lambda     K       H
55    %.2f     %.3f      %.2f
56
57 Matrix: %s matrix:%s %s
58 Gap Penalties: Existence: %s, Extension: %s
59 Number of Sequences: %s
60 Length of database: %s
61 </PRE>
62 </BODY>
63 </HTML>''' %(fr.database,fr.num_letters_in_database,
64              fr.num_sequences_in_database,
65              fr.ka_params[0],fr.ka_params[1],
66              fr.ka_params[2],fr.application,
67              fr.sc_match,fr.sc_mismatch,
68              fr.gap_penalties[0],fr.gap_penalties[1],
69              fr.num_sequences_in_database,
70              nldb))
71      fo.close()
72      return None
73
74 def prettyalign(fo,q,qs,qe,m,s,ss,se):
75      """ Format the alignment in slices of 60 characters
76      """
77      #fo file handler
78      #q query sequence
79      #qs query_start (or query_from)
80      #qe query_end (or query_to)
81      #m match sequence
82      #s, ss and se are the equivalent for subject/hit
83      pos = 0
84      qr = range(qs,qe-1,-1) if qs>qe else range(qs,qe+61,1)
85      qini = qs
86      qend = qe
87      sr=range(ss,se-1,-1) if ss>se else range(ss,ss+len(s),1)
88      mx = max(len(str(qr[-1])),len(str(sr[-1])))
89      q_desp = 0
90      s_desp = 0
91      if max(len(q),len(s))>=60:
```

```
92              finant_u = (pos+1 if ss>se else pos-1)
93              finant_d = (pos+1 if ss>se else pos-1)
94              while pos<max(len(q)-(len(q)%60),len(s)-(len(s)%60)):
95                  q_desp += (q[pos:pos+60].count('-')
96                          if '-' in q[pos:pos+60] else 0)
97                  s_desp += (s[pos:pos+60].count('-')
98                          if '-' in s[pos:pos+60] else 0)
99                  fo.write('Query: %-*s %s %s\n'%
100                         (mx,qr[finant_u-1 if ss>se else finant_u+1],
101                          q[pos:pos+60],qr[pos+59-q_desp]))
102                 fo.write('       '+' '*mx+' '+m[pos:pos+60]+'\n')
103                 fo.write('Sbjct: %-*s %s %s\n\n'%
104                         (mx,sr[finant_d-1 if ss>se else finant_d+1],
105                          s[pos:pos+60],sr[pos+59-s_desp]))
106                 finant_u = pos+59-q_desp
107                 finant_d = pos+59-s_desp
108                 pos += 60
109             if len(q)%60!=0:
110                 q_desp += (q[pos:pos+60].count('-') if
111                         '-' in q[pos:pos+60] else 0)
112                 s_desp += (s[pos:pos+60].count('-') if
113                         '-' in s[pos:pos+60] else 0)
114                 fo.write('Query: %-*s %s %s\n'%(mx,qr[pos-q_desp],
115                         q[pos:pos+60],qend))
116                 fo.write('       '+' '*mx+' '+m[pos:pos+60]+'\n')
117                 fo.write('Sbjct: %-*s %s %s\n\n'%(mx,sr[pos-s_desp],
118                         s[pos:pos+60],sr[-1]))
119         else:
120             fo.write('Query: %-*s %s %s\n'%(mx,qini,
121                     q[pos:pos+60],qend))
122             fo.write('       '+' '*mx+' '+m[pos:pos+60]+'\n')
123             fo.write('Sbjct: %-*s %s %s\n\n'%(mx,sr[pos],
124                     s[pos:pos+60],sr[-1]))
125         return None
126
127 def blastconv(rec,fo,fr,de=None,al=None):
128     """ Get a blast record in XML en saves the same
129         record in HTML
130     """
131     fo.write('\n<b>Query=</b> %s' %(rec.query))
132     fo.write('\n          (%s letters)' %(rec.query_letters))
133     fo.write('\n<b>Database:</b> %s' %(rec.database))
134     try:
135         fo.write('\n          %s sequences; %s total letters\n'
136         %(fr.num_sequences_in_database,
```

```
137                fr.num_letters_in_database))
138     except:
139         # For Biopython > 1.45
140         fo.write('\n         %s sequences; %s total letters\n'
141         %(fr.num_sequences_in_database,
142             fr._num_letters_in_database))
143     fo.write('''Searching.................................\
144 ...done
145 <PRE>
146
147
148                                                               \
149 Score    E
150 Sequences producing significant alignments:                   \
151 (bits) Value
152
153 ''')
154     for d in rec.descriptions[:de]:
155         k = d.accession
156         desc = d.title
157         bs = d.bits
158         sc = d.e
159         if 'gi|' in k:
160             m = k.index('gi|')+3
161             gi = k[m:k[m:].index('|')+3]
162             fo.write('<a href="http://www.ncbi.nlm.nih.gov\
163 /entrez/query.fcgi?cmd=Retrieve&db=Nucleotide&list_uids\
164 =%s&dopt=GenBank" >%s</a> %s <a href = #%s>%.1f</a>%s%s\n'
165 %(gi,k.replace('gi|'+gi+'|',''),desc,k,bs,4*" ",sc))
166         else:
167             fo.write('%s ... <a href = #%s>%.1f</a>%s%s\n'
168             %(desc[:60],k,bs,4*" ",sc))
169     fo.write('</PRE>\n')
170     for alig in rec.alignments[:al]:
171         fo.write('<PRE>\n')
172         k = alig.hit_id
173         desc = alig.hit_def
174         if 'gi|' in k:
175             m = k.index('gi|')+3
176             gi = k[m:k[m:].index('|')+3]
177             fo.write('><a name=%s></a><a href="http://\
178 www.ncbi.nlm.nih.gov/entrez/query.fcgi?\
179 cmd=Retrieve&db=Nucleotide&list_uids=%s&dopt=GenBank" >\
180 %s</a> %s \n' %(alig.accession,gi,k.replace('gi|'+gi+'|',''),
181                 desc))
```

```
182            else:
183                fo.write('><a name=%s></a>%s'%(alig.accession,desc))
184            fo.write('\n Length = %s\n' %(alig.length))
185            # Walk over all the hsps
186            for hsp in alig.hsps:
187                bs = hsp.bits
188                hsc = hsp.score
189                sc = hsp.expect
190                h_id = hsp.identities
191                h_pos = hsp.positives
192                h_alen = hsp.align_length
193                q_frame = hsp.frame[0]
194                try:
195                    h_frame = hsp.frame[1]
196                except IndexError:
197                    h_frame = q_frame
198                q_from = hsp.query_start
199                q_to = hsp.query_end
200                h_from = hsp.sbjct_start
201                h_to = hsp.sbjct_end
202                qseq = hsp.query
203                hseq = hsp.sbjct
204                mid = hsp.match
205                qf = 'Plus' if q_frame>0 else 'Minus'
206                hf = 'Plus' if h_frame>0 else 'Minus'
207                fo.write('\n\nScore = %s bits (%s), Expect = %s\n'
208                        %(bs,hsc,sc))
209                fo.write('Identities = %s/%s (%.0f%%)\n'
210                    %(h_id,h_alen,float(int(h_id))/int(h_alen)*100))
211                fo.write('Strand = %s/%s\n\n' %(qf,hf))
212                prettyalign(fo,qseq,q_from,q_to,mid,hseq,h_from,h_to)
213            fo.write('</PRE>\n')
214    return fo
215
216 def doconvert(f_in,outfile,desc,align):
217    # fr is the first record, where 'Parameters' are stored.
218    fr = NCBIXML.parse(open(f_in)).next()
219    f_out = htmlhead(fr,outfile,f_in)
220    for b_rec in NCBIXML.parse(open(f_in)):
221        f_out = blastconv(b_rec,f_out,fr,desc,align)
222    htmlfoot(f_out,fr)
223    return None
224
225 (opts, args) = parser.parse_args()
226 if len(args)<1:
```

```
227     errmsg = "Bad or missing option in input."
228     errmsg += " This program requires an input file"
229     errmsg += " Please see the help with -h or --help"
230     parser.error(errmsg)
231 elif len(args)==1:
232     f = args[0]
233     if opts.o_file is None:
234         opts.o_file = f[:-3]+'html'
235     doconvert(f, opts.o_file, opts.desc_n, opts.align_n)
236 elif len(args)>1:
237     for f in args:
238         outfile = f[:-3]+'html'
239         doconvert(f, outfile, opts.desc_n, opts.align_n)
```

20.1.3   Biopython Version for Multiple BLAST Commented Source Code

Listing 20.3: From XML to HTML using Biopython (py3.us/78)

```
 1 #!/usr/bin/env python
 2
 3 # From BLAST XML to HTML. By Sebastian Bassi.
 4 # Tested with BLASTN xml files from 2.2.16 to 2.2.18.
 5 # BLASTN xml files < 2.2.16 are not properly formatted.
 6
 7 import os
 8 from optparse import OptionParser
 9
10 from Bio.Blast import NCBIXML
11
12 helpstr = '''XML2HTML converts a BLAST XML file into one or
13 multiple HTML files. This version requires Biopython 1.45
14 with CSV fixes or higher.
15
16 Author: Sebastian Bassi (sbassi@genesdigitales.com)
17 Thanks to Yoan Jacquemin for help in testing.
18 License: GPL 3.0 (http://www.gnu.org/licenses/gpl-3.0.txt)'''
19 usage = helpstr + '\n\nusage: %prog input_file [options]'
20 parser = OptionParser(usage=usage)
21 parser.add_option("-o", "--output", dest="o_dir", default='.',
22                 help="name of the output directory")
23 parser.add_option("-v", '--descriptions', dest="desc_n",
24                 default=None, type="int",
25                 help="descriptions to keep in output file")
```

```
26 parser.add_option("-b", '--alignments', dest="align_n",
27                       default=None, type="int",
28                       help="alignments keep in output file")
29 parser.add_option("-V", '--verbose', dest="verb",
30                       action="store_true", default=False,
31                       help="prints output filename(s)")
32 parser.add_option("-t", '--title', dest="title",
33                       action="store_true", default=False,
34                       help="use sequence title as filename")
35
36 def htmlhead(fr,oid,f_in,odir,rec):
37     if oid:
38         fo = open(os.path.join(odir,rec.query_id+'.html'),'w')
39     else:
40         fo = open(os.path.join(odir,rec.query+'.html'),'w')
41     fo.write('''<HTML>
42 <TITLE>BLAST Search Results</TITLE>
43 <BODY BGCOLOR="#FFFFFF" LINK="#0000FF" \
44 VLINK="#660099" ALINK="#660099">
45 <!-- Generated from %s by XML2HTML (Sebastian Bassi) -->
46 <PRE>''' %(f_in))
47     fo.write('<b>%s %s %s</b>' %(fr.application,fr.version,
48                                    fr.date))
49     fo.write('\n<b><a href="http://www.ncbi.nlm.nih.gov/entrez/\
50 query.fcgi?db=PubMed&cmd=Retrieve&list_uids=9254694&dopt=\
51 Citation">Reference</a>:</b>'+fr.reference.replace('~',' ')
52 +'\n')
53     return fo
54
55 def htmlfoot(fo,fr):
56     try:
57         nldb = fr.num_letters_in_database
58     except:
59         nldb = fr._num_letters_in_database
60     fo.write('''<PRE>
61 Database: %s
62 Number of letters in database: %s
63 Number of sequences in database:  %s
64
65 Lambda     K        H
66    %.2f    %.3f     %.2f
67
68 Matrix: %s matrix:%s %s
69 Gap Penalties: Existence: %s, Extension: %s
70 Number of Sequences: %s
```

```
71 Length of database: %s
72 </PRE>
73 </BODY>
74 </HTML>''' %(fr.database,fr.num_letters_in_database,
75                 fr.num_sequences_in_database,
76                 fr.ka_params[0],fr.ka_params[1],
77                 fr.ka_params[2],fr.application,
78                 fr.sc_match,fr.sc_mismatch,
79                 fr.gap_penalties[0],fr.gap_penalties[1],
80                 fr.num_sequences_in_database, nldb))
81     fo.close()
82     return None
83
84 def prettyalign(fo,q,qs,qe,m,s,ss,se):
85     """ Format the alignment in slices of 60 characters
86     """
87     #fo=file handler
88     #q query sequence
89     #qs query_start (or query_from)
90     #qe query_end (or query_to)
91     #m match sequence
92     #s, ss and se are the equivalent for subject/hit
93     pos = 0
94     qr=range(qs,qe-1,-1) if qs>qe else range(qs,qe+61,1)
95     qini = qs
96     qend = qe
97     sr = range(ss,se-1,-1) if ss>se else range(ss,ss+len(s),1)
98     mx = max(len(str(qr[-1])),len(str(sr[-1])))
99     q_desp = 0
100    s_desp = 0
101    if max(len(q),len(s))>=60:
102        finant_u = (pos+1 if ss>se else pos-1)
103        finant_d = (pos+1 if ss>se else pos-1)
104        while pos<max(len(q)-(len(q)%60),len(s)-(len(s)%60)):
105            q_desp += (q[pos:pos+60].count('-')
106                       if '-' in q[pos:pos+60] else 0)
107            s_desp += (s[pos:pos+60].count('-')
108                       if '-' in s[pos:pos+60] else 0)
109            fo.write('Query: %-*s %s %s\n'%
110                     (mx,qr[finant_u-1 if ss>se else finant_u+1],
111                      q[pos:pos+60],qr[pos+59-q_desp]))
112            fo.write('      '+' '*mx+' '+m[pos:pos+60]+'\n')
113            fo.write('Sbjct: %-*s %s %s\n\n'%
114                     (mx,sr[finant_d-1 if ss>se else finant_d+1],
115                      s[pos:pos+60],sr[pos+59-s_desp]))
```

```
116                 finant_u = pos+59-q_desp
117                 finant_d = pos+59-s_desp
118                 pos += 60
119         if len(q)%60!=0:
120             q_desp+=(q[pos:pos+60].count('-') if
121                     '-' in q[pos:pos+60] else 0)
122             s_desp+=(s[pos:pos+60].count('-') if
123                     '-' in s[pos:pos+60] else 0)
124             fo.write('Query: %-*s %s %s\n'%(mx,qr[pos-q_desp],
125                         q[pos:pos+60],qend))
126             fo.write('        '+' '*mx+' '+m[pos:pos+60]+'\n')
127             fo.write('Sbjct: %-*s %s %s\n\n'%(mx,sr[pos-s_desp],
128                         s[pos:pos+60],sr[-1]))
129     else:
130         fo.write('Query: %-*s %s %s\n'%(mx,qini,
131                     q[pos:pos+60],qend))
132         fo.write('        '+' '*mx+' '+m[pos:pos+60]+'\n')
133         fo.write('Sbjct: %-*s %s %s\n\n'%(mx,sr[pos],
134                     s[pos:pos+60],sr[-1]))
135     return None
136
137 def blastconv(rec,fo,odir,de=None,al=None,oid='T'):
138     """ Get a blast record in XML en saves the same
139         record in HTML
140     """
141     fo.write('\n<b>Query=</b> %s' %(rec.query))
142     fo.write('\n          (%s letters)' %(rec.query_letters))
143     fo.write('\n<b>Database:</b> %s' %(rec.database))
144     try:
145         fo.write('\n           %s sequences; %s total letters\n'
146         %(fr.num_sequences_in_database,
147            fr.num_letters_in_database))
148     except:
149         # For Biopython > 1.45
150         fo.write('\n           %s sequences; %s total letters\n'
151         %(fr.num_sequences_in_database,
152            fr._num_letters_in_database))
153     fo.write('''Searching...............................done
154
155 <PRE>
156
157
158 ''')
159     fo.write(' '*65+'Score    E\n')
160     fo.write('Sequences producing significant alignments:'+\
```

```
161                    ' '*22+'(bits) Value\n')
162        for d in rec.descriptions[:de]:
163            k = d.accession
164            desc = d.title
165            bs = d.bits
166            sc = d.e
167            if 'gi|' in k:
168                m = k.index('gi|')+3
169                gi = k[m:k[m:].index('|')+3]
170                fo.write('<a href="http://www.ncbi.nlm.nih.gov\
171    /entrez/query.fcgi?cmd=Retrieve&db=Nucleotide&list_uids\
172    =%s&dopt=GenBank" >%s</a> %s <a href = #%s>%.1f</a>%s%s\n'
173    %(gi,k.replace('gi|'+gi+'|',''),desc,k,bs,4*" ",sc))
174            else:
175                fo.write('%s ... <a href = #%s>%.1f</a>%s%s\n'
176                %(desc[:60],k,bs,4*" ",sc))
177        fo.write('</PRE>\n')
178        for alig in rec.alignments[:al]:
179            fo.write('<PRE>\n')
180            k = alig.hit_id
181            desc = alig.hit_def
182            if 'gi|' in k:
183                m = k.index('gi|')+3
184                gi = k[m:k[m:].index('|')+3]
185                fo.write('><a name=%s></a><a href="http://\
186    www.ncbi.nlm.nih.gov/entrez/query.fcgi?\
187    cmd=Retrieve&db=Nucleotide&list_uids=%s&dopt=GenBank" >\
188    %s</a> %s \n'
189    %(alig.accession,gi,k.replace('gi|'+gi+'|',''),desc))
190            else:
191                fo.write('><a name=%s></a>%s'%(alig.accession,desc))
192            fo.write('\n Length = %s\n' %(alig.length))
193            # Walk over all the hsps
194            for hsp in alig.hsps:
195                bs = hsp.bits
196                hsc = hsp.score
197                sc = hsp.expect
198                h_id = hsp.identities
199                h_pos = hsp.positives
200                h_alen = hsp.align_length
201                q_frame = hsp.frame[0]
202                try:
203                    h_frame = hsp.frame[1]
204                except IndexError:
205                    h_frame = q_frame
```

```
206                    q_from = hsp.query_start
207                    q_to = hsp.query_end
208                    h_from = hsp.sbjct_start
209                    h_to = hsp.sbjct_end
210                    qseq = hsp.query
211                    hseq = hsp.sbjct
212                    mid = hsp.match
213                    qf = 'Plus' if q_frame>0 else 'Minus'
214                    hf = 'Plus' if h_frame>0 else 'Minus'
215                    fo.write('\n\nScore = %s bits (%s), Expect = %s\n'
216                             %(bs,hsc,sc))
217                    fo.write('Identities = %s/%s (%.0f%%)\n'
218                        %(h_id,h_alen,float(int(h_id))/int(h_alen)*100))
219                    fo.write('Strand = %s/%s\n\n' %(qf,hf))
220                    prettyalign(fo,qseq,q_from,q_to,mid,hseq,h_from,h_to)
221            fo.write('</PRE>\n')
222       return fo
223
224 (opts, args) = parser.parse_args()
225 if len(args)<1:
226     errmsg = "Bad or missing option in input."
227     errmsg += " This program requires an input file"
228     errmsg += " Please see the help with -h or --help"
229     parser.error(errmsg)
230 else:
231     title = opts.title
232     desc = opts.desc_n
233     align = opts.align_n
234     for f in args:
235         if opts.o_dir=='.':
236             o_dir = ""
237         else:
238             o_dir = opts.o_dir
239         fr = NCBIXML.parse(open(f)).next()
240         for rec in NCBIXML.parse(open(f)):
241             f_out = htmlhead(fr,title,f,o_dir,rec)
242             f_out = blastconv(rec,f_out,o_dir,desc,align,title)
243             htmlfoot(f_out,fr)
244             if opts.verb:
245                 print(rec.query_id if title else rec.query+
246                     '.html')
```

Chapter 21

Infering Splicing Sites

21.1 Problem Description

An expressed sequence tag or EST is a short sub-sequence of a transcribed spliced nucleotide sequence (either protein-coding or not). They may be used to identify gene transcripts, and are instrumental in gene discovery and gene sequence determination.[1] The identification of ESTs has proceeded rapidly, with approximately 52 million ESTs now available in public databases (e.g. GenBank 5/2008, all species).

An EST is produced by one-shot sequencing of a cloned mRNA (i.e. sequencing several hundred base pairs from an end of a cDNA clone taken from a cDNA library). The resulting sequence is a relatively low quality fragment whose length is limited by current technology to approximately 500 to 800 nucleotides. Because these clones consist of DNA that is complementary to mRNA, the ESTs represent portions of expressed genes. They may be present in the database as either cDNA/mRNA sequence or as the reverse complement of the mRNA, the template strand.

Preparatory program,

Listing 21.1: Convert data for entering into a SQLite database (py3.us/79)

```
1  """
2  Convert TAIR fasta file in a CSV file for making a
3  SQLite database.
4  """
5  from Bio import SeqIO
6
7  seqfile = open('TAIR8_seq_20080412')
8  cdsfile = open('TAIR8_cds_20080412')
9  f_out = open('TAIR.csv','w')
10 atD = {}
11 # Get all sequences from TAIR sequences file.
12 for record in SeqIO.parse(seqfile, "fasta"):
13     sid = record.id
14     seq = record.seq.data
```

```
15     atD[sid] = [seq]
16 # Get all sequences from TAIR CDS file.
17 for record in SeqIO.parse(cdsfile, "fasta"):
18     sid = record.id
19     seq = record.seq.data
20     atD[sid].append(seq)
21 # Write to a CSV file only the entries of the dictionary that
22 # has data from both sources
23 for x in atD:
24     if len(atD[x])==2:
25         # Write in this order: Seq. ID, CDS, SEQ.
26         f_out.write('%s,%s,%s\n' %(x,atD[x][1],atD[x][0]))
27 f_out.close()
```

Program 21.1 generates the CSV file used in the SQLite database. Here are the steps to create and populate the database:

```
$ ./sqlite3-3.5.9.bin AT.db
SQLite version 3.5.9
Enter ".help" for instructions
sqlite> create table seqs(ID, CDS, Seq);
sqlite> .separator ,
sqlite> .import TAIR.csv seqs
sqlite> CREATE INDEX IDidx on seqs (ID);
```

Command to format the TAIR cds database to BLAST use:

```
$ formatdb -t ATCDS -i TAIR8_seq_20080412 -p F -n TAIR8seq
```

21.1.1 Infer Splicing Sites with Commented Source Code

Listing 21.2: Estimate introns (py3.us/80)

```
 1 #!/usr/bin/env python
 2
 3 import sys
 4 import os
 5 import sqlite3
 6 from tempfile import NamedTemporaryFile
 7 from Bio import SeqIO, SeqRecord, Seq, Clustalw
 8 from Bio.Blast import NCBIStandalone
 9 from Bio.Blast import NCBIXML
10 from Bio.Clustalw import MultipleAlignCL
11
12 dbpath = 'AT.db'
13 blast_exe ='/home/sb/blast-2.2.20/bin/blastall'
```

```
14 blast_db = '/home/sb/blast-2.2.20/data/TAIR8cds'
15
16 def allgaps(seq):
17     """Return a list with tuples containing all gap positions
18        and length. seq is a string."""
19     i = 0
20     gaps = []
21     indash = False
22     for c in seq:
23         if indash is False and c=='-':
24             c_ini = i
25             indash = True
26             dashn = 0
27         elif indash is True and c=='-':
28             dashn += 1
29         elif indash is True and c!='-':
30             indash = False
31             gaps.append((c_ini,dashn+1))
32         i += 1
33     return gaps
34
35 def iss(record):
36     """Infer Splicing Sites from a FASTA file full of EST
37     sequences"""
38     usersid = record.id
39     userseq = record.seq
40     tf = NamedTemporaryFile()
41     fth = tf.file
42     fth.write(record.format("fasta"))
43     tfn = tf.name
44     fth.flush()
45     # Note: expectation, descriptions, alignments are passed as
46     # strings due to Biopython bug#2538. It is fixed in Biopython
47     # 1.48.
48     result, err = NCBIStandalone.blastall(blast_exe, "blastn",
49                     blast_db, tfn, expectation='1e-10',
50                     descriptions='1', alignments='1')
51     fth.close()
52     b_record = NCBIXML.read(result)
53     if len(b_record.alignments) > 0:
54         title = b_record.alignments[0].title
55         sid = title[title.index(' ')+1:title.index(' |')]
56         # Polarity information of returned sequence.
57         # 1 = normal, -1 = reverse.
58         frame = b_record.alignments[0].hsps[0].frame[1]
```

```
59          db = sqlite3.connect(dbpath)
60          # Run the SQLite query
61          t = (sid,)
62          c = db.cursor()
63          c.execute("SELECT CDS, Seq from seqs WHERE ID=?", t)
64          cds,seq = c.fetchone()
65          if cds=='':
66              print 'There is no matching CDS'
67              exit()
68          # Check sequence polarity.
69          if frame==1:
70              seqCDS = SeqRecord.SeqRecord(Seq.Seq(cds),
71                          id=sid+'-CDS',name="",description="")
72              fullseq = SeqRecord.SeqRecord(Seq.Seq(seq),
73                          id=sid+'-SEQ',name="",description="")
74          else:
75              seqCDS = SeqRecord.SeqRecord(
76              Seq.Seq(cds).reverse_complement(),id=sid+'-CDS',
77                  name="",description="")
78              fullseq = SeqRecord.SeqRecord(
79                  Seq.Seq(seq).reverse_complement(),id=sid+'-SEQ',
80                  name="",description="")
81          # Create a tuple with the user sequence and both AT seque
82          allseqs = (record,seqCDS,fullseq)
83          tf = NamedTemporaryFile()
84          trifh = tf.file
85          # Write the file with the three sequences.
86          SeqIO.write(allseqs,trifh,"fasta")
87          tfn = tf.name
88          trifh.flush()
89          # Do the alignment:
90          cline = MultipleAlignCL(tfn)
91          cline.set_output(usersid+".aln")
92          alignment = Clustalw.do_alignment(cline)
93          trifh.close()
94          # Walk over all aligned sequences and look for query sequ
95          for seq in alignment.get_all_seqs():
96              if usersid in seq.id:
97                  seqstr = str(seq.seq)
98                  gaps = allgaps(seqstr.strip('-'))
99                  break
100         print "Original sequence:",usersid
101         print "\nBest match in AT CDS:",sid
102         i = 0
103         acc = 0
```

```
104          for gap in gaps:
105              i += 1
106              print "Intron #%s: Start at position %s, length %s"\
107                  %(i,gap[0]-acc,gap[1])
108              acc += gap[1]
109          print '\n'+seqstr.strip('-')
110          print '\nAlignment file: %s.aln\n'%usersid
111      return None
112
113 try:
114     f_name = sys.argv[1]
115 except:
116     print "Run this program from command line as:"
117     print "iss.py file_in"
118     exit()
119 records = SeqIO.parse(open(f_name), "fasta")
120 for record in records:
121     iss(record)
```

21.1.2   Sample Run of Estimate Intron Program

```
$ iss.py /mnt/hda2/bio/t3.txt
Original sequence: secu3

Best match in AT CDS: AT1G14990.1
Putative Intron #1: Start at position 171, length 95
Putative Intron #2: Start at position 250, length 153


CTAGCCACTTCCAACGAGTTGGCCTTGAGATAGAAGGTGAGCCATGTATTGGGAGTGGTAAA<=
CGTATGGAGATTTTCCCTGGCGATCAAAATGCTTAGCCATTATGCAGAATTCAACAGGACCG<=
GAATCTTCAGATTCATAGCCTTTCCCAAGCGCCGCTTTGTACAGCTT---------------<=
------------------------------------------------------------<=
-----------------AGCTGTGTCGGTCAAAAGTTCGGTGCCAGCAGTCGAAGATGCAT<=
AAAACTGATCTCCCCTGGAATATCCTGCTCTTGTT-------------------------<=
------------------------------------------------------------<=
------------------------------------------------------------<=
--GTGTTGTTTGTATAGAAGAATGTGAGGGCAGCAGTGAAGCAGTAGAATCCGGCGTAAGAG<=
ACAGCCCGTCGTAGCTTCTGGATAATTATAACCTCTGAGCGGTCATCCAAGATCATCAT

Alignment file: secu3.aln
```

Chapter 22

DNA Mutations with Restrictions

22.1 Problem Description

A researcher needs to design a DNA sequence based on a coding sequence. This sequence codes for a polypeptide that the researcher is interested in. The new sequence has to code for the same polypeptide, so all nucleotide changes in the new sequence must be "silent mutations." A silent mutation is one that does not result in a change to the amino acid sequence of the resulting protein. This is done by taking advantage of the DNA code redundancy. For example AAG codes for "Lysine" (K), as AAA, so changing G for A doesn't change the resulting polypeptide.

The point of generating different DNA sequences is to be able to sort them by using restriction enzymes.

Given a DNA sequence, the program must first convert it into a polypeptide and then generate all possible DNA sequences that code such a polypeptide. The next step is to calculate which enzymes cut the original sequence and each new generated DNA sequence and compare them. The program should print the names of those enzymes that are exclusive to each sequence.

22.1.1 Introduce Point Mutations and Get Restriction Profile

Listing 22.1: Introduce point mutations (py3.us/81)

```python
1 #!/usr/bin/env python
2
3 from Bio import Translate
4 from Bio import Seq
5 from Bio.Alphabet import IUPAC
6 from Bio import Restriction
7 from Bio.Data import CodonTable
8
9 def backtrans(ori_pep,TableID=1):
10     # Function to make backtranslation (from peptide to DNA)
11     # This function needs the peptide sequence and the code of
```

```
12      # translation table. Code number is the same as posted in:
13      # http://www.ncbi.nlm.nih.gov/Taxonomy/Utils/wprintgc.cgi
14      def recurs(order, pos):
15          for letter in bt[order[pos]]:
16              if pos == len(order) - 1:
17                  yield letter
18                  continue
19              for prox in recurs(order, pos+1):
20                  yield (letter+prox)
21      def combine(order):
22          ordened = set()
23          for frase in recurs(order, 0):
24              ordened.add(frase)
25          return ordened
26      t = CodonTable.generic_by_id[TableID]
27      bt = dict()
28      for a1 in "ATCG" :
29          for a2 in "ATCG" :
30              for a3 in "ATCG" :
31                  codon = a1+a2+a3
32                  try :
33                      amino = t.forward_table[codon]
34                  except KeyError :
35                      assert codon in t.stop_codons
36                      continue
37                  try :
38                      bt[amino].append(codon)
39                  except KeyError :
40                      bt[amino] = [codon]
41      return list(combine(ori_pep))
42
43 def seqcomp(s1,s2):
44     # Compares 2 sequences and returns a value with
45     # how many differents elements they have.
46     p = len(s1)
47     for x,y in zip(s1,s2): # Walk through 2 sequences.
48         if x==y:
49             p -= 1
50     return p
51
52 n_mut = 1  # Number or allowed mutations.
53 trans = Translate.unambiguous_dna_by_id[1]
54 builtin_seq = "ATGggtaaTtgcaacggggCATCCAAG".upper()
55 dna = Seq.Seq(builtin_seq, IUPAC.unambiguous_dna)
56 # Translate DNA sequence.
```

```
57 ori_pep = str(trans.translate(dna))
58 # Get all backtranslations.
59 bakpeps = backtrans(ori_pep)
60 print 'builtin_seq: %s\nPeptide: %s\n' %(builtin_seq,ori_pep)
61 print "ORIGINAL SEQUENCE:"
62 # Make a restriction analysis for the orignal sequence.
63 anal = Restriction.Analysis(Restriction.CommOnly, dna)
64 anal.print_as("map")
65 anal.print_that()
66 # Store the enzymes that cut in the original sequence.
67 enzORI = anal.with_sites().keys()
68 enzORIset = set(enzORI)
69 # Get a string out of the enzyme list, only for
70 # printing purposes.
71 oname = str(enzORI)[1:-1]
72 # Note: str(enzORI)[1:-1] == ", ".join(str(n) for n in enzORI)
73 print "========================="
74
75 for x in bakpeps:
76     if x not in builtin_seq:
77         # Make a restriction analysis for each sequence.
78         anal = Restriction.Analysis(Restriction.CommOnly, \
79             Seq.Seq(x, IUPAC.unambiguous_dna))
80         # Store the enzymes that cut in this sequence.
81         enzTMP = anal.with_sites().keys()
82         enzTMPset = set(enzTMP)
83         # Get the number of mutations in backpep sequence.
84         y = seqcomp(builtin_seq,x)
85         if enzTMPset!=enzORIset and enzORI!=None and y<=n_mut:
86             print 'Original sequence enzymes: %s' % oname
87             # Get a string out of the enzyme list, only for
88             # printing purposes.
89             pames = str(enzTMP)[1:-1]
90             print 'Proposed sequence enzymes: %s' % pames
91             anal.print_as("map")
92             anal.print_that()
93             # o: Only in original sequences, p: proposed seq.
94             o = str(list(enzORIset.difference(enzTMPset)))[1:-1]
95             p = str(list(enzTMPset.difference(enzORIset)))[1:-1]
96             print "Enzimes only in original sequence: %s\n" % o
97             print "Enzimes only in proposed sequence: %s" % p
98             print "========================="
```

22.1.2 Sample Run of Introduce Point Mutations Program

```
builtin_seq: ATGGGTAATTGCAACGGGGCATCCAAG
Peptide: MGNCNGASK

ORIGINAL SEQUENCE:

        7 FokI Tsp509I TspEI Sse9I
        |
        |    12 HpyCH4V CviRI
        |     |
        |     |         20 BseGI BstF5I
        |     |          |
ATGGGTAATTGCAACGGGGCATCCAAG
|||||||||||||||||||||||||||
TACCCATTAACGTTGCCCCGTAGGTTC
1                         27


    Enzymes which do not cut the sequence.

AccII     AciI      AfaI      AluI      AspLEI    BfaI   ...
BshFI     BsiSI     Bsp143I   BspANI    BstFNI    BstHHI ...
(...cut...)
=========================
Original sequence enzymes: BstF5I, Tsp509I, TspEI, FokI, <=
HpyCH4V, BseGI, Sse9I, CviRI
Proposed sequence enzymes: FokI, BseGI, HpyCH4V, BstF5I, <=
CviRI, MaeIII

      5 MaeIII
      |
      | 7 FokI
      | |
      | |    12 HpyCH4V CviRI
      | |     |
      | |     |         20 BseGI BstF5I
      | |     |          |
ATGGGTAACTGCAACGGGGCATCCAAG
|||||||||||||||||||||||||||
TACCCATTGACGTTGCCCCGTAGGTTC
1                         27


    Enzymes which do not cut the sequence.
```

```
AccII    AciI     AfaI     AluI     AspLEI    BfaI    ...
BshFI    BsiSI    Bsp143I  BspANI   BstFNI    BstHHI ...
(...cut...)
```

Enzimes only in original sequence: TspEI, Sse9I, Tsp509I

Enzimes only in proposed sequence: MaeIII
=========================
(...cut...)

22.2 Additional Resources

- Roberts, R.J., Vincze, T., Posfai, J., Macelis, D. (2007). "REBASE–enzymes and genes for DNA restriction and modification." Nucleic Acids Res. 35: D269-D270.
  http://rebase.neb.com/rebase/rebase.html

- Bickle TA, Kruger DH (June 1993). "Biology of DNA restriction." Microbiol. Rev. 57 (2): 434-50.
  http://mmbr.asm.org/cgi/reprint/57/2/434?view=long&pmid=8336674

Chapter 23

Web Server for Multiple Alignment

23.1 Problem Description

DNA sequences of different organisms are often related. The closer the species, the more similar are their genomes. Some genes are highly conserved while others have extensive arrangement and mutations. Sequence multiple alignment (msa) help to evidence the relationship between sequences and infer an evolutionary history.

There are several programs to perform msa. It is out of the scope of this book to review them, but there are pointers to several papers in "Additional Resources" for those interested in msa software.

One of these programs is **muscle** (**Mu**ltiple **S**equence alignment by log-expectation), that is characterized by its improved speed and accuracy over currently available programs. Since **muscle** has no graphical interface (it is a command line application), we will build a GUI using a web server.

The advantage of the use of a web server it is not only the GUI, but the ability to use it from several computers.

23.1.1 Web Interface: Front-End. HTML Code

The first step is to make the GUI in HTML. Before reinventing the wheel, I searched for a Muscle Web server and found one at the EBI website (`http://www.ebi.ac.uk/Tools/muscle`). Inspired on this site, I made the form displayed in figure refmusclewi. The HTML code for this form is shown in listing 23.1.

Listing 23.1: Web Interface to Muscle: Front End (py3.us/82)

```
1 <html>
2 <head>
3 <title>Muscle Web Interface</title>
4 </head>
5 <body bgcolor="#eef5f5">
6 <h2>Muscle Web Interface</h2>
7 <form action='musclewi.py' method='post' enctype="multipart<=
```

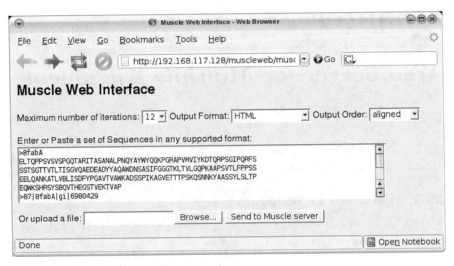

FIGURE 23.1: Muscle Web interface.

```
/form-data">
 8 Maximum number of iterations:
 9 <select name="iterat" style="width: 45px" >
10  <option value="1" selected="selected">1</option>
11  <option value="4">4</option>
12  <option value="8">8</option>
13  <option value="10">10</option>
14  <option value="12">12</option>
15  <option value="14">14</option>
16  <option value="14">16</option>
17 </select>
18 Output Format:
19 <select name="output" style="width: 140px" >
20  <option value="fasta" selected="selected">FASTA</option>
21  <option value="clw">ClustalW2</option>
22  <option value="clwstrict">ClustalW2 (Strict)</option>
23  <option value="html">HTML</option>
24  <option value="msf">MSF</option>
25 </select>
26 Output Order:
27 <select name="outorder" style="width: 90px">
28  <option value="group" selected="selected">aligned</option>
29  <option value="stable">input</option>
30 </select>
31 <p>Enter or Paste a set of Sequences in any supported format:
32 <br/><textarea name="seq" rows="5" cols="90"></textarea><p>
```

```
33 Or upload a file: <input type="file" name="upfile" />
34 <input type='submit' value='Send to Muscle server'></form>
35 </body></html>
```

23.1.2   Web Interface: Server Side Script. Commented Source Code

Listing 23.2: Web Interface to Muscle (py3.us/83)

```
 1 #!/usr/bin/env python
 2
 3 import cgi
 4 import cgitb
 5 import subprocess
 6 import sys
 7 import os
 8 from tempfile import mkstemp
 9
10 # Uncomment the following line when debugging
11 #cgitb.enable()
12
13 def badrequest(bad):
14     """ Display an error message """
15     print("<h1>Bad Request</h1>\n")
16     print("Use the options provided in the form: %s"%bad)
17     print("</body></html>")
18     # Get out of here:
19     return sys.exit()
20
21 print("Content-Type: text/html\n")
22
23 form = cgi.FieldStorage()
24 iterat = form.getvalue("iterat","4")
25 output = form.getvalue("output","html")
26 outorder = form.getvalue("outorder","group")
27 # Get sequence data from text area
28 seqs = form.getvalue("seq")
29 if not seqs:
30     # Since the textarea is empty, check the uploaded file
31     seqs = form.getvalue("upfile")
32
33 # Verify that the user entered valid information.
34 if iterat not in set(('1','4','8','10','12','14','16')):
```

```
35      badrequest(iterat)
36 valid_output = set(('html','fasta','msf','clw','clwstrict'))
37 if output not in valid_output:
38      badrequest(output)
39 if outorder not in set(('group', 'stable')):
40      badrequest(outorder)
41
42 print "<html><head><title>A CGI script</title></head><body>"
43
44 # Make a random filename for user entered data
45 fi_name=mkstemp('.txt','userdata_',"/var/www/muscleweb/")[1]
46 # Open this random filename
47 fi_fh = open(fi_name,'w')
48 # Write the user entered sequences into this temporary file
49 fi_fh.write(seqs)
50 fi_fh.close()
51
52 # Make a random filename for program output
53 fo_name=mkstemp('.txt','outfile_',"/var/www/muscleweb/")[1]
54
55 erfh = open('err.log','w')
56 cmd = ['./muscle', '-in', fi_name, '-out', fo_name,
57          '-quiet', '-maxiters', iterat, '-%s'%output,
58          '-%s'%outorder]
59
60 # Uncomment to check the generated command
61 #print ' '.join(cmd)
62 # Run the program with user provided parameters
63 p = subprocess.Popen(cmd, stderr=erfh, cwd='/var/www/muscleweb')
64 # Wait until finished
65 p.communicate()  # Same result as os.waitpid(p.pid,0)
66 erfh.close()
67 # Remove the input file since the it is not needed anymore.
68 os.remove(fi_name)
69
70 fout_fh = open(fo_name)
71 if output=='html':
72      print(fout_fh.read())
73 else:
74      print('<pre>%s</pre>'%fout_fh.read())
75 fout_fh.close()
76
77 # Remove the output file
78 os.remove(fo_name)
79
```

```
80 print("</body></html>")
```

Code explanation: The code is widely commented, but I think it worth some explanations. Note line 33 to 40, data entered by the user is checked before feeding the script with data originated from external sources (as commented in Chapter 11). In lines 45 and 53, two files with random names are created. One file is for storing the data entered by the user and the other is the filename of the **muscle** output. You may be wondering why I choose to use a function to generate files with random names instead of using a fixed name for each file. The problem with fixed files is that a web program can be used simultaneously by several users and there is a risk of data override. Another feature of this program is the use **subprocess.Popen** in line 63 and 65. The **subprocess** module replaces **os.system**, **os.spawn** and others. This module allows you to spawn new processes and have complete control over it. Temporary files are removed (lines 68 and 78).

23.2 Additional Resources

- Edgar, Robert C. (2004), "MUSCLE: multiple sequence alignment with high accuracy and high throughput." Nucleic Acids Research 32(5), 1792-97.

- Edgar, Robert C. (2004), "MUSCLE: a multiple sequence alignment method with reduced time and space complexity." BMC Bioinformatics 2004, 5:113doi:10.1186/1471-2105-5-113.

- Sellis Diamantis and Charissi Anna. "Comparison of Multiple Sequence Alignment programs."
  http://www.ceng.metu.edu.tr/~tcan/ceng465/Spring2006/Schedule/MSAComparison.pdf

- Notredame, C (2007). "Recent evolutions of multiple sequence alignment algorithms." PLOS Computational Biology 8(3):e123 doi:10.1371/journal.pcbi.0030123.

Chapter 24

Drawing Marker Positions Using Data Stored in a Database

24.1 Problem Description

This program makes a graphical representation of a selected locus in five chromosomes of *Arabidopsis thaliana*.[1] The position data of the locus are stored in a relational database, so the program has to connect such a database and retrieve the data before plotting it. The first implementation uses a MySQL database and the second one uses SQLite, just to exemplify the use of both databases covered in the book. The drawing part is made by using the **BasicChromosome** class from Biopython.

In Figure 24.1 at the end of this chapter there is an example of what the output looks like.

24.1.1 Preliminary Work on the Data

The raw data used by this program is provided by the Arabidopsis Information Resource[2] (TAIR). The file is located at their FTP server and can be retrieved with any web browser from `ftp://ftp.arabidopsis.org/home/tair/Genes/TAIR7_genome_release/TAIR7_Transcripts_by_map_position.gz`. It is a gzipped compressed CSV file with more information than the locus position into the chromosome. From this file we need four fields: Locus, Chromosome, `Map_start_coordinate` and `Map_end_coordinate`.

Here is a brief sample of TAIR data:

```
Locus Locus_orientation_is_5 Genbank_acc external_id <=
Type(1=cDNA 2=EST) Chromosome Transcript_orientation_is_5 <=
Map_start_coordinate Map_end_coordinate
(... cut ...)
AT1G01280  1  BX814827  42472162  1  1  1  112263  113195
```

---

[1] *Arabidopsis thaliana* (wall cress or mouse-ear cress) is a small flowering plant that is widely used as a model organism in plant biology. It is used in this example because there are completed physical and genetic maps available.

[2] TAIR website is available at `http://www.arabidopsis.org`.

```
AT1G01280  1  BX814827  42472162  1  1  1  113279  113861
AT1G01280  1  AA720028  2733638   2  1  1  112341  112589
AT1G01280  1  AV535036  8695319   2  1  1  112300  112919
AT1G01280  1  AV532990  8693273   2  1  0  113720  113947
AT1G01280  1  BT022023  63003811  1  1  1  112283  113195
AT1G01280  1  BT022023  63003811  1  1  1  113279  113944
(... cut ...)
```

Another nuisance in this dataset is that the lowest and highest positions
are not properly marked. In the small text snip displayed above, the lower
position is 112263 and the highest is 113947, so this text should be translated
into the following line:

```
AT1G01280,1,112263,113947
```

Therefore a custom made script is needed to convert the data for entering
into a database:

Listing 24.1: Convert the format of a csv for entering into a database
(py3.us/84)

```python
1 #!/usr/bin/env python
2
3 import csv
4 import sys
5 import gzip
6
7 f_name = 'TAIR7_Transcripts_by_map_position.gz'
8
9 # Get a file handler of an uncompressed file:
10 f_unzip = gzip.GzipFile(f_name)
11
12 lines = csv.reader(f_unzip, delimiter='\t')
13 lines.next() # Ignore the header
14
15 # Dictionary for storing markers and associated information:
16 atD = {}
17 # Load the dictionary using the data in the file:
18 for line in lines:
19     if line[0] in atD:
20         tup = atD[line[0]]
21         l7 = int(line[7])
22         left = l7 if l7<int(tup[1]) else tup[1]
23         l8 = int(line[8])
24         right = l8 if l8>int(tup[2]) else tup[2]
25         atD[line[0]] = (tup[0],left,right)
```

```
26      else:
27          atD[line[0]] = (line[5],int(line[7]),int(line[8]))
28
29 # Prints the contend of the dictionary to a CSV file:
30 out_fname= 'TAIR7.csv'
31 o_fh = open(out_fname,'w')
32 for x in atD:
33     chrom = atD[x][0] # Chromosome number
34     s_pos = atD[x][1] # Start position
35     e_pos = atD[x][2] # End position
36     o_fh.write('%s,%s,%s,%s\n' %(x,chrom,s_pos,e_pos))
37 o_fh.close()
```

24.1.2 MySQL and SQLite Database Creation

To have a database where you can retrieve marker position, you need to create it first. This is a two step procedure. It begins by creating the database with an appropriate table for this data. A MySQL command line session would look like this:

```
$ mysql -uroot -p
Enter password:
Welcome to the MySQL monitor.  Commands end with ; or \g.
Your MySQL connection id is 29
Server version: 5.0.45-Debian_1ubuntu3 Debian etch distribution

Type 'help;' or '\h' for help. Type '\c' to clear the buffer.

mysql> CREATE DATABASE at;
Query OK, 1 row affected (0.20 sec)
mysql> USE at2;
Database changed
mysql> CREATE TABLE 'pos' (
    ->    'Locus' varchar(9) NOT NULL,
    ->    'Chrom' tinyint(4) NOT NULL,
    ->    'LStart' int(11) NOT NULL,
    ->    'LEnd' int(11) NOT NULL
    -> ) ENGINE=MyISAM DEFAULT CHARSET=latin1;
Query OK, 0 rows affected (0.71 sec)
```

Alternatively, the following procedure should be used to create a SQLite database and import the data generated by the code in 24.1:

```
$ sqlite3 AT.db
SQLite version 3.4.2
```

```
Enter ".help" for instructions
sqlite> CREATE TABLE pos (Locus TEXT, Chrom INTEGER, LStart <=
INTEGER, LEnd INTEGER);
sqlite> .separator ,
sqlite> .import TAIR7.csv pos
sqlite> .quit
```

With these commands the SQLite database *AT.db* with the table *pos* is
created. This file should be accessible from the program that uses SQLite
(listing 24.3). The advantage of using SQLite is that you only need this file,
instead of setting up a database server.

24.1.3   MySQL Version with Commented Source Code

With the database in place, we finally can make a program to retrieve the
marker information from the MySQL database and plot the PDF document
with the graphic.

The program asks for a list of loci, it checks if each locus conforms to a
specific pattern (lines 142 and 175 show how to check a pattern using regex)
and then retrieves the data from the database. This program also has two
"test modes": **DBDEMO** and **NODBDEMO**. These modes are used to
test the program without entering all loci by hand. The first mode uses a
predefined list of loci (starting at line 156) and then retrieves them from the
database. The second mode uses a built-in list of loci with its positions (from
line 161) to test the program without a database connection.[3]

Listing 24.2: Draw markers in chromosomes from data extracted from a
MySQL database (py3.us/85)

```
1 #!/usr/bin/env python
2
3 # standard library
4 import os
5 import sys
6 import re
7
8 # local stuff
9 import MySQLdb
10 from Bio.Graphics import BasicChromosome
11
12 # reportlab
```

---

[3]Having the data inside the program is calling *hardcoded*. In most cases it is not recommended since it is a better idea to have the data in an easy to change external file. In this
case the data is hardcoded since this data is only for debugging purposes.

```
13 from reportlab.lib import colors
14
15 def sortmarkers(crms,end):
16     """ Sort markers into chromosomes
17     """
18     i = 0
19     crms_o = [[] for r in range(len(end))]
20     crms_fo = [[] for r in range(len(end))]
21     for crm in crms:
22         for marker in crm:
23             # add the marker start position at each chromosome.
24             crms_fo[i].append(marker[1])
25         crms_fo[i].sort() # Sort the marker positions.
26         i += 1
27     i = 0
28     for order in crms_fo:
29         # Using the marker order set in crms_fo, fill crms_o
30         # with all the marker information
31         for pos in order:
32             for mark in crms[i]:
33                 try:
34                     if pos==mark[1]:
35                         crms_o[i].append(mark)
36                 except:
37                     pass
38         i += 1
39     return crms_o
40
41 def getchromo(crms_o,end):
42     """ From an ordered list of markers, generate chromosomes.
43     """
44     chromo = [[] for r in range(len(end))]
45     i = 0
46     for crm_o in crms_o:
47         j = 0
48         if len(crm_o)>1:
49             for mark in crm_o:
50                 if mark==crm_o[0]: #first marker
51                     chromo[i].append(('',None,mark[1]))
52                     chromo[i].append((mark[0],colors.red,
53                                        mark[2]-mark[1]))
54                     ant = mark[2]
55                 elif mark==crm_o[-1]: #last marker
56                     chromo[i].append(('',None,mark[1]-ant))
57                     chromo[i].append((mark[0],colors.red,
```

```
58                                        mark[2]-mark[1]))
59                     chromo[i].append(('',None,end[i]-mark[2]))
60                 else:
61                     chromo[i].append(('',None,mark[1]-ant))
62                     chromo[i].append((mark[0],colors.red,
63                                        mark[2]-mark[1]))
64                     ant=mark[2]
65           elif len(crm_o)==1: # For chromosomes with one marker
66               chromo[i].append(('',None,crm_o[0][1]))
67               chromo[i].append((crm_o[0][0],colors.red,
68                                  crm_o[0][2]-crm_o[0][1]))
69               chromo[i].append(('',None,end[i]-crm_o[0][2]))
70           else:
71               # For chromosomes without markers
72               # Add 3% of each chromosome.
73               chromo[i].append(('',None,int(0.03*end[i])))
74               chromo[i].append(('',None,end[i]))
75               chromo[i].append(('',None,int(0.03*end[i])))
76           i += 1
77           j += 1
78       return chromo
79
80  def addends(chromo):
81       """ Adds a 3% of blank region at both ends for better
82           graphic output.
83       """
84       # get length:
85       size = 0
86       for x in chromo:
87           size += x[2]
88       # get 3% of size of each chromosome:
89       endsize = int(float(size)*.03)
90       # add this size to both ends in chromo:
91       chromo.insert(0,('', None, endsize))
92       chromo.append(('', None, endsize))
93       return chromo
94
95  def load_chrom(chr_name):
96       """ Generate a chromosome with information
97       """
98       cur_chromosome = BasicChromosome.Chromosome(chr_name[0])
99       chr_segment_info = chr_name[1]
100
101      for seg_info_num in range(len(chr_segment_info)):
102          label, color, scale = chr_segment_info[seg_info_num]
```

```
103                  # make the top and bottom telomeres
104                  if seg_info_num == 0:
105                      cur_segment = BasicChromosome.TelomereSegment()
106                  elif seg_info_num == len(chr_segment_info) - 1:
107                      cur_segment = BasicChromosome.TelomereSegment(1)
108                  ## otherwise, they are just regular segments
109                  else:
110                      cur_segment = BasicChromosome.ChromosomeSegment()
111                  if label != "":
112                      cur_segment.label = label
113                      cur_segment.label_size = 12
114                  if color is not None:
115                      cur_segment.fill_color = color
116                  cur_segment.scale = scale
117                  cur_chromosome.add(cur_segment)
118
119          cur_chromosome.scale_num = max(end) + (max(end)*.04)
120          return cur_chromosome
121
122  def dblookup(atgids):
123          """ Code to retrieve all marker data from name using mysql.
124          """
125          db = MySQLdb.connect(host="localhost", user="root",
126                              passwd="12345", db="at")
127          markers = []
128          cur = db.cursor()
129          for x in atgids:
130              cur.execute("SELECT * from pos WHERE Locus = '%s'"%x)
131              # Check if the requested marker is on the DB.
132              mrk = cur.fetchone()
133              if mrk:
134                  markers.append((mrk[0],(mrk[1],mrk[2],mrk[3])))
135              else:
136                  print "Marker %s is not in the DB" %x
137          return markers
138
139  # Size of each chromosome:
140  end=(30427563,19696817,23467989,18581571,26986107)
141  gids = []
142  rx_rid = re.compile('^AT[1-5]G\d{5}$')
143  print '''Enter AT ID or press 'enter' to stop entering IDs.
144  Valid IDs:
145  AT2G28000
146  AT3G03020
147
```

```
148 Also you can enter DBDEMO to use predefined set of markers
149 fetched from a MySQL database. Enter NODBDEMO to use a
150 predefined set of markers without database access.'''
151 while True:
152     rid = raw_input("Enter Gene ID: ")
153     if not rid:     global z
154         break
155     if rid=="DBDEMO":
156         gids = ['AT3G14890','AT1G66160','AT3G55260','AT5G59570',
157                 'AT4G32551','AT1G01430','AT4G26000','AT2G28000',
158                 'AT5G21090','AT5G10470']
159         break
160     elif rid=="NODBDEMO":
161         samplemarkers=[('AT3G14890', ('3', 5008749, 5013275)),
162                        ('AT1G66160', ('1', 24640827, 24642411)),
163                        ('AT3G55260', ('3', 20500225, 20504056)),
164                        ('AT1G10960', ('1', 3664385, 3665040)),
165                        ('AT5G23350', ('5', 7857646, 7859280)),
166                        ('AT5G15250', ('5', 4950414, 4952780)),
167                        ('AT1G55700', ('1', 20825263, 20827306)),
168                        ('AT5G21090', ('5', 7164583, 7167257)),
169                        ('AT5G10470', ('5', 3289228, 3297249)),
170                        ('AT2G28000', ('2', 11933524, 11936523)),
171                        ('AT3G03020', ('3', 680920, 682009)),
172                        ('AT4G26000', ('4', 13197255, 13199845)),
173                        ('AT4G32551', ('4', 15707516, 15713587))]
174         break
175     if rx_rid.match(rid):
176         gids.append(rid)
177     else:
178         print "Bad format, please enter it again"
179
180 if rid!="NODBDEMO":
181     samplemarkers = dblookup(gids)
182
183 crms = [[] for r in range(len(end))]
184 for x in samplemarkers:
185     crms[int(x[1][0])-1].append((x[0],x[1][1],x[1][2]))
186
187 crms_o = sortmarkers(crms,end)
188 chromo = getchromo(crms_o,end)
189 all_chr_info = [("I",chromo[0]),("II",chromo[1]),
190                 ("III",chromo[2]), ("IV",chromo[3]),
191                 ("V",chromo[4])]
192
```

```
193 pdf_organism = BasicChromosome.Organism()
194 for x in all_chr_info:
195     newcrom = (x[0],addends(x[1]))
196     pdf_organism.add(load_chrom(newcrom))
197
198 pdf_organism.draw('at.pdf','Arabidopsis thaliana')
```

24.1.4 SQLite Version with Commented Source Code

Note that the following code is almost a verbatim copy of code 24.2, the only differences are in the lines related to the database connection (lines 9 and 125).

Listing 24.3: Draw markers in chromosomes from data extracted from a SQLite database (py3.us/86)

```
 1 #!/usr/bin/env python
 2
 3 # standard library
 4 import os
 5 import sys
 6 import re
 7
 8 # local stuff
 9 import sqlite3
10 from Bio.Graphics import BasicChromosome
11
12 # reportlab
13 from reportlab.lib import colors
14
15 def sortmarkers(crms,end):
16     """ Sort markers into chromosomes
17     """
18     i = 0
19     crms_o = [[] for r in range(len(end))]
20     crms_fo = [[] for r in range(len(end))]
21     for crm in crms:
22         for marker in crm:
23             # add the marker start position at each chromosome.
24             crms_fo[i].append(marker[1])
25         crms_fo[i].sort() # Sort the marker positions.
26         i += 1
27     i = 0
28     for order in crms_fo:
```

```
29              # Using the marker order set in crms_fo, fill crms_o
30              # with all the marker information
31              for pos in order:
32                  for mark in crms[i]:
33                      try:
34                          if pos==mark[1]:
35                              crms_o[i].append(mark)
36                      except:
37                          pass
38              i += 1
39      return crms_o
40
41 def getchromo(crms_o,end):
42      """ From an ordered list of markers, generate chromosomes.
43      """
44      chromo = [[] for r in range(len(end))]
45      i = 0
46      for crm_o in crms_o:
47          j = 0
48          if len(crm_o)>1:
49              for mark in crm_o:
50                  if mark==crm_o[0]: #first marker
51                      chromo[i].append(('',None,mark[1]))
52                      chromo[i].append((mark[0],colors.red,
53                                        mark[2]-mark[1]))
54                      ant = mark[2]
55                  elif mark==crm_o[-1]: #last marker
56                      chromo[i].append(('',None,mark[1]-ant))
57                      chromo[i].append((mark[0],colors.red,
58                                        mark[2]-mark[1]))
59                      chromo[i].append(('',None,end[i]-mark[2]))
60                  else:
61                      chromo[i].append(('',None,mark[1]-ant))
62                      chromo[i].append((mark[0],colors.red,
63                                        mark[2]-mark[1]))
64                      ant=mark[2]
65          elif len(crm_o)==1: # For chromosomes with one marker
66              chromo[i].append(('',None,crm_o[0][1]))
67              chromo[i].append((crm_o[0][0],colors.red,
68                                crm_o[0][2]-crm_o[0][1]))
69              chromo[i].append(('',None,end[i]-crm_o[0][2]))
70          else:
71              # For chromosomes without markers
72              # Add 3% of each chromosome.
73              chromo[i].append(('',None,int(0.03*end[i])))
```

```
74              chromo[i].append(('',None,end[i]))
75              chromo[i].append(('',None,int(0.03*end[i])))
76          i += 1
77          j += 1
78      return chromo
79
80  def addends(chromo):
81      """ Adds a 3% of blank region at both ends for better
82          graphic output.
83      """
84      # get length:
85      size = 0
86      for x in chromo:
87          size += x[2]
88      # get 3% of size of each chromosome:
89      endsize = int(float(size)*.03)
90      # add this size to both ends in chromo:
91      chromo.insert(0,('', None, endsize))
92      chromo.append(('', None, endsize))
93      return chromo
94
95  def load_chrom(chr_name):
96      """ Generate a chromosome with information
97      """
98      cur_chromosome = BasicChromosome.Chromosome(chr_name[0])
99      chr_segment_info = chr_name[1]
100
101     for seg_info_num in range(len(chr_segment_info)):
102         label, color, scale = chr_segment_info[seg_info_num]
103         # make the top and bottom telomeres
104         if seg_info_num == 0:
105             cur_segment = BasicChromosome.TelomereSegment()
106         elif seg_info_num == len(chr_segment_info) - 1:
107             cur_segment = BasicChromosome.TelomereSegment(1)
108         ## otherwise, they are just regular segments
109         else:
110             cur_segment = BasicChromosome.ChromosomeSegment()
111         if label != "":
112             cur_segment.label = label
113             cur_segment.label_size = 12
114         if color is not None:
115             cur_segment.fill_color = color
116         cur_segment.scale = scale
117         cur_chromosome.add(cur_segment)
118
```

```
119      cur_chromosome.scale_num = max(end) + (max(end)*.04)
120      return cur_chromosome
121
122 def dblookup(atgids):
123      """ Code to retrieve all marker data fom name using mysql.
124      """
125      db = sqlite3.connect('TAIR.db')
126      markers = []
127      cur = db.cursor()
128      for x in atgids:
129          cur.execute("SELECT * from pos WHERE Locus = '%s'"%x)
130          # Check if the requested marker is on the DB.
131          mrk = cur.fetchone()
132          if mrk:
133              markers.append((mrk[0],(mrk[1],mrk[2],mrk[3])))
134          else:
135              print "Marker %s is not in the DB" %x
136      return markers
137
138 # Size of each chromosome:
139 end=(30427563,19696817,23467989,18581571,26986107)
140 gids = []
141 rx_rid = re.compile('^AT[1-5]G\d{5}$')
142 print '''Enter AT ID or press 'enter' to stop entering IDs.
143 Valid IDs:
144 AT2G28000
145 AT3G03020
146
147 Also you can enter DBDEMO to use predefined set of markers
148 fetched from a SQLite database. Enter NODBDEMO to use a
149 predefined set of markers without database access.'''
150 while True:
151      rid = raw_input("Enter Gene ID: ")
152      if not rid:
153          break
154      if rid=="DBDEMO":
155          gids = ['AT3G14890','AT1G66160','AT3G55260','AT5G59570',
156                  'AT4G32551','AT1G01430','AT4G26000','AT2G28000',
157                  'AT5G21090','AT5G10470']
158          break
159      elif rid=="NODBDEMO":
160          samplemarkers=[('AT3G14890', ('3', 5008749, 5013275)),
161                         ('AT1G66160', ('1', 24640827, 24642411)),
162                         ('AT3G55260', ('3', 20500225, 20504056)),
163                         ('AT1G10960', ('1', 3664385, 3665040)),
```

```
164                            ('AT5G23350', ('5', 7857646, 7859280)),
165                            ('AT5G15250', ('5', 4950414, 4952780)),
166                            ('AT1G55700', ('1', 20825263, 20827306)),
167                            ('AT5G21090', ('5', 7164583, 7167257)),
168                            ('AT5G10470', ('5', 3289228, 3297249)),
169                            ('AT2G28000', ('2', 11933524, 11936523)),
170                            ('AT3G03020', ('3', 680920, 682009)),
171                            ('AT4G26000', ('4', 13197255, 13199845)),
172                            ('AT4G32551', ('4', 15707516, 15713587))]
173            break
174        if rx_rid.match(rid):
175            gids.append(rid)
176        else:
177            print "Bad format, please enter it again"
178
179 if rid!="NODBDEMO":
180     samplemarkers = dblookup(gids)
181
182 crms = [[] for r in range(len(end))]
183 for x in samplemarkers:
184     crms[int(x[1][0])-1].append((x[0],x[1][1],x[1][2]))
185
186 crms_o = sortmarkers(crms,end)
187 chromo = getchromo(crms_o,end)
188 all_chr_info = [("I",chromo[0]),("II",chromo[1]),
189                 ("III",chromo[2]), ("IV",chromo[3]),
190                 ("V",chromo[4])]
191
192 pdf_organism = BasicChromosome.Organism()
193 for x in all_chr_info:
194     newcrom = (x[0],addends(x[1]))
195     pdf_organism.add(load_chrom(newcrom))
196
197 pdf_organism.draw('at.pdf','Arabidopsis thaliana'
```

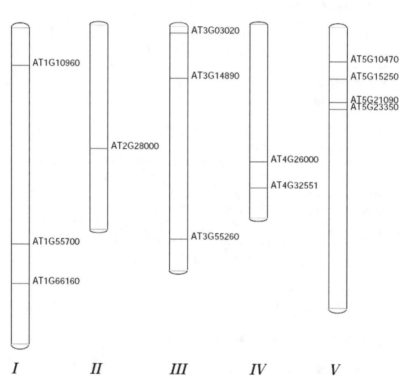

FIGURE 24.1: Product of code 24.2, using the demo dataset (`NODBDEMO`).

Appendix A

Python and Biopython Installation

As was mentioned on page 19, Python installation is straightforward. This chapter shows Python installation on most systems used (Windows, Mac OS X and Linux). A special case is taken into account: Installing more than one version of Python on the same machine. Developers sometimes need to try their programs in more than one Python version. Users with no administrative rights on their machine can install in their home directory a Python version that is different from the one that came installed in the system (if any). Another advantage in installing a Python version in your directory is that you can install modules even if the system administrator can't do it system-wide.

A.1 Python Installation

A.1.1 Windows

Python is not included in any version of Windows, so if you want to do Python programming under Windows, you will have to install it. This is a nonissue since a full Python installation is only a few clicks away.

Tip: Vista Note. Installing Python in Vista for All Users.

Administrators installing Python for all users on Windows Vista either need to be logged in as Administrator or use the *runas* command, as in,

```
runas /user:Administrator "msiexec /i <path>\<file>.msi"
```

A.1.2 Mac OS X

OS X comes with Python preinstalled. The problem with the preinstalled version is that it tends not to be the last Python version.

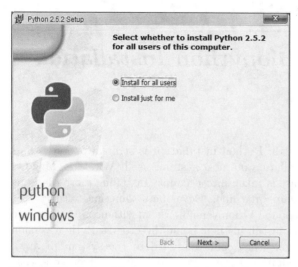

FIGURE A.1: First screen of the Python installer for Windows: This is the first screen you should get when you double click the Python installer. It gives you the option to install Python for all users or just for you. If this is not your machine, it is best to ask a system administrator which to choose first.

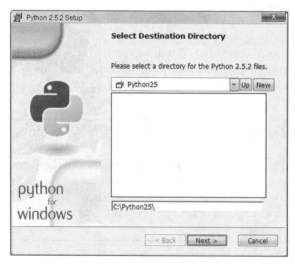

FIGURE A.2: Second screen of the Python installer for Windows: In this second screen, you have the option of which location to install Python. The default is C:\Python25\. To continue, click Next.

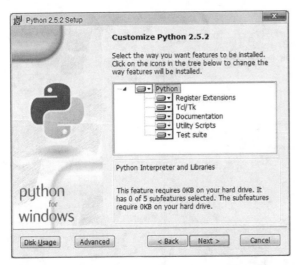

FIGURE A.3: Customize Python installation: This screen lets you customize the installation of Python 2.5.2. It has a list of the components of Python, along with hard drive icons. If you click the drive icon beside a component, you can choose to install on the local hard drive, or leave out that particular feature. A disk usage button allows you to see if there is free space, and an Advanced button lets you elect to compile the .py files to bytecode after installation. Click Next to continue. (See also Advanced and disk usage screen).

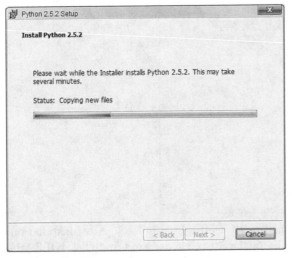

FIGURE A.4: Installation progress: After continuing, this screen shows the progress of the install.

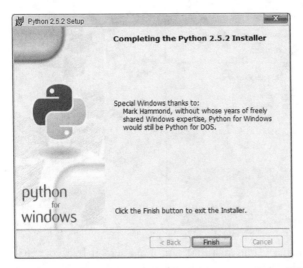

FIGURE A.5:   Last installation screen under Windows: This is the install finished page. Click Finish to close the installer.

FIGURE A.6:   Python shell ready to use in Windows: After installation, the Python25 folder should be in the Start Menu. Find and click IDLE (the Python shell and editor) to open it.

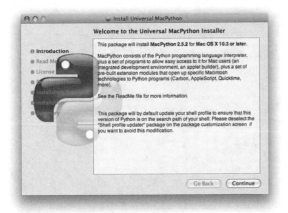

FIGURE A.7: First screen of the Python installer for Mac: After double-clicking and mounting the installer .dmg, then double-clicking MacPython.mpkg, this is the first screen that should appear. Read it carefully and hit Continue.

This version can be downloaded from `http://www.python.org/download/mac`.

Here is the step-by-step guide for Mac OS X:

A.1.3 Linux

As stated in section 2.1 (Installing Python), Python comes preinstalled in most Linux distributions. Installation and update are done with available package management software (like apt-get or rpm). The next subsection deals with the particular case of installation without administrative rights.

A.1.4 Installing Python with No Administrative Permissions

On your own machine you are supposed to have full administrative rights, so you can perform a system wide installation without asking anyone else. On a shared PC, you should ask the system administrator to have your program installed for all the users. This is typical for a university cluster, a corporate server or for a shared hosting. Sometimes there could be reasons where the system administrator will not be able to process your request. Limited resources, restrictive security policies, lack of packages for a specific Linux version, compatibility problems and other (unknown) reasons could prevent you from getting the latest Python version installed into your system.

This can be solved by installing Python into your own directory. The first step is to download the latest Python source code from `www.python.org`:

```
$ wget http://www.python.org/ftp/python/2.5.2/Python-2.5.2.tgz
```

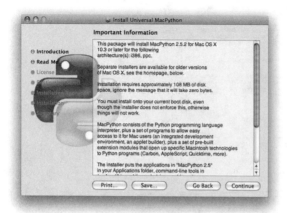

FIGURE A.8: Second screen of the Python installer for Mac: This screen displays important information also contained in the ReadMe.txt file. Do note that it says you must only install Python on your boot disk. It also explains where on your hard drive you can find the Python applications.

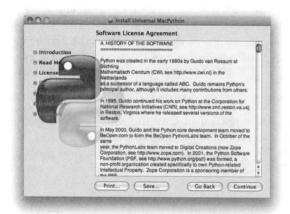

FIGURE A.9: Python software agreement in Mac: This is the Python software agreement. It explains the terms of using Python. After reading press Continue to accept or decline the agreement.

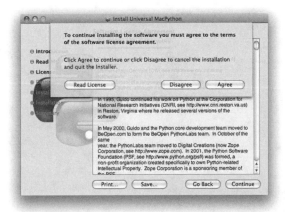

FIGURE A.10: Accept to continue in Mac: If you understand and accept the license from screen three, press Agree to continue.

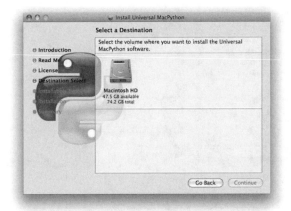

FIGURE A.11: Select where to install Python: Select the volume where you would like to install Python. This should be your boot disk. After selection, press Continue.

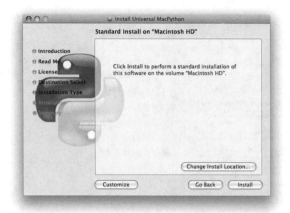

FIGURE A.12:   Ready to install Python: On this screen, pressing Install will perform a Standard install of Python. If you need to install only certain part of Python, you can choose selected components by clicking Customize(See Custom Screen). Once you click install, the installer will move on to the next screen.

FIGURE A.13:   Enter your user name and password to continue (Mac).

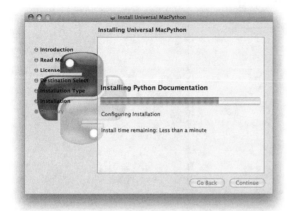

FIGURE A.14:   Installation progress: Now you should see a screen with a progress bar that indicates the progress of the installation

FIGURE A.15:   Python is successfully installed! You can close the Installer by pressing Close.

FIGURE A.16: Python shell in Mac: After install, you can find the Python Editor IDLE inside the MacPython 2.5 folder located in Applications. Double-click it to bring up the shell, and you can now write your own Python scripts!.

Uncompress the file[1] and change to this directory:

```
$ tar xfz Python-2.5.2.tgz
$ cd Python-2.5.2
```

Configure it with the path of the directory where you want to install it:

```
$ ./configure --prefix=/home/sb/py25
checking MACHDEP... linux2
checking EXTRAPLATDIR...
checking for --without-gcc... no
checking for gcc... gcc
checking for C compiler default output file name... a.out
(... output deleted ...)
creating Modules/Setup
creating Modules/Setup.local
creating Makefile
$ make
gcc -pthread -c -fno-strict-aliasing -DNDEBUG -g  -O3 -Wall <=
-Wstrict-prototypes -I. -IInclude -I./Include   <=
```

---

[1]Use **tar xfj** if you downloaded the bz2 version.

```
-DPy_BUILD_CORE -o Modules/python.o ./Modules/python.c
(... output deleted ...)
$ make install
/usr/bin/install -c -m 644 ./Include/abstract.h <=
/mnt/hda2/bio/py252/include/python2.5
(... output deleted ...)
```

Check if the program was successfully installed:

```
sb@xubuntu:~/python25/bin$ ./python2.5
Python 2.5.2 (r252:60911, Mar 23 2008, 17:03:19)
[GCC 4.0.3 (Ubuntu 4.0.3-1ubuntu5)] on linux2
Type "help", "copyright", "credits" or "license" for more <=
information.
>>>
```

A.2 Biopython Installation

A.2.1 Windows

Software requirements for Biopython are Python and Numpy.[2]

Requirements

Biopython needs Numpy. It is available from `http://www.scipy.org/` `Download`. The file for Windows is `numpy-1.2.1-win32-superpack-python2-` `.5.exe`.[3] It is installed as any Windows program.

If you are installing an old Biopython version (before 1.49), you need Numeric (also called Numerical Python). Its development is halted at version 24.2 and you can find a version for Python 2.3 and Python 2.4 in `http://` `sourceforge.net/project/showfiles.php?group_id=1369&package_id=1351`. Numeric for Python 2.5 is available at `http://biopython.org/DIST/Numeric-24.` `2.win32-py2.5.exe`.

Installing Biopython

The current version of Biopython, 1.49 at the moment of writing this, works with Python 2.4, 2.5 and 2.6.[4]

---

[2] Biopython up to version 1.48 used Numeric instead of Numpy.

[3] A Numpy for Python 2.6 is not available at the moment, this limits Biopython to Python 2.5 in Windows.

[4] Biopython 1.49 can't be installed in Python 2.6 with a graphical installer.

Windows users should download the Biopython installer for their Python version. For Python 2.5 the installer is biopython-1.49.win32-py2.5.exe and it is available from the Biopython download page: `http://biopython.org/wiki/Download`.

Biopython installation is a three step process: Click next first setup windows, select the Python version which Biopython will be installed (or leave the default option) and click next until the program is installed.

Tip: Installing Biopython for All Users.

If you installed Python for all users, you have to install Numpy as an Administrator. To do this right-click on the Numpy executable file and select "Run as administrator."

A.2.2   Linux

Installing with Package Manager

Biopython is available in most Linux repositories and can be installed with one command. The advantage of this method is that with a single program (the package manager) you can keep track, update or remove every package in your system. The main drawback is that the Biopython package available in the repository usually is not the latest version. Installation is a one step process:

For Debian/Ubuntu based systems,

```
$ sudo apt-get install python-biopython
```

For RedHat based systems:

```
$ sudo yum install python-biopython
```

Installing Biopython from Sources

When the Biopython version in the repository is not the same as the one in the Biopython website, there is the need to install it from sources. The first step is to install **NumPy**, the only external program needed by Biopython. Numpy is available in all Linux repositories, so you can install it using your package manager:

For Debian/Ubuntu based systems:

```
$ sudo apt-get install python-numpy
```

For RedHat based systems:

```
$ sudo yum install python-numpy
```

Before installing Biopython, check if you need to install some of the optional software: Reportlab and MySQLdb.

ReportLab

ReportLab is a library used for generating PDF documents. It is used in the BasicChromosome module. If you don't plan to draw chromosomes with this module, you don't need to install it. In Debian based Linux (like Ubuntu) it is available at program repositories with the name `python-reportlab` and can be installed as usual:

```
$ sudo apt-get install python-reportlab
```

It can be installed in a particular Python installation if needed: Download the sources from `http://www.reportlab.org/downloads.html` and copy the file contents into a directory reachable for that Python version:

```
$ tar xfz ReportLab_2_1.tgz
$ cd reportlab_2_1/
$ cp -R reportlab /home/sb/py25/lib/python2.5/site-packages
```

MySQLdb

MySQLdb is used to access MySQL databases. Biopython uses it for BioSQL, a database schema for storing sequence data.[5] This module was covered in section 13.6.1 (page 289). It is also available in all Linux repositories with the name `python-mysqldb`, so it is installed with:

```
$ sudo apt-get install python-mysqldb
```

Biopython

Once requirements are met, install Biopython from sources:

```
$ setup.py build
$ setup.py test
$ setup.py install
```

---

[5]See `http://www.biopython.org/wiki/BioSQL` for more information on BioSQL.

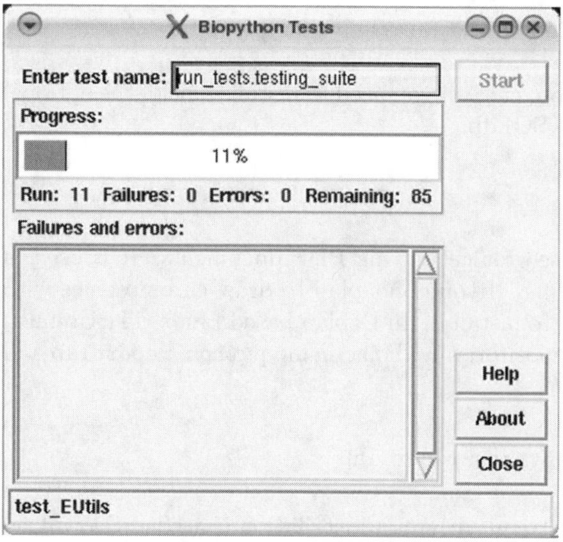

FIGURE A.17:   Biopython testing dialog.

A.2.3   Installing Old Biopython Versions

Biopython up to version 1.48 required **mx-texttools** and **Numeric**. Note that in Biopython 1.49 **mx-texttools** is not needed and **Numeric** was replaced by **Numpy**. Since each Biopython version fixes bugs and adds functionality, it is not recommended to install an old version. But if you need to install an old Biopython version, here are the instructions to install its dependencies.

Installing mx-texttools

mx-texttools is a Python package that provides high-performance text manipulation and searching algorithms and it is used by the Martel parser. It can be installed with the package manager with the name `python-egenix-` `-mxtexttools`. It also can be installed in a particular directory by doing a manual installation: Download the egenix-mx-base package from `http:` `//www.egenix.com/products/python/mxBase`, select the package for your system (for this example I choose "Source Code") and press "download." Unpack the file and go to the created directory:

```
$ tar xfz egenix-mx-base-3.1.0.tar.gz
$ cd egenix-mx-base-3.1.0
```

Install the program by calling the Python version you want to install it into.[6] If this custom Python installation is located at "/home/sb/py25/bin/", you should run:

```
$ /home/sb/py25/bin/python2.5 setup.py install
```

Check that **mx-texttools** is installed:

```
>>> from mx.DateTime import now
>>> print now()
2008-03-20 10:51:49.04
```

Numeric (or Numerical Python)

Numeric is a Python package that provides support for large, multidimensional arrays and matrices, and a large library of high-level mathematical functions. Numeric is needed for Biopython up to version 1.48 and it is now replaced with **NumPy**. It is available at Linux repositories with the name `python-numeric`. To make a custom installation, download it from `http://numpy.scipy.org/#older_array`. Uncompress, compile (build), and install:

```
$ tar xfz Numeric-24.2.tar.gz
$ cd Numeric-24.2
$ /home/sb/py25/bin/python2.5 setup.py build
$ /home/sb/py25/bin/python2.5 setup.py install --prefix=/<=
home/sb/py25
```

Testing **Numeric** installation:

```
>>> import Numeric
>>>
```

A.3   Biopython with Easy Install

If you have easy_install installed on your computer, you can download and install the latest biopython distribution by executing this command:

```
# easy_install biopython
```

---

[6]It is important not to just call the first "python" installation you have in your path. You need to call the python version that is installed in your directory.

You will have to have administrator's rights to do this. Eventually, on a *nix system, you could use **sudo**:

```
$ sudo easy_install biopython
```

To install Biopython in a user directory (without affecting the rest of the system), use **virtualenv** as shown on page 118.

Appendix B

Selected Papers

B.1 Python All a Scientist Needs

By: Julius B. Lucks. Open writing projects/Python all a scientist needs. (2008, December 3). OpenWetWare.

Abstract

Any cutting-edge scientific research project requires a myriad of computational tools for data generation, management, analysis and visualization. Python is a flexible and extensible scientific programming platform that offered the perfect solution in our recent comparative genomics investigation [1]. In this paper, we discuss the challenges of this project, and how the combined power of Biopython [2], Matplotlib [3] and SWIG [4] were utilized for the required computational tasks. We finish by discussing how Python goes beyond being a convenient programming language, and promotes good scientific practice by enabling clean code, integration with professional programming techniques such as unit testing, and strong data provenance.

The Scientist's Dilemma

A typical scientific research project requires a variety of computational tasks to be performed. At the very heart of every investigation is the generation of data to test hypotheses. An experimental physicist builds instruments to collect light scattering data; a crystallographer collects X-ray diffraction data; a biologist collects flouresence intensity data for reporter genes, or DNA sequence data for these genes; and a computational researcher writes programs to generate simulation data. All of these scientists use computer programs to control instruments or simulation code to collect and manage data in an electronic format.

Once data is collected, the next task is to analyze it in the context of hypothesis-driven models that help them understand the phenomenon they are studying. In the case of light, or X-ray scattering data, there is a well-proven physical theory that is used to process the data and calculate the observed structure function of the material being studied [5]. This structure function is then compared to predictions made by the hypotheses being tested. In the case of biological reporter gene data, light intensity is matched up with phenotypic traits or DNA sequences, and statistically analyzed for trends that might explain the observed patterns.

As these examples illustrate, across science, the original raw data of each investigation is extensively processed by computational programs in an effort to understand the underlying phenomena. Visualization tools to create a variety of scientific plots are often a preferred tool for both troubleshooting ongoing experiments, and creating publication-quality scientific plots and charts. These plots and charts are often the final product of a scientific investigation in the form of data-rich graphics that demonstrate the truth of a hypothesis compared to its alternatives [6].

Unfortunately, all too often scientists resort to a grab-bag of tools to perform these varied computational tasks. For physicists and theoretical chemists, it is common to use C or Fortran to generate simulation data, and C code is used to control experimental apparatus; for biologists, perl is the language of choice to manipulate DNA sequence data [7]. Data analysis is performed in separate, external software packages such as MATLAB or Mathematica for equation solving [8, 9], or Stata, SPSS or R for statistical calculations [10, 11, 12]. Furthermore, separate data visualization packages can be used, making the scientific programming toolset extremely varied.

Such a mixed bag of tools is an inadequate solution for a variety of reasons. From a computational perspective, most of these tools cannot be pipelined easily which necessitates many manual steps or excessive glue code that most scientists are not trained to write. Far more important than just an inconvenience associated with gluing these tools together is the extreme burden placed on the scientist in terms of data management. In such a complicated system there are often a plethora of different data files in several different formats residing at many different locations. Most tools do not produce adequate metadata for these files, and scientists typically fall back on cryptic file naming schemes to indicate what type of data the files contain and how it was generated. Such complications can easily lead to mistakes. This in turn provides poor at best data provenance when it is in fact of utmost importance in scientific studies where data integrity is the foundation of every conclusion reached and every fact established.

Furthermore, when data files are manually moved around from tool to tool, it is not clear if an error is due to program error, or human error in using the wrong file. Analyses can only be repeated by following work flows that have to be manually recorded in a paper or electronic lab notebook. This practice makes steps easily forgotten, and hard to pass on to future generations of scientists, or current peers trying to reproduce scientific results.

The Python programming language and associated community tools [13] can help scientists overcome some of these problems by providing a general scientific programming platform that allows scientists to generate, analyze, visualize and manage their data within the same computational framework. Python can be used to generate simulation data, or control instrumentation to capture data. Data analysis can be accomplished in the same way, and there are graphics libraries that can produce scientific charts and graphs. Furthermore python code can be used to glue all of these python solutions together so that visualization code resides alongside the code that generates the data it is applied to. This allows streamlined generation of data and its analysis, which makes data management feasible. Most importantly, such a uniform tool set allows the scientist to record the steps used in data work flows to be written down in Python code itself, allowing automatic provenance tracking.

In this paper, we outline a recent comparative genomics case study where Python and associated community libraries were used as a complete scientific programming platform. We introduce several specific Python libraries and tools, and how they were used to facilitate input of standardized biological data, create scientific plots, and provide solutions to speed bottle-necks in the code. Throughout, we provide detailed tutorial-style examples of how these tools were used, and point to resources for further reading on these topics. We conclude with ideas about how Python promotes good scientific programming practices, and tips for scientists interested in learning more about Python.

Comparative Genomics Case Study

Recently we performed a comparative genomics study of the genomic DNA sequences of the 74 sequenced bacteriophages that infect *E. coli*, *P. aeruginosa*, or *L. lactis* [1]. Bacteriophages are viruses that infect bacteria. The DNA sequences of these bacteriophages contain important clues as to how the relationship with their host has shaped their evolution.

Each virus that we examined has a DNA genome that is a long strand of four nucleotides called Adenine (A), Threonine (T), Cytosine (C), and Guanine (G). The specific sequences of As, Ts, Cs, and Gs encode for proteins that the virus uses to take over the host bacteria and create more copies of itself. Each protein is encoded in a specific region of the genomic DNA called a gene.

Proteins are made up of linear strings of 20 amino acids. There are 4 bases encoding for 20 amino acids, and the translation table that governs the encoding, called the genetic code, is comprised of 3 base triplets called codons. Each codon encodes a specific amino acid. Since there are 64 possible codons, and only 20 amino acids, there is a large degeneracy in the genetic code. For more information on the genetic code, and the biological process of converting DNA sequences into proteins, see [14].

Because of this degeneracy, each protein can be "spelled" as a sequence of codons in many possible ways. The particular sequence of codons used to spell a given protein in a gene is called the gene's "codon usage." As we found in [1], bacteriophages genomes favor certain codon spellings of genes over the other possibilities. The primary question of our investigation was does the observed spellings of the bacteriophage genome shed light onto the relationship between the bacteriophage and its host [1]?

To address this question, we examined the codon usage of the protein coding genes in these bacteria for any non-random patterns compared to all the possible spellings, and performed statistical tests to associate these patterns with certain aspects about the proteins.

The computational requirements of this study included:

- Downloading and parsing the genome files for viruses from GenBank in order to get the genomic DNA sequence, the gene regions and annotations: GenBank [15] is maintained by the National Center of Biotechnology Information (NCBI), and is a data warehouse of freely available DNA sequences. For each virus, we needed to obtain the genomic DNA sequence, the parts of the genome that code for genes, and the annotated function of these genes. Listing B.1 displays this information for lambda phage, a well-studied bacterophage that infects E. coli [14], in GenBank format, obtained from NCBI. Once these files were downloaded and stored, they were parsed for the required information.

- Storing the genomic information: The parsed information was stored in a custom genome python class that also included methods for retrieving the DNA sequences of specific genes.

- Drawing random genomes to compare to the sequenced genome: For each genome, we drew random genomes according to the degeneracy rules of the genetic code so that each random genome would theoretically encode the same proteins as the sequenced genome. These genomes were then visually compared to the sequenced genome through zero-mean cumulative sum plots discussed below.

- Visualize the comparisons through "genome landscape" plots: Genome landscapes are zero-mean cumulative sums, and are useful visual aids when comparing nucleotide frequency properties of the genomes they are constructed from (see [1] for more information). Genome landscapers were computed for both the sequenced genome, and each drawn genome. The genome landscape of the sequenced genome was compared to the distribution of genome landscapes generated from the random genomes to detect regions of the genomes that have extremely non-random patterns in codon usage.

- Statistically analyzing the non-random regions with annotation and host information: To understand the observed trends, we performed analysis of variance (ANOVA) [16] analysis to detect correlations between protein function annotation or host lifestyle information with these regions.

Python was used in every aspect of this computational work flow. Below we discuss in more detail how Python was used in several of these areas specifically, and provide illustrative tutorial-style examples. For more information on the details of the computational work flow, and the biological hypotheses we tested, see [1]. For specific details on the versions of software used in this paper, and links to free downloads, see Materials and Methods.

Listing B.1: Lambda phage GenBank file snippet. The full file can be found online, see [17].

```
LOCUS       NC_001416 48502 bp DNA   linear   PHG 28-NOV-2007
DEFINITION  Enterobacteria phage lambda, complete genome.
ACCESSION   NC_001416
VERSION     NC_001416.1  GI:9626243
PROJECT     GenomeProject:14204
KEYWORDS    .
SOURCE      Enterobacteria phage lambda
  ORGANISM  Enterobacteria phage lambda
            Viruses; dsDNA viruses, no RNA stage; <=
Caudovirales; Siphoviridae;
            Lambda-like viruses.
REFERENCE   1  (sites)
  AUTHORS   Chen,C.Y. and Richardson,J.P.
  TITLE     Sequence elements essential for rho-dependent<=
            transcription termination at lambda tR1
  JOURNAL   J. Biol. Chem. 262 (23), 11292-11299 (1987)
  PUBMED    3038914
...
FEATURES             Location/Qualifiers
     source          1..48502
                     /organism="Enterobacteria phage lambda"
                     /mol_type="genomic DNA"
                     /specific_host="Escherichia coli"
                     /db_xref="taxon:10710"
     gene            191..736
                     /gene="nu1"
                     /locus_tag="lambdap01"
```

```
                        /db_xref="GeneID:2703523"
        CDS             191..736
                        /gene="nu1"
                        /locus_tag="lambdap01"
                        /codon_start=1
                        /transl_table=11
                        /product="DNA packaging protein"
                        /protein_id="NP_040580.1"
                        /db_xref="GI:9626244"
                        /db_xref="GeneID:2703523"
                        /translation="MEVNKKQLADIFGASIRTIQNWQEQGMPV<=
LRGGGKGNEVLYDSA
                        AVIKWYAERDAEIENEKLRREVEELRQASEADLQPGTIEYERH<=
RLTRAQADAQELKNA
...
ORIGIN
        1 gggcggcgac ctcgcgggtt ttcgctattt atgaaaattt tccgg<=
tttaa ggcgtttccg
       61 ttcttcttcg tcataactta atgttttat ttaaaatacc ctctg<=
aaaag aaaggaaacg
      121 acaggtgctg aaagcgaggc tttttggcct ctgtcgtttc ctttc<=
tctgt ttttgtccgt
...
```

Biopython

Biopython is an open-source suite of bioinfomatics tools for the Python language [2]. The suite is comprehensive in scope, and offers Python modules and routines to parse bio-database files, facilitate the computation of alignments between biological sequences (DNA and protein), interact with biological web-services such as those provided by NCBI, and examine protein crystallographic data to name a few.

In this project, Biopython was used both to download and parse genomic viral DNA sequence files from the NCBI Genbank database [15] as outlined in Listing B.2.

Listing B.2: Downloading and parsing the GenBank genome file for lambda phage (refseq number NC_001416).

```
# genbank.py - utilities for downloading
# and parsing GenBank files

from Bio import GenBank #(1)
from Bio import SeqIO
```

```
def download(accession_list):
    """Download and save all GenBank records in
       accession_list.
    """

    try:
        handle = GenBank.download_many(accession_list) #(2)
    except:
        print "Are you connected to the internet?"
        raise

    genbank_strings = handle.read().split('//\n') #(3)
    for i in range(len(accession_list)):
        #Save raw file as .gb
        gb_file_name = accession_list[i]+'.gb'
        f = open(gb_file_name,'w')
        f.write(genbank_strings[i]) #(4)
        f.write('//\n')
        f.close()


def parse(accession_list):
    """ Parse all records in accession_list. """

    parsed = []
    for accession_number in accession_list:
        gb_file_name = accession_number+'.gb'
        print 'Parsing ... ',accession_number
        try:
            gb_file = file(gb_file_name,'r')
        except IOError:
            print 'Is the file %s downloaded?' % gb_file_name
            raise

        gb_parsed_record = SeqIO.parse(gb_file,
                                        "genbank").next() #(5)
        gb_file.close()

        print gb_parsed_record.id #(6)
        print gb_parsed_record.seq

        parsed.append(gb_parsed_record) #(7)

    return parsed
```

Example of use of the above code:

Listing B.3: Using `genbank` parser.

```
import genbank # (8)
genbank.download(['NC_001416'])
genbank.parse(['NC_001416'])
```

1. The Biopython module is called Bio. The Bio.Genbank module is used to download records from GenBank, and the Bio.SeqIO module provides a general interface for parsing a variety of biological formats, including GenBank.

2. The Bio.GenBank.download_many method is used in the genbank.download method to download GenBank records over the Internet. It takes a list of GenBank accession numbers identifying the records to be downloaded.

3. GenBank records are separated by the character string '\n'. Here we manually separate GenBank files that are part of the same character string.

4. When we save the GenBank records as individual files to disk, we include the '\n' separator again.

5. The Bio.SeqIO.parse method can parse a variety of formats. Here we use it to parse the GenBank files on our local disk using the "genbank" format parameter. The method returns a generator, whose next() method is used to retrieve an object representing the parsed file.

6. The object representing the parsed GenBank file has a variety of methods to extract the record id and sequence. See Example 2 for more details.

7. The genbank.parse method returns a listed of parsed objects, one for each input sequence file.

8. To run the code in genbank.py, Biopython 1.44 must first be installed (see Materials and Methods). Executing the following code should create a file called 'NC_001416.gb' on the local disk (see listing B.1), as well as produce the following output:

```
Parsing ...   NC_001416
NC_001416.1
Seq('GGGCGGCGACCTCGCGGGTTTTCGCTATTTATGAAAATTTTCCGGTTTAAGG<=
CGTTTCCG ...', IUPACAmbiguousDNA())
```

The benefits of using Biopython in this project are several, including:

1. Not having to write or maintain this code ourselves. This is an important point as the number of web-available databases and services grows. These often change rapidly, and require rigorous maintenance to keep up with tweaks to API's and formats - a monumental task that is completed by an international group of volunteers for the Biopython project.

2. The Biopython parsing code can be wrapped in custom classes that make sense for a particular project. Example B.4 illustrates the latter by outlining a custom genome class used in this project to store the location of coding sequences for genes (CDS_seq).

Listing B.4: Using genbank parser.

```python
# genome.py - a custom genome class which wraps
# biopython parsing code

import genbank # (1)
from Bio import Seq
from Bio.Alphabet import IUPAC

class Genome(object):
    """Genome - representing a genomic DNA sequence with genes

    Genome.genes[i] returns the CDS sequences for each gene i."""

    def __init__(self, accession_number):

        genbank.download([accession_number]) # (2)
        self.parsed_genbank = genbank.parse([
                              accession_number])[0]

        self.genes = []

        self._parse_genes()


    def _parse_genes(self):
        """Parse out the CDS sequence for each gene."""

        for feature in self.parsed_genbank.features: # (3)
            if feature.type == 'CDS':

                #Build up a list of (start,end) tuples that will
                #be used to slice the sequence in
```

```
#self.parsed_genbank.seq
#
#Biopython locations are zero-based so can be
#directly used in sequence splicing

locations = []
if len(feature.sub_features): # (4)
    # If there are sub_features, then this gene
    # is made up of multiple parts.  Store the
    # start and end positins for each part.
    for sf in feature.sub_features:
        locations.append((sf.location.start.position,
                          sf.location.end.position))
else:
    # This gene is made up of one part.  Store
    # its start and end position.
    locations.append((feature.location.start.position,
                      feature.location.end.position))


# Store the joined sequence and nucleotide
# indices forming the CDS.
seq = '' # (5)
for begin,end in locations:
    seq += self.parsed_genbank.seq[
            begin:end].tostring()

# Reverse complement the sequence if the CDS is on
# the minus strand
if feature.strand == -1:  # (6)
    seq_obj = Seq.Seq(seq,IUPAC.ambiguous_dna)
    seq = seq_obj.reverse_complement().tostring()

# append the gene sequence
self.genes.append(seq) # (7)
```

---

1. Here we import the genbank module outlined in Example B.2, along with two more biopython modules. The Bio.Seq module has methods for creating DNA sequence objects used later in the code, and the Bio.Alphabet module contains definitions for the types of sequences to be used. In particular, we use the Bio.Alphabet.IUPAC definitions.

2. We use the genbank methods to download and parse the GenBank record for the input accession number.

3. The parsed object stores the different parts of the GenBank file as a list of features. Each feature has a type, and in this case, we are looking for features with type 'CDS', which stores the coding sequence of a gene.

4. For many organisms, genes are not contiguous stretches of DNA, but rather are composed of several parts. For GenBank files, this is indicated by a feature having sub-features. Here we gather the start and end positions of all sub features, and store them in a list of 2-tuples. In the case that the gene is a contiguous piece of DNA, there is only one element in this list.

5. Once the start and end positions of each piece of the gene are obtained, we use them to slice the seq of the parsed_genbank object, and collect the concatenated sequence into a string.

6. Since DNA has polarity, there is a difference between a gene that is encoded on the top, plus strand, and the bottom, minus strand. The strand that the gene is encoded in is stored in feature.strand. If the strand is the minus strand, we need to reverse compliment the sequence to get the actual coding sequence of the gene. To do this we use the Bio.Seq module to first build a sequence, then use the reverse_complement() method to return the reverse compliment.

7. We store each gene as an element of the Genome.genes list. The CDS of the ith gene is then retrievable through Genome.genes[i].

For a more detailed introduction to the plethora of Biopython features, as well as introductory information on Python, see [18] .

MatPlotLib

Matplotlib [3] is a suite of open-source Python modules that provide a framework for creating scientific plots similar to the Matlab [8] graphical tools. In this project, matplotlib was used to create genome landscape plots both to have a quick look at data as it was generated, and to produce publication quality figures. Genome landscapes are cumulative sums of a zero-mean sequence of numbers, and are useful visualization tools for understanding the distribution of nucleotides across a genome (see [1] for more information).

Example B.5 outlines how matplotlib was used to quickly generate graphics to test raw simulation data as it was being generated.

Listing B.5: Sample matplotlib script that calculates and plots the zero-mean cumulative sum of the numbers listed in a single column of an input file.

```
# landscape.py - plotting a zero-mean cumulative sum of numbers.
```

```
import fileinput # (1)
import numpy
from matplotlib import pylab

def plot(filename):
    """Read single-column numbers in filename and plot zero-mean
    cumulative sum"""

    numbers = []
    for line in fileinput.input(filename): # (2)
        numbers.append(float(line.split('\n')[0]))

    mean = numpy.mean(numbers) # (3)
    cumulative_sum = numpy.cumsum([number -
                                 mean for number in numbers])

    pylab.plot(cumulative_sum[0::10],'k-') # (4)
    pylab.xlabel('i')
    pylab.title('Zero Mean Cumulative Sum')

    pylab.savefig(filename+'.png') # (5)
    pylab.show()
```

---

1. We use several Python community modules to plot the zero-mean cumulative sum. As part of the Python standard library, fileinput can be used as a quick and easy solution to reading in a file containing a column of entries. numpy is a comprehensive Python project aimed at providing numerical routines for scientific applications [19]. Finally we import the matplotlib.pylab module which provides a Matlab-like plotting environment.

2. Here we use fileinput to read successive lines of the input file, which takes care of opening and closing the input file automatically. Notice that we split each line by the newline character '\n', and take everything to the left of it, assuming that each line contains a single number.

3. The numpy module provides many convenient methods such as mean to compute the mean of a list of numbers, and cumsum which computes the cumulative sum. To shift the input numbers by the mean, we use a Python list comprehension to subtract the mean from each number, and then input the shifted list to numpy.cumsum.

4. The pylab module presents a Matlab-like plotting environment. Here we use several methods to create a basic line plot with an xlabel and title.

5. To view the plot, we use pylab.show(), after we have saved the figure as a PNG file using pylab.savefig. The following script uses the genome class outlined in Example B.4, along with the landscape class to plot the GC-landscape for the lambda phage genome. The genome class is used to download and parse the GenBank file for lambda phage. Each gene sequence is then scanned for 'G' or 'C' nucleotides. For every 'G' or 'C' nucleotide encountered, a 1 is appended to the list GC; for every 'A' or 'T' encountered, a 0 is appended. This sequence of 1's and 0's representing the GC-content of the lambda phage genome is saved in a file, and input into the landscape.plot method. A plot corresponding to executing this script is shown in Figure B.1.

Listing B.6: Plots the zero-mean cumulative sum.

```
import genome,landscape
lambda_phage = genome.Genome('NC_001416')
GC = []
for gene_sequence in lambda_phage.genes:
    for nucleotide in gene_sequence:
        if nucleotide == 'G' or nucleotide == 'C':
            GC.append(1)
        else:
            GC.append(0)

f = file('NC_001416.GC','w')
for num in GC:
    f.write('%i\n' % num)
f.close()


landscape.plot('NC_001416.GC')
```

Matplotlib was also used to make custom graphics classes for creating publication-quality plots. To do this, we used the object oriented interface to matplotlib plotting routines to inherit functionality in our classes.

The benefits of using matplotlib in this project were several:

1. The code that produced the scientific plots resided alongside the code that produced the underlying data that was used to produce the plots. The importance of this cannot be stressed enough as having the code structured in this way removed many opportunities for human error involved in manually shuffling raw data files into separate graphical programs. Moreover, the instructions for producing the plots from the underlying raw data was Python code, which not only described these instructions, but could be executed to produce the plots. Imagine instead the often practiced use of spreadsheets to create plots from raw data - in these spreadsheets, formulas are hidden by the results of the

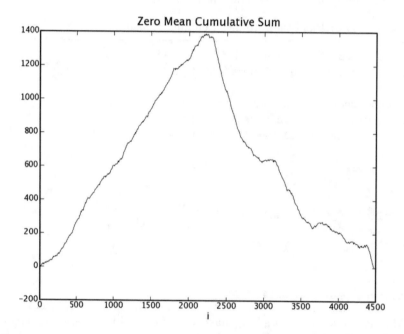

FIGURE B.1:   The lambda phage GC-landscape generated by the sample code in Example B.5.

calculations, and it is often very confusing to construct a picture of the computational flow used to produce a specific plot.

2. Having the graphics instructions in code allowed for quick trouble shooting when creating the plots, or evaluating raw data as it was generated.

3. Complicated plots were easily regenerated by tweaking the code for particular graphical plots.

SWIG

The Simple Wrapper and Interface Generator (SWIG) [4], is an easy-to-use system for extending Python. In particular, it allows the speed up of selected parts of an application by writing these routines in another more low-level language such as C or C++. Furthermore, SWIG implements the use of this low-level code using the standard Python module importing structure. This allows developers to first prototype code in Python, then re-implement the code in C and SWIG causing no change in the Python code that uses the re-implemented module.

This project relied heavily on drawing random numbers from an input discrete distribution. For example, we often needed to draw a sequence of As, Ts, Cs, or Gs corresponding to the nucleotide sequence of the genome, but preserving the genomic distribution of these four nucleotide bases. For some viruses, the distribution might look like: $PA = 0.2, PT = 0.2, PC = 0.3, PG = 0.3$, with $PA + PT + PC + PG = 1.0$. Example B.7 illustrates the outline of a Python module that has methods to draw numbers according to a discrete distribution with 4 possible outcomes. It also illustrates how this module could be implemented in C, and included in a Python module with SWIG.

Listing B.7: Drawing random numbers from a specified discrete distribution with four possibilities. The Python code to do this is shown first, followed by a re-implementation in C and inclusion in a Python module with SWIG. The procedure for using SWIG is described below.

```
# module discrete_distribution.py - drawing numbers from
# a discrete probability distribution

import random # (1)

def seed(): # (2)
    random.seed()

def draw(distribution): # (3)
    """ Drawing an index according to distribution.

    distribution is a list of floating point numbers,
```

one for each index number, representing the probability
of drawing that index number.

Example: [0.5, 0.5] would represent equal probabilities
of returning a 0 or 1.
    """

```
sum = 0 # (4)
r = random.random()
for i in range(0,len(distribution)):
    sum += distribution[i]
    if r < sum:
        return i
```

Listing B.8: Discrete Distribution

```
import discrete_distribution # (5)
discrete_distribution.seed()
print sum([discrete_distribution.draw([0.2,0.2,0.3,0.3])<=
for x in range(10000)])/10000.
```

---

1. Import the random number generator.

2. We use the discrete_distribution.seed() method to seed the random number generator. If no arguments are supplied to random.seed(), the system time is used to seed the number generator [20].

3. The draw function takes an argument distribution, which is a list of floating point numbers.

4. The algorithm for drawing a number according to a discrete distribution is to draw a number, r, from a uniform distribution on $[0, 1]$; compute a cumulative sum of the probabilities in the discrete distribution for successive indices of the distribution; when r is less than this cumulative sum, return the index that the cumulative sum is at.

5. To test this code, plug in a distribution $[0.2, 0.2, 0.3, 0.3]$, draw 10000 numbers from this distribution, and compute the mean, which theoretically should be $0 * 0.2 + 1 * 0.2 + 2 * 0.3 + 3 * 0.3 = 1.7$. In this case, when this code was executed, the result 1.7013 was returned.

   In the rest of the example, we implement this routine using C, and use SWIG to create a Python module of the C implementation.

Listing B.9: Discrete distribution in C

```
//c_discrete_distribution.c -
//A C implementation of the discrete_distribution.py module

#include <stdlib.h> // (6)
#include <stdio.h>
#include <time.h>

void seed() {
    srand((unsigned) time(NULL) * getpid());
}

int draw(float distribution[4]) { // (7)
    float r= ((float) rand() / (float) RAND_MAX);
    float sum = 0.;
    int i = 0;
    for(i = 0; i < 4; i++) {
        sum += distribution[i];
    if (r < sum) {
            return i;
    }
    }
}
```

6. Here we define two functions, seed and draw, which correspond to the Python methods in discrete_distribution.py. Note that the Python implementation of discrete_distribution.draw() worked with distributions of arbitrary numbers of elements. For simplicity, we are restricting the C implementation to work with distributions of length 4.

7. The draw routine is implemented using the same algorithm as in the Python implementation. For simplicity, we use the C standard library rand() routine, although there are more advanced random number generators that would be more appropriate for scientific applications [21].

Listing B.10: Draw routine in C

```
// c_discrete_distribution.i - A Swig interface file for the <=
c_discrete_distribution module // (8)


// Grab a 4 element array as a Python 4-list // (10)
 a local variable
```

```
    int i;
    if (PyList_Check($input)) {
        PyObject* input_to_tuple = PyList_AsTuple($input);
        if (!PyArg_ParseTuple(input_to_tuple,"ffff",temp,temp+1,<=
temp+2,temp+3)) {
            PyErr_SetString(PyExc_TypeError,"tuple must have 4 <=
elements");
            return NULL;
        }
        $1 = &temp[0];
    } else {
        PyErr_SetString(PyExc_TypeError,"expected a tuple.");
        return NULL;
    }
}

void seed(); // (11)
int draw(float distribution[4]);
```

8. To use SWIG, we create a SWIG interface file that describes how to translate Python inputs to the C code, and C outputs to the Python code.

9. SWIG directives are preceded by the % sign. Here we declare that the module we are going to make is called c_discrete_distribution. In general, the module name, the C source name, and the interface file name should all be the same outside of the file extension.

10. SWIG will automatically handle the conversion of many data-types from Python to C and C to Python. For illustration purposes, we create an explicit typemap which converts a 4-element Python list into a 4 element C list of floats. Since we are using the typemap(in) directive, SWIG knows that we are converting Python to C. The rest of the code checks that a list was passed from Python to C, and the list has 4 elements. If these conditions are not met, Python errors are thrown. If they are met, an array of floats called temp is called, and passed to C. This conversion is adapted from the SWIG reference manual [4].

11. The last thing to do in the SWIG interface file is to declare the function signatures of the C implementation.

    To use this module then, we have to call SWIG to generate wrapper code, then compile and link our code with the wrapper code. With SWIG installed, the procedure would look something like

    ```
    swig -python -o c_discrete_distribution_wrap.c c_discr<=
    ete_distribution.i
    ```

We first use SWIG to generate the wrapper code. Using the c_discrete_-distribution.i interface file, SWIG will generate c_discrete_distribution_-wrap.c using the Python C API, since we specified the -python flag. In addition, SWIG will also generate c_discrete_distribution.py, which we will use to import the module into our code.

```
gcc -c c_discrete_distribution.c c_discrete_distributi<=
on_wrap.c -I/usr/include/python2.5 -I/usr/lib/python2.5
```

Next we use a C compiler to compile each of the C files (our C source, and the SWIG generated wrapper). We have to include the python header files and libraries for the python version we are using. In our case, we used python 2.5. After this procedure completes, we should have two additional files: c_discrete_distribution.o and c_discrete_distribution-_wrap.o.

```
gcc -bundle -flat_namespace -undefined suppress -o _c_<=
discrete_distribution.so c_discrete_distribution.o c_d<=
iscrete_distribution_wrap.o
```

The final step is to link them all together. The linking options are platform dependent, and the official SWIG documentation should be consulted [4]. For Mac OS X, we use the "-bundle -flat_namespace -undefined suppress" options for gcc. When this step is done, the file _c_discrete_distribution.so is created.

The Python module file c_discrete_distribution.py can be used in the same way as in (5) above,

```
import c_discrete_distribution as discrete_distribution
discrete_distribution.seed()
print sum([discrete_distribution.draw([
          0.2,0.2,0.3,0.3]) for x in range(10000)])/10000.
```

This code produces the number 1.6942.

The benefits of using SWIG in this project were several:

1. We used all the benefits of Python with the increased speed for critical bottlenecks of our simulation code.

2. The parts that were sped up were used in the exact same context through the Python module import structure, removing the need for glue code to tie in external C-programs.

More generally, SWIG allows scientists using Python to leverage experience in other languages that they typically have, while staying within the Python framework with all its benefits outlined above. This promotes a scientific work flow which consists of prototyping simulation code using the more simple Python, then profiling the Python code to identify the speed bottlenecks. These can then be re-implemented in C or C++ and wrapped into the existing Python code using SWIG. This is a much preferred methodology than writing unnecessarily complicated and error-prone C programs, and using glue code to integrate them within the larger simulation methodology.

Conclusions

There are several practical conclusions to draw for scientists. The first is that Python, and its associated modules supported by the Python community, offer a general platform for computing that is useful across a broad range of scientific disciplines. We have only outlined several such tools in this article, but there exist many more relevant to scientists [22]. The second is that Python and its community modules can easily be used by scientists. The clean nature of the code is quick to learn, and its high-level features make complicated tasks quick to accomplish. We have not discussed the interactive programming environments offered by Python[23, 24], which when combined with the power of the language makes prototyping ideas and algorithms extremely easy.

The bigger picture conclusion is that Python promotes good scientific practice. The code readability and package structure enables code to be easily understood by different researchers working on the same project. In fact, Python code is often self-documenting which allows researchers to go back to code they wrote in the past and easily understand it. Python and its community modules provide a consistent framework to generate data, and shuttle it to the various analysis tasks. This in turn promotes data provenance through a written record in code of every step used to analyze specific data, which removes many manual steps, and thus many errors.

Finally, by using Python, scientists can start to use other community tools and practices originally designed for professional programmers, but also useful to scientists. The most important of these, but not discussed in this article, is unit testing, whereby test code is written alongside scientific code that tests to see if that code is working properly. This allows scientists to re-write aspects of the code, perhaps using a different algorithm, and to rerun the tests to see if it still works as they think it should. For large projects this is critical, and removes the need for often-used adhoc practices of looking at some sample data by eye, which is not only tedious, but not guaranteed to uncover subtle numerical bugs that could cause crucial misinterpretation of scientific data.

Since Python is a well-established language and has a large and active community, the resources available for beginners can be overwhelming. For the scientist interested in learning more about scientific programming in Python, we recommend visiting the web page and mailing lists of the SciPy project for an introduction to scientific modules [22], and [25, 26] for excellent introductory Python tutorials.

Materials and Methods

All code examples in this paper were written by the author. The particular versions of the relevant software used were: Python 2.5, Biopython 1.44, MatPlotLib 0.91.2, and SWIG 1.3.33. Documentation and free downloads of this software are available at the following URLs:

- Python - http://python.org

- Biopython - http://biopython.org

- MatPlotLib - http://matplotlib.sourceforge.net

- SWIG - http://www.swig.org/

Acknowledgments

The author would like to thank Adrian Del Maestro, Joao Xavier, David Thompson, and Stanley Qi for helpful comments during the preparation of this manuscript. The author also thanks the Miller Institute for Basic Research in Science at the University of California, Berkeley for support.

References/Resources

1. J. B. Lucks, D. R. Nelson, G. Kudla, J. B. Plotkin. Genome landscapes and bacteriophage codon usage, PLoS Computational Biology, 4, .1000001, 2008. (doi:10.1371/journal.pcbi.1000001) [LucksJB-PlOSCompBio-2008].

2. The Biopython project homepage is at http://biopython.org, and the documentation can be found at http://biopython.org/wiki/Documentation.

3. The Matplotlib project homepage is at http://matplotlib.sourceforge.net, where the documentation can also be found.

4. The Simple Wrapper and Interface Generator (SWIG) project homepage is at http://www.swig.org, and the documentation can be found at http://www.swig.org/doc.html.

5. N. Ashcroft and N. Mermin. Solid State Physics. New York. Holt, Reinhart and Winston, 1976.

6. E. Tufte. The Visual Display of Quantitative Information. 2nd ed. Cheshire, CT. Graphics Press, 2001.

7. L. Stein. How Perl Saved the Human Genome Project (`http://www.bioperl.org/wiki/How_Perl_saved_human_genome`).

8. The Matlab programming environment is developed by Mathworks - `http://www.mathworks.com`.

9. The Mathematica software is developed by Wolfram Research - `http://www.wolfram.com`.

10. The Stata statistical software is developed by StataCorp - `http://www.stata.com`.

11. The SPSS statistical software is developed by SPSS - `http://www.spss.com`.

12. The R statistical programming project homepage is at `http://www.r-project.org`.

13. The Python project homepage is at `http://python.org`.

14. Alberts, Johnson, Lewis, Raff, Roberts, Walter. Molecular Biology of the Cell, 4th ed. New York: Garland Science, 2002.

15. The GenBank data repository can be found at `http://www.ncbi.nlm.nih.gov/Genbank`.

16. For information on the Analysis of Variances Statistical Method, see Julian Faraway, Practical Regression and Anova using R, which can be found at `http://cran.r-project.org/other-docs.html`.

17. The GenBank file for lambda phage can be downloaded from `http://www.ncbi.nlm.nih.gov/nuccore/9626243` [Lambda-GenBank].

18. Bassi, S. (2007). A Primer on Python for Life Science Researchers. PLoS Comput Biol 3(11): e199. ([http://dx.doi.org/10.1371/journal.pcbi.0030199 doi:10.1371/journal.pcbi.0030199).

19. The numpy project provides a numerical back-end for scientific applications. The project homepage is at `http://numpy.scipy.org/`.

20. The Python random module documentation can be found at `http://docs.python.org/lib/module-random.html`.

21. For an extensive discussion of random numbers, see Numerical Recipes in C Chapter 7, which can be found at `http://www.nrbook.com/a/bookcpdf.php`.

22. The SciPy project is aimed at collecting and developing scientific tools for Python. The project homepage is at `http://www.scipy.org/`.

23. The ipython project can be found at `http://ipython.scipy.org`.

24. F. Perez and B. Granger. IPython: A System for Interactive Scientific Computing, Computing in Science and Engineering, 2007. `http://ieeexplore.ieee.org/iel5/5992/4160244/04160251.pdf?arnumber=4160251`.

25. Swaroop, C. H. A Byte of Python. `http://www.ibiblio.org/swaroopch/byteofpython/read/`.

26. Pilgrim, M. Dive Into Python. `http://www.diveintopython.org/`.

B.2   Diving into the Gene Pool with Biopython

Zachary Voase. Published in Volume 2 Issue 4 of **Python Magazine**. (`http://pymag.phparch.com`)

Bioinformatics is on the rise in the world of science. More and more computer scientists have begun to gravitate towards this exciting field, and as tools and libraries such as Biopython evolve, this number will only increase. In this article I introduce developers and dabblers who are familiar with Python to the biology powering this exciting subject, and the utilities available for Pythonistas to work with biological data.

Introduction

DNA: The blueprint of life. Surely an oversimplified description of such a complex and important molecule, but a necessity nonetheless. What most people overlook is the effort that goes into determining, analyzing and using this blueprint to gain even more information about the world, both around us and within us. The creation of phylogenetic trees, for example, is one pursuit of the field of bioinformatics. These trees show, with a high probability, how organisms evolved from the thriving soup of the Earth into amazingly intricate beings and expose genetic relationships between different organisms or proteins.

Let's begin with the basics. The DNA molecule is essentially a string of small units known as *nucleotides*, held together by a sugar-phosphate backbone. Nucleotides have attached to them chemical groups known as bases that distinguish the different types of nucleotide. In each of the nucleotides that make up DNA, the base can be one of four possibilities: Adenine (A), Guanine (G), Thymine (T) or Cytosine (C) (see Figure B.2). So, we can look at DNA as being a string of information in base-4 notation. Figure B.3 shows

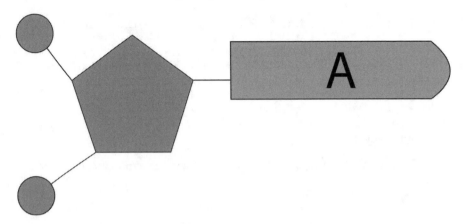

FIGURE B.2: Schema of an Adenine nucleotide.

a strand of nucleotides with the sugar-phosphate backbone as a pentagon with two circles in the left and the bases at the right. The strands of bases join up along their length in order to make a complete molecule of DNA, in the famous double-helix structure. Given one strand, the second strand's composition is predictable because A will only join up with T, and C with G. The name given to this other strand is the *complementary strand* or sequence. Figure B.4 illustrates a strand of DNA along with its complementary strand. There is no real chemical difference between the two strands; the strand on the left is complementary to that on the right, and vice versa. The difference comes in the interpretation of the information held by the DNA. Some information is encoded in a strand-dependent way, and other information is not; the factors affecting this are too numerous and complex to describe here.

Another feature of the DNA molecule is its directionality. The structure of the molecule is such that a strand has two chemically distinct endpoints that tell the DNA-reading machinery in the cell which end to start from. These two ends are called 5′ and 3′ (pronounced "five prime" and "three prime"), and conventionally DNA sequences are written from 5′ to 3′. In Figures B.3 and B.4, the arrows run in the direction the sequences are read and written. The complementary strands run in opposite directions, so in order to represent data in a conventional manner it is necessary to give complementary sequences in reverse order, the so-called *reverse complement*. This can be seen in Figure B.4 also; the strands running opposite to each other run in different directions, because their 5′ and 3′ ends have been swapped.

Essentially, DNA provides a code for creating proteins, the machinery of our cells. There are other regions of DNA which perform several other complex and, at the moment, not entirely understood functions, but the main concern of this article is in the protein-coding regions. This code is understood by each and every cell in our body. Different sequences of three bases, known as *codons*, correspond to one of 20 different chemical subunits called *amino*

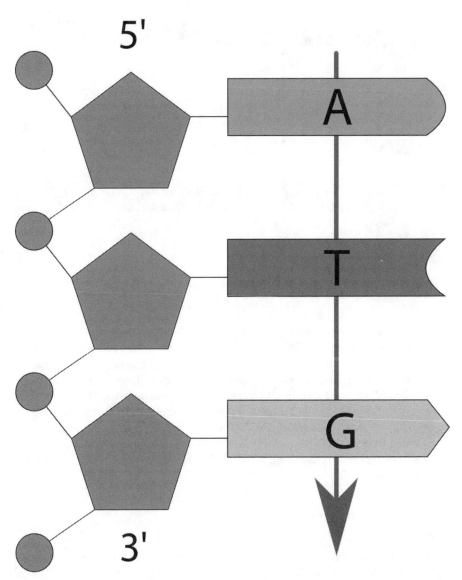

FIGURE B.3: Nucleotides forming a DNA strand of Adenine, Thymine, and Guanine (ATG).

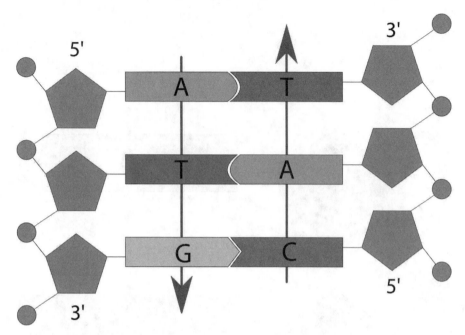

FIGURE B.4:   DNA strand and its complementary sequence.

acids. The amino acids string together to form *proteins.* A long sequence of DNA, therefore, codes for many of these units. The units are strung together to form the more complex molecules making up proteins via a process called *translation.* During translation a sequence of codons is literally translated into a sequence of amino acid subunits. These proteins carry out specific functions in our cells. Some, called *enzymes,* allow particular chemical reactions to occur in our cells. Others, known as *hormones,* send messages to other cells in our body. Still others perform one of a host of other unique abilities.

That's enough background material on DNA coding; it's GATC time to get back into the Python coding.

Basics of Sequences

Let's look at how easy it is to manipulate a sequence of nucleotides. Initialize your environment by importing the Seq class, then create a sequence by instantiating a Seq with a sequence as the only argument.

```
>>> from Bio.Seq import Seq
>>> sequence = Seq('AAACAACTTCGTAAGTAAGTATA')
>>> print sequence
Seq('AAACAACTTCGTAAGTAAGTATA',
SingleLetterAlphabet())
```

REQUIREMENTS

PYTHON: 2.3+
Other Software:

- Biopython v1.4.4

- C compiler (if compiling Biopython from source)

- mxTextTools v2.0

- Numerical Python v24.2

Useful/Related Links:

- AMY1.fa - `http://tinyurl.com/2cgzrg`

- AMY1.gb - `http://tinyurl.com/yppgow`

- Biopython Home Page - `http://biopython.org`

- SwissProt Protein Database - `http://www.expasy.ch/sprot`

- FASTA Format Wikipedia Entry - `http://en.wikipedia.org/wiki/FASTA_format`

- PubMed Home Page - `http://www.pubmed.gov`

- Biopython SeqIO Wiki Page - `http://www.biopython.org/wiki/SeqIO`

The reverse complement of a sequence can be found by calling *reverse- complement()*.

```
>>> print sequence.reverse_complement()
Seq('TATACTTACTTACGAAGTTGTTT',
SingleLetterAlphabet())
```

To reiterate, the reverse complement is a representation of the complement as it would be read from the 5′ to 3′ end, due to the opposite directionality of the complementary strand of DNA.

Alphabets

The problem with just using simple Seq objects as a way of storing sequences is that there is no meaning to the data. One way of adding metadata to a sequence is to associate a particular *alphabet* with it. The alphabet specifies which characters should be present in a sequence, and protects it from several illegal operations which we will see later. A lot of very commonly used alphabets are held in `Bio.Alphabet`, with the ones we want located in `Bio.Alphabet.IUPAC`.

```
>>> from Bio.Alphabet import IUPAC
>>> print IUPAC.unambiguous_dna
IUPACUnambiguousDNA()
>>> print IUPAC.unambiguous_dna.letters
GATC
```

From this, it's clear that unambiguous_dna is an instance of the `IUPACUn- ambiguousDNA` class, which is held inside the same module. It has a set of letters, which represent the valid characters forming a sequence of unambiguous DNA. There is also an alphabet for ambiguous DNA, in which there are several more letters to represent ambiguity in bases - for example, 'R' indicates that a base may be either 'A' or 'G', and 'Y' represents 'T' or 'C'. The whole specification may be found at the Wikipedia entry for the FASTA format. To add an alphabet to a sequence, simply give the alphabet instance as the second positional argument when initializing the sequence. Alternatively, this may be done by changing the alphabet attribute of the sequence object.

```
>>> sequence.alphabet = IUPAC.unambiguous_dna
>>> print sequence
Seq('AAACAACTTCGTAAGTAAGTATA',
IUPACUnambiguousDNA())
```

Having access to the alphabet when working with a lot of sequences helps because you can quickly see what sequence is what.

Using Sequences with Alphabets

The Seq object is very versatile. It supports slices, concatenation, and the majority of operations which may be performed on strings. Using the usual slicing syntax, and the sequence object we created previously, let's cut out the first five characters of the sequence.

```
>>> sub_seq1 = sequence[:5]
>>> print sub_seq1
Seq('AAACA', IUPACUnambiguousDNA())
```

The alphabet of the parent sequence is preserved in the newly created sequence. Create another sub-sequence of the last five characters, and stick the two together.

Alphabets also act to validate sequence data and can help prevent illegal operations between incompatible molecules. For example, RNA molecules are similar to DNA molecules except that they do not form double stranded molecules normally, and that the grey section in Figure B.2 varies between the two. Another major difference is that Thymine (T) on the DNA molecule is a counterpart to Uracil (U) on the RNA molecule. Due to their differing structures the two are incompatible with one another; you cannot have a mixed strand of DNA and RNA. We can make a new sequence using the IUPAC.unambiguous_rna alphabet, the alphabet for sequences of RNA, and try to concatenate the two.

```
>>> rna_seq = Seq('GUAAGUAUA',
... IUPAC.unambiguous_rna)
>>> print rna_seq
Seq('GUAAGUAUA', IUPACUnambiguousRNA())
>>> new_seq + rna_seq
...
<type 'exceptions.TypeError'>: ...
```

An error message is returned: it is impossible to concatenate a DNA sequence with an RNA sequence. In this way, using alphabets helps to avoid errors which may arise from trying to do something which is physically impossible.

Mutable Sequences

Sequences are like primitive strings or lists in that they allow slicing and access to particular positions along the sequence. On the other hand, sequences do not allow you to replace slices or give positions within the sequence different values; in fact, they don't actually have a __setitem__() method.

```
>>> sequence[5] = 'G'
```

```
Traceback (most recent call last):
...
AttributeError: Seq instance has no attribute
'__setitem__'
```

Objects like this are said to be *immutable*. So, if you want to make changes to slices of a sequence, it is necessary to change the sequence so that it becomes mutable, using Biopython's `MutableSeq` class. Seq instances can be transformed into `MutableSeq` instances by calling their `tomutable()` method.

```
>>> mut_seq = sequence.tomutable()
>>> print mut_seq
MutableSeq('AAACAACTTCGTAAGTAAGTATA',
IUPACUnambiguousDNA())
```

Using a mutable sequence, you can reassign whole slices of a sequence, or just single characters, and the newly changed mutable sequence can be made immutable again by calling `toseq()`.

```
>>> mut_seq[:5] = 'TCAGG'
>>> print mut_seq
MutableSeq('TCAGGACTTCGTAAGTAAGTATA',
IUPACUnambiguousDNA())
>>> changed_seq = mut_seq.toseq()
>>> print changed_seq
Seq('TCAGGACTTCGTAAGTAAGTATA',
IUPACUnambiguousDNA())
```

The functionality offered by the MutableSeq class is very useful, as it allows sequences to be modified in order to carry out mutations, as would happen in the body.

Reading Sequences from Files

Of course, all of this manipulation of genetic sequences would be useless if you couldn't store your data after manipulating it. One of the most popular sequence storage formats is FASTA, which was originally developed for use with the FASTA sequence alignment algorithm (a method for finding similarity in sequences). It's a very simple format, essentially consisting of a line beginning with a > character, followed by information about a sequence, and then several lines containing the sequence itself. This allows many sequences to be placed in the same file, and it is trivial to implement a parser and writer for this format. Biopython contains tools for working with FASTA files.

We can begin by getting an exemplar sequence file from the Internet. In the related links for this article, I have provided a TinyURL version of a link to a query on a public database called PubMed. Download that file and save it as `AMY1.fa` so you can use it with the following examples.

As an example, we'll write this sequence's reverse complement to an output file. The first step is to import the necessary Biopython resources. Bio.SeqIO contains even more tools to read and write a variety of formats, but I'm only going to demonstrate FASTA now. The IUPAC module contains needed alphabets, and the SeqRecord class will be used later to write the sequence to a file.

```
>>> from Bio.SeqIO import FastaIO
>>> from Bio.Alphabet import IUPAC
>>> from Bio.SeqRecord import SeqRecord
```

The generator function **FastaIO.FastaIterator()** reads the sequence data. Biopython makes extensive use of generators, because they allow procedures to be carried out on streams and provide more intuitive interfaces. We can assign an alphabet to the data returned by this generator by specifying an alphabet keyword argument. Invoke the generator using the sequence file as the argument and specifying an alphabet.

```
>>> seq_file = open('AMY1.fa')
>>> seq_reader = FastaIO.FastaIterator(seq_file,
    alphabet=IUPAC.unambiguous_dna)
```

Records are returned by successive calls to the generator's next() method. Get the first (and only) record from the file, and close the file.

```
>>> seq_record = seq_reader.next()
>>> seq_file.close()
```

The returned objects are instances of Bio.SeqRecord.SeqRecord. In addition to the sequence itself, sequence records contain metadata about a particular sequence, as parsed by the FASTA reader, including the name, ID, and description of the sequence. In the case of the FASTA format, the name and ID are the same when read. Other formats provide different names and IDs, with IDs primarily being used for accessing records in large databases.

```
>>> print seq_record.__class__
Bio.SeqRecord.SeqRecord
```

Writing Sequences to Files

The sequence held in a SeqRecord instance is available through its seq attribute. We can now get the reverse complement by extracting seq from the original record, finding its reverse complement, and assigning it to a new variable.

```
>>> seq = seq_record.seq
>>> seq_rc = seq.reverse_complement()
```

In order to write sequences out, you need to package them in a SeqRecord object. SeqRecord.__init__() accepts the sequence as the first positional argument, with a host of other keyword arguments usually used for more complex things such as database cross-references and annotations to the sequence. Sequence records to be written to FASTA-formatted files only need name, ID, and description. Here I've created some variables to hold the information before creating the record, in the interest of clean code.

```
>>> name = seq_record.name
>>> id = seq_record.id
>>> description = seq_record.description + \
...         ' (Reverse Complement)'
>>> seq_record_rc = SeqRecord(seq_rc,
...         id=id, name=name,
...         description=description)
```

In addition to passing them to the constructor, you can assign these attributes to the SeqRecord object after instantiating the object, like this:

```
>>> seq_record_rc.name = name
>>> seq_record_rc.id = id
>>> seq_record_rc.description = description
```

The next step is to write this information in the FASTA format using SeqIO.write. It accepts iterable containing several records, a file handle and the format name ("fasta" in this case). Create a writable output file called AMY1_RC.fa to which we will write both the original sequence and its reverse complement.

```
>>> seq_file_rc = open('AMY1_RC.fa','w')
```

Now import SeqIO and use its write method:

```
>>> from Bio import SeqIO
>>> SeqIO.write([seq_record, seq_record_rc], seq_file_rc, 'fasta')
```

You now have a new FASTA-formatted file containing both the original and reverse complement of the genetic sequence of the human alpha-amylase 1 enzyme.

Proteins

The last Biology section touched on proteins lightly, but it is relevant to go slightly more in-depth now before doing more work with them.

Proteins, like DNA, are strands of units which are each different but link together in a similar fashion. The units of proteins are known as *amino acids*, and several of these join end-to-end to form a large molecule known as a *polypeptide*. Often, proteins will be made up of only one polypeptide, but some are composed of several polypeptides that combine in a very specific way. Hemoglobin, the pigment that allows our blood to transport oxygen, is one such protein; it is composed of four polypeptides known as *hem groups*, with an atom of iron trapped between the molecular entanglement.

DNA has many functions in our cells. In this article, we are concentrating on how it acts as a blueprint for the creation of proteins. As mentioned before, there is a code that maps 3-base-long codons of DNA onto particular amino acids in a process known as *translation*. Translation is carried out by a structure in cells called the *ribosome*, which joins this string of amino acids up to form a polypeptide. Because there are 64 possible codons and only 20 amino acids, the code is called *degenerate*, as several different codons will code for the same amino acid. Nevertheless, due to the structure of polypeptides it is possible to write them as sequences, albeit using 20 letters instead of 4. If you take a look at the `Bio.Alphabet.IUPAC module`, you will notice an alphabet named protein. This is the alphabet of amino acids, the specification for which may also be found at the FASTA format Wikipedia entry. It is also worth noting that because polypeptides do not join up with another complementary molecule, the concepts of complements and reverse complements do not apply. If you try calling the `reverse_complement()` method of a protein sequence, you will receive an error.

Querying from Public Databases

Due to the rapid expansion of the field of bioinformatics, and the necessity for information interchange between research parties around the world, several resources have been created which contain quickly growing repositories of information for bioinformaticists. PubMed, mentioned earlier, is a database of articles, genes, whole genomes (the gene sequences of entire organisms), and protein sequences run by the US National Center for Biotechnology Information. Swiss-Prot, part of the Swiss Institute of Bioinformatics' ExPASy server, is a publicly available database of protein structures and information, complete with large amounts of annotation and references. Biopython offers ways to query these (and other) resources from within your Python programs, using several modules included with the main distribution. In this section, we're going to grab a protein sequence from the Swiss-Prot protein knowledge-base.

The first step is to import Bio.ExPASy and SwissIterator.

```
>>> from Bio import ExPASy
```

```
>>> from Bio.SeqIO.SwissIO import SwissIterator
```

Then, using ExPASy.get_sprot_raw(), create a handle, that returns the data fetched from Swiss-Prot on the protein specified by the identifier 'P04745', which is essentially the index of a record in the database.

```
>>> connection = ExPASy.get_sprot_raw('P04745')
```

Instantiate the SwissIterator class around this connection. Because the connection acts like a file object, it is perfectly safe to iterate over it, instead of reading data into a file or StringIO buffer and then using that. The SwissIterator is a generator function which returns SeqRecord objects.

```
>>> sprot_reader = SwissIterator(connection)
>>> sprot_record = sprot_reader.next()
>>> connection.close()
```

Due to the large amount of annotative data held in the Swiss-Prot database, this sequence record will have its other possible attributes filled; annotations, dbxrefs and features, for example, will all contain a lot of information.

```
>>> print sprot_record.name
AMY1_HUMAN
>>> print sprot_record.annotations['organism']
Homo Sapiens (Human)
```

Protein P04745's name is 'AMY1_HUMAN', and it comes from the *Homo Sapiens* organism, otherwise known as the human. ExPASy's entries have a lot of metadata bundled with them, including references to papers which talk about the protein, and the location of the protein on other databases. You have successfully downloaded and parsed the polypeptide sequence for the alpha-amylase 1 enzyme, the genetic sequence of which we just worked with a short while ago.

Proteins are not simply mapped from the gene to the polypeptide sequence. Before being translated, DNA is copied onto a strand of RNA in a process called "transcription." This messenger RNA (mRNA) then moves from the nucleus of the cell, where the DNA is stored, out into the cell's cytoplasm. There a process called splicing occurs. Splicing is essentially the removal of particular fragments of code from the sequence, which has the effect of changing the polypeptide sequence produced. The fragments which remain, known as exons, join together to form a new sequence, and those that were removed, called introns, move off and are recycled by the cell. The spliced mRNA strand is used as the template to make the polypeptide.

Splicing is still not fully understood. Biologists spend a lot of time and effort figuring out what parts of a gene are spliced, and why. It is useful to have the gene and the protein sequences available at the same time because by using the protein sequence it is possible to figure out what parts of the gene were spliced out, and hence which parts of the gene are introns and which are exons. Splicing is responsible for the large size difference between the human genome (the collective set of all human genes) and the human proteome (the set of all human proteins). This difference in size is because it is possible for one gene to code for several different polypeptides by alternative splicing. New evidence also suggests that some segments of RNA have a catalytic action which causes them to splice themselves out of the mRNA strand, without external help from proteins. The rapid advances occurring in this field every day, combined with its incredible and exciting complexity, are what is making it so attractive to computer scientists around the world.

Data Included with Biopython

In addition to a large collection of procedures, Biopython comes with a rich set of useful data. The Bio.Data module includes data such as the molecular weights of bases for both DNA and RNA and the codon mappings used by several different organisms to translate mRNA into polypeptide sequences. The specification for ambiguous DNA is included in the dictionary Bio.Data.IUPACData.ambiguous_dna_values. Individual character keys are mapped onto strings of several letters. For example, 'R' is mapped onto 'AG' and 'Y' onto 'CT', with a whole host of other ambiguous letters.

As an example of the other features of Bio.Data, we're going to write a small application to calculate the molecular weight of a specific protein, given its SwissProt identifier. The program will have to download a SwissProt entry, parse it, and then use the molecular weight data included in Biopython to calculate the weight of the protein. For this, it will need the **Bio.ExPASy.get_sprot_raw()**, **Bio.SeqIO.SwissIO.SwissIterator** and **Bio.Data.IUPACData.protein_weights** respectively. Listing B.11 shows the whole program, lines 3-7 of which contain the necessary imports.

Listing B.11: Using data included in Biopython

```
1   #!/usr/bin/env python
2
3   import sys
4
5   from Bio.Data.IUPAaCData import protein_weights
6   from Bio.ExPASy import get_sprot_raw
7   from Bio.SeqIO.SwissIO import SwissIterator
8
9   def get_prot_record(prot_id):
10      connection = get_sprot_raw(prot_id)
11      reader = SwissIterator(connection)
12      prot_record = reader.next()
```

```
13      connection.close()
14      return prot_record
15
16 def calc_weight(prot_seq):
17      weight_list = map(protein_weights.get, prot_seq)
18      return sum(weight_list)
19
20 def id_to_weight(prot_id):
21      prot_record = get_prot_record(prot_id)
22      weight = calc_weight(prot_record.seq)
23      return weight
24
25 if __name__ == '__main__':
26      args = sys.argv[1:]
27      for arg in args:
28          print '%s: %.3f' % (arg, id_to_weight(arg))
```

A good starting point is the small function to download the sequence record for a specified protein identifier (lines 9-14). It is merely a repetition of the procedure used earlier for the alpha-amylase 1 enzyme. It accepts a protein ID, connects to SwissProt using get_sprot_raw(), and uses SwissIterator() to grab the sequence record. It then closes the connection and returns the record. The reason I chose not to make it return only the sequence, which is really all that is needed, is so this function can be used again by other programs. Next, the program needs another function to calculate the weight of a protein given a sequence (lines 16-18). Because calc_weight() only requires that its argument implements iteration, it can accept strings, unicode objects, Seq, and MutableSeq instances. The protein letter-to-weight mappings are held in a simple dictionary, called protein_weights, located in Bio.Data.IUPACData. This dictionary allows calc_weight() to simply map the weight of each amino acid to a list, and then return the sum of this list.

Again, fragmenting this rather simple program up into several functions enables other programs to use various parts of this one. Finally, the wrapper function id_to_weight() cements it all together (lines 20-23). If id_to_weight()is fed the protein identifier for alpha-amylase 1, 'P04745', it gives back a floating point number of approximately equal to 66955.810. I've also added a small section which will allow the function to be run from the command line, and called with several identifiers. As you can see, Biopython comes with a real treasure trove of data, and I recommend you have a look around at the Bio.Data module in the interpreter and online documentation as much as you can to see what can be done with it.

Sequence Features

The sequences we have examined so far have all been pretty simple; just strings of letters, with optional metadata if wrapped in a SeqRecord instance. But, of course, different segments of sequences often have different functions or roles. Sometimes you will have a sequence of pre-mRNA (i.e. mRNA that has not yet been spliced) and you will want to know which parts are introns and which parts are exons, etc. It is easy to do this with Biopython's SeqFeature module. Sequence Features are annotations to sequences which can optionally span a segment of the sequence. In this section we will use sequence features to separate out a DNA sequence into several different parts. Each of these parts will correspond to an exon on the DNA sequence. For us to do this, the FASTA format will not be enough; we're going to have to use a format known as GenBank. This format allows for verbose detail and annotation of a sequence, and Biopython can parse the GenBank format's annotations into sequence features. I've made available a link to the data we will need. Download the AMY1.gb file as you did for the FASTA-formatted sequence, and make sure to save it under that name.

Listing B.12: Reading Sequence Features

```
1 from Bio.SeqIO import parse, write
2
3 gb_file = open('AMY1.gb')
4 gb_iterator = parse(gb_file, 'genbank')
5 gb_record = gb_iterator.next()
6 gb_file.close()
7 exons = [ feat for feat in gb_record.features
8          if feat.type == 'exon' ]
9
10 seq_records = []
11 for exon in exons:
12     start = exon.location.start.position
13     end = exon.location.end.position
14     number = exon.qualifiers['number'][0]
15     sequence = gb_record.seq[start:end]
16     record = SeqRecord(sequence)
17     record.name = 'AMY1.%s' % (number,)
18     record.id = record.namev
19     record.description =
20         'Human amylase gene, exon %s' % (number,)
21     seq_records.append(record)
22 outfile = open('AMY1_exons.fa','w')
23 write(seq_records, outfile, 'fasta')
24 outfile.close()
```

Listing B.12 starts by importing the necessary modules and functions. In this case, we need to import only two functions that are held in the Bio.SeqIO module: parse() and write(). These functions offer a general I/O ability, with the type of file specified as the second positional argument. For example, on line 4 we open the file using the 'genbank' format string, meaning it will read this as a GenBank file. Other file types include 'fasta' (the file format of which should be obvious) and 'swiss' (for SwissProt files). The list of supported file formats, along with their capabilities (with respect to reading and writing) can be found at the Biopython wiki page for the SeqIO module.

Once the file is open, we grab the only record in the file, and close the file. Lines 7-8 extract all the exon data from each feature in the sequence record.

SeqRecord instances contain a features attribute which is a list of SeqFeature instances. Each feature has a type attribute; for example, a type of 'gene' means that the feature contains information on where that particular sequence can be found within an organism's genome. Features of type 'exon' contain information on a particular exon within that sequence, including the exon number for that exon within that gene and its location, given as a range of positions along the sequence. The location attribute of the first exon shows that it is made up of the bases between positions 0 and 168 along that sequence, with 0 being the first base (as with all iterables in Python).

The next step is to create a list of the sequence records cut out of the master sequence so we can write them to a file afterwards. This list is called seq_records. The loop processes the exons. First, it gets the start and end positions by accessing its location attribute. Then, it retrieves the exon's number, or index on the strand, from the 'qualifier' with title 'number'. The first item in the list is used because the values are held in lists so keys can be specified several times for a particular feature. Qualifiers are essentially key/value pairs of metadata associated with features, which means you can specify your own qualifiers without having to adhere to a globally defined standard.

The sequence for the exon is then obtained by slicing out the region between the start and end positions previously obtained (line 15). A new record is created around this sequence, and its name is set to our custom format, which will be different for each exon. Because we are writing this out to the FASTA format, the id and name attributes should be the same, so we can set the record's id to be its name. In addition, we can also give each record a description, which we've set to give a basic summary of the whole gene's function and the particular exon number. Finally, each newly created record is appended to the list.

Now that it has a list of fully created sequence records, the program can write them all out to 'AMY1_exons.fa' (lines 22-25). This example uses the standard form of write(), which is usually used for writing output in a quick-and-dirty way. For more precise control, it is often necessary to use the more specialized writers such as the FastaWriter class used before. There you have it: a FASTA-formatted file which contains all of the exons which will go on to make the alpha-amylase 1 enzyme.

Conclusion

Biopython is an incredibly versatile and well documented tool for bioinformaticists. In this article I have barely scraped the surface of all that it has to offer. There are some very well-written tutorials out there that go into a much greater depth. Biopython is constantly adapting to both the large and the small discoveries being made every day, and it will continue to do so with time. I can only wish you good luck getting to know it in the future.

Appendix C

Included DVD: Virtual Machine Installation and Use

C.1 General Overview

The DVD includes a virtual machine (VM) with a Linux distribution with all the software included in this book. A VM is a software container that can run its own operating systems and applications as if it were a physical computer. A VM is composed of two elements: A virtual machine monitor (VMM), and the VM data. The VMM is the software that allows one or multiple operating systems (guest systems) to run on a physical computer (host). The VM data are the files where the guest operating system is stored.

The VMM included in the DVD is called VM Player. This program can run on top of most used operating systems.[1] With a VM you can run a pre-configured system in minutes with no hassle. In this particular case, the included system is a special edition of DNALinux. In order to run DNALinux, you have to install VM player and load the included VM.

In case you can't or don't want to install the VM, all programs and files used in the book are available in the **software** directory in the DVD.

Installation from included DVD takes three steps:

1. Uncompress virtual machine files.

2. Install VM player.

3. Load virtual machine into the player.

C.1.1 Uncompress

Since the VM disk size is larger than DVD available space, it is compressed with the **7zip utility**. You must first install the uncompress utility (unless you already have one) to be able to uncompress the virtual disk. This program is included in the DVD. Find the apropiate version in your Operating System

---

[1]Windows and Linux players are freely available, but Mac OSX virtual machine (called VMWare Fusion) is not free, but there is a 30-day free trial.

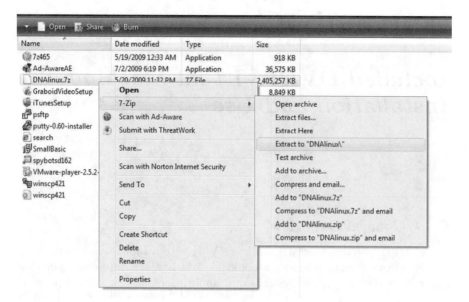

FIGURE C.1: Select the compressed virtual disk and extract it contents with 7zip.

directory on the DVD. Windows version installation is straightforward. Linux users may want to install `xarchive` since it provides a GUI front end for `7zip`.

In any case, copy the VM disk file (`DNALinux.7z`) from the DVD to the harddisk. In Windows, select the file and press the secondary mouse button. From the popup menu, select 7-Zip and then Extract to `DNALinux` as shown in Figure C.1.

C.1.2 VMWare Player Installation

The next step is to install the VM player. It is installed as any standard Windows program (rebooting included). Double-click the "VMWare Player" installer, `VMware-player-2.5.2-156735.exe`, and you will see a screen like C.2.

Press "Install" and follow the prompts until you get a "Finished" screen (see Figure C.3 in page 451).

Installing VMWare in Linux has almost the same procedure as Windows installation, albeit without the rebooting stage.

Before installing VMWare, check if you have the kernel headers and a build environment. In Ubuntu these packages are installed with

```
$ sudo apt-get install linux-headers-'uname -r' build-essential
```

The executable file is called `VMware-Player-2.5.2-156735.i386.bundle` and it should be called from the command line with:

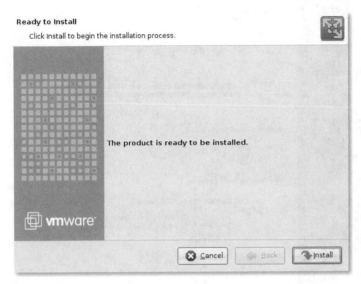

FIGURE C.2:   VMWare First Installation Screen.

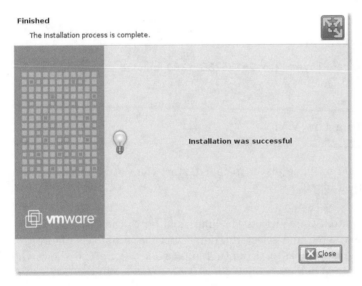

FIGURE C.3:   VMWare Finish Screen.

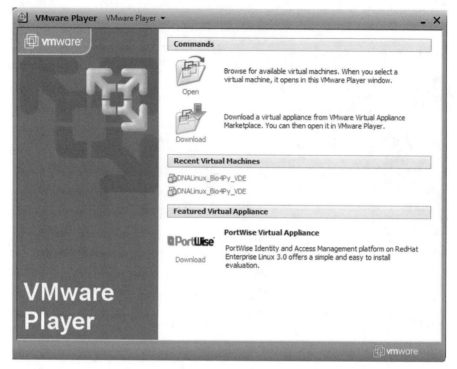

FIGURE C.4:   Starting VMWare Player.

```
$ sudo sh VMware-Player-2.5.2-156735.i386.bundle
```

C.1.3   Loading the VM

With the virtual machine player installed, start the virtual machine. You should see a screen like C.4.

Press "Open an existing Virtual Machine" and go to the directory where VM files were extracted and select the only .vmx file available. Figure C.5 shows how this dialog looks in Ubuntu Linux while Figure C.6 shows the equivalent dialog in Windows.

DNALinux should boot. There will be a Xubuntu screen during boot-up since Xubuntu is the Linux distribution taken as a base to make DNALinux.

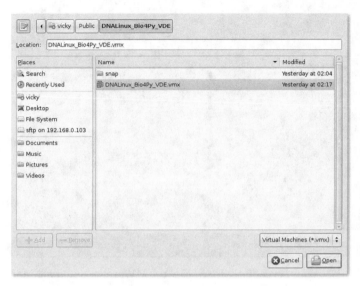

FIGURE C.5: Open File Dialog Box in Linux.

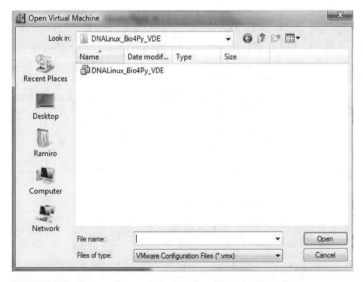

FIGURE C.6: Open File Dialog Box in Windows.

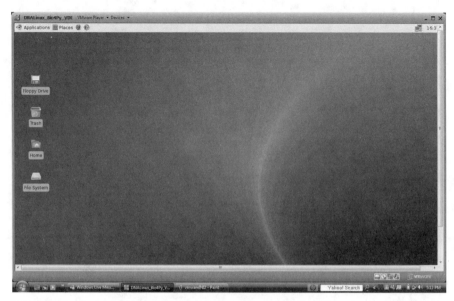

FIGURE C.7: DNALinux ready to use (inside a VMWare virtual machine).

C.2   Instructions for Mac Users

VMWare free player is not available for Mac. Although there are two options: Use "VMWare Fusion"[2] or use the .py programs included in the software directory in the DVD. All programs are named with the same number of the py3.us URL. For example, program 10.1 (page 182) has as URL, py3.us/36, so the program name is 36.py.

There is a Mac directory with the 7zip program. Use this program to uncompress the DNAlinux.7z file if you want to use "VMWare Fusion."

C.3   Accessing the Virtual Machine

Once the VMWare is installed and the DNALinux virtual machine is loaded, you should see a screen like Figure C.7.

The DNALinux desktop is ready to use. There is no need to log-in since it login automatically with the user sb. The password for this user is dnalinux.

---

[2]Available from http://www.vmware.com/products/fusion/.

FIGURE C.8: Loading a page from the DNALinux Web server.

There is no root user. To execute a command with root privileges, use `sudo`
command. To issue the command `top` as root, type

```
$ sudo top
```

When prompt for a password, use your password (`dnalinux`).

To run the Python IDLE, go to **Applications**, **Development** and the
click on **IDLE**. From the IDLE menu, you can open any Python program.
All programs featured in the book are available in the `/home/sb/bioinfo`
directory. As mentioned in "Instructions for Mac Users," programs are named
with the same number of the `py3.us` URL.

To run a terminal, go to **Applications**, **Accessories**, and **Terminal**.

C.3.1   Using DNALinux as a Server

An Apache Web server is preconfigured in DNALinux. It can be accessed
from your host computer. You need the IP address of the virtual machine to
point your browser in your host computer at the Web server. To find out the
IP address, open a terminal and type: `ifcongif`. Look at the second line of
the output and find a line starting with `inet addr:`. The IP I get is always
192.168.117.133, but this may change on other setup. Enter this IP in the
address bar on the browser of the host machine, and it should load a page like
Figure C.8.

The contents of this page is being served from the directory `/var/www/`.
Take this into account to test programs from Chapter 11 (Web Applications).

C.3.2   Using Databases in DNALinux

A MySQL server is installed and loaded in DNALinux. The password is `root` and can be accessed from the command line with the MySQL client or from the Web server using the installed **PHPMyAdmin**.

There is a SQLite viewer installed. It is called **SQLite Database Browser** and it is available in **Applications**, **Development** menu.

C.4   Additional Resources

- VMWare Player. "Run Virtual Machines on Linux or Windows PCs for Free." `http://www.vmware.com/products/player`

- VMWare Player. Getting started guide (PDF). `http://www.vmware.com/pdf/vmware_player250.pdf`

- DNALinux homepage.
  `http://www.dnalinux.com`

- Citing DNALinux:
  Bassi, Sebastian and Gonzalez, Virginia. DNALinux Virtual Desktop Edition. Available from Nature Precedings <`http://dx.doi.org/10.1038/npre.2007.670.1`> (2007).

- DNALinux forum.
  `http://groups.google.com/group/dnalinux-forum/`

- Ubuntu Support
  `http://www.ubuntu.com/support/communitysupport`

Appendix D

Python Language Reference

D.1 Python 2.5 Quick Reference

D.1.1 Invocation Options

python[w] [-dEhimOQStuUvVWxX?] [-c *command* | *scriptFile* | -] [*args*]
See Table D.1 on page 458 for a complete list.
(python**w** does not open a terminal/console; python does)

- Available **IDEs** in std distrib: **IDLE** (tkinter based, portable), **Python-win** (on Windows). Other free IDEs: IPython (enhanced interactive Python shell), Eric, SPE, BOA constructor, PyDev (Eclipse plugin).

- Typical python **module header** :

```
#!/usr/bin/env python
# -*- coding: latin1 -*-
```

Since 2.3, the encoding of a Python source file must be declared as one of the two first lines (or defaults to **7 bits ASCII**) [PEP-0263], with the format:

TABLE D.1: Invocation Options

Option	Effect
-d	Output parser debugging information (also PYTHONDE-BUG=x)
-E	Ignore environment variables (such as PYTHONPATH)
-h	Print a help message and exit (formerly -?)
-i	Inspect interactively after running script (also PYTHONIN-SPECT=x) and force prompts, even if stdin appears not to be a terminal.
-m *module*	Search for *module* on `sys.path` and runs the module as a script. (Implementation improved in 2.5: module `runpy`)
-O	Optimize generated bytecode (also PYTHONOPTIMIZE=x). Asserts are suppressed.
-OO	Remove doc-strings in addition to the -O optimizations.
-Q *arg*	Division options: -Qold (default), -Qwarn, -Qwarnall, -Qnew
-S	Don't perform `import site` on initialization.
-t	Issue warnings about inconsistent tab usage (-tt: issue errors).
-u	Unbuffered binary stdout and stderr (also PYTHONUN-BUFFERED=x).
-U	Force Python to interpret all string literals as Unicode literals.
-v	Verbose (trace import statements) (also PYTHONVER-BOSE=x).
-V	Print the Python version number and exit.
-W *arg*	Warning control (arg is action:message:category:module:lineno)
-x	Skip first line of source, allowing use of non-unix forms of `#!cmd`
-c *command*	Specify the command to execute (see next section). This terminates the option list (following options are passed as arguments to the command).
scriptFile	The name of a python file (.py) to execute. Read from stdin.
-	Program read from stdin (default; interactive mode if a tty).
args	Passed to script or command (in `sys.argv[1:]`)

```
# -*- coding: encoding -*-
```

Std encodings are defined here, e.g. ISO-8859-1 (aka latin1), iso-8859-15 (latin9), UTF-8... Not all encodings supported, in particular UTF-16 is not supported.

- It's now a **syntax error** if a module contains string literals with 8-bit characters but doesn't have an encoding declaration (was a warning before).

- Since 2.5, from __future__ import feature statements must be declared at **beginning** of source file. item Site customization: File sitecustomize.py is automatically loaded by Python if it exists in the Python path (ideally located in **$PYTHONHOME/lib/site-packages**).

D.2 Environment Variables

See Table D.2 on page 460.

D.3 Notable Lexical Entities

D.3.1 Keywords

and	del	for	is	raise
assert	elif	from	lambda	return
break	else	global	not	try
class	except	if	or	while
continue	exec	import	pass	with
def	finally	in	print	yield

- (List of keywords available in std module: **keyword**)

- Illegitimate Tokens (only valid in strings): $? (plus @ before 2.4)

- A statement must all be on a single line. To break a statement over multiple lines, use "\", as with the C preprocessor.

- Exception: can always break when inside any (), [], or {} pair, or in triple-quoted strings.

- More than one statement can appear on a line if they are separated with semicolons (";").

TABLE D.2: Environment Variables

Variable	Effect
PYTHONHOME	Alternate prefix directory (or prefix;exec_prefix). The default module search path uses prefix/lib
PYTHONPATH	Augments the default search path for module files. The format is the same as the shell's $PATH: one or more directory pathnames separated by ':' or ';' without spaces around (semi-) colons ! On Windows Python first searches for Registry key HKEY_LOCAL_MACHINE\-Software\Python\PythonCore\x.y\PythonPath (default value). You can create a key named after your application with a default string value giving the root directory path of your appl. Alternatively, you can create a text file with a .pth extension, containing the path(s), one per line, and put the file somewhere in the Python search path (ideally in the `site-packages` directory). It's better to create a .pth for each application, to make it easy to uninstall them.
PYTHONSTARTUP	If this is the name of a readable file, the Python commands in that file are executed before the first prompt is displayed in interactive mode (no default).
PYTHONDEBUG	If non-empty, same as -d option
PYTHONINSPECT	If non-empty, same as -i option
PYTHONOPTIMIZE	If non-empty, same as -O option
PYTHONUNBUFFERED	If non-empty, same as -u option
PYTHONVERBOSE	If non-empty, same as -v option
PYTHONCASEOK	If non-empty, ignore case in file/module names (imports)

Literal
"a string enclosed by double quotes"
'another string delimited by single quotes and with a " inside'
"'a string containing embedded newlines and quote (') marks, can be delimited with triple quote*s*."'
""" may also use 3- double quotes as delimiters """
u'a <u>unicode</u> string'
U"Another <u>unicode</u> string"
r'a <u>raw</u> string where are kept (literalized): handy for regular expressions and windows paths!'
R"another raw string" – raw strings cannot end with a
ur'a <u>unicode</u> raw string'
UR"another raw <u>unicode</u>"

- Comments start with "#" and continue to end of line.

D.3.2   Identifiers

```
(letter | "_") (letter | digit | "_")*
```

- Python identifiers keywords, attributes, etc. are **case-sensitive**.

- Special forms: _ident (not imported by 'from module import *'); __ident__ (system defined name); __ident (class-private name mangling).

D.3.3   String Literals

Two flavors: `str` (standard 8 bits locale-dependent strings, like ASCII, iso 8859-1, utf-8, ...) and `unicode` (16 or 32 bits/char in utf-16 mode or 32 bits/char in utf-32 mode); one common ancestor `basestring`.

- Use \ at end of line to continue a string on next line.

- Adjacent strings are concatenated, e.g. `'Monty ' 'Python'` is the same as `'Monty Python'`.

- `u'hello' + ' world' --> u'hello world'` (coerced to unicode)

String Literal Escapes

See Table D.3 on page 462.

- NUL byte (\000) is **not** an end-of-string marker; NULs may be embedded in strings.

- Strings (and tuples) are immutable: they cannot be modified.

TABLE D.3: String Literal Escapes

Escape	Meaning
newline	Ignored (escape newline)
\ \	Backslash (\)
\e	Escape (ESC)
\v	Vertical Tab (VT)
\'	Single quote (')
\f	Formfeed (FF)
\ooo	char with octal value ooo
\"	Double quote (")
\n	Linefeed (LF)
\a	Bell (BEL)
\r	Carriage Return (CR)
\xhh	char with hex value hh
\b	Backspace (BS)
\t	Horizontal Tab (TAB)
\uxxxx	Character with 16-bit hex value xxxx (unicode only)
\Uxxxxxxxx	Character with 32-bit hex value xxxxxxxx (unicode only)
\N{name}	Character named in the Unicode database (unicode only), e.g. `u'\NGreek Small Letter Pi' <=> u'\u03c0'`. (Conversely, in module `unicodedata`, `unicodedata.name(u'\u03c0') == 'GREEK SMALL LETTER PI'`)
\AnyOtherChar	left as-is, including the backslash, e.g. `str('\z') == '\ \z'`

D.3.4 Boolean Constants (Since 2.2.1)

- True

- False

In 2.2.1, True and False are integers 1 and 0. Since 2.3, they are of new type `bool`.

D.3.5 Numbers

- **Decimal integer**: 1234, 1234567890546378940L (or l)

- **Octal** integer: 0177, 0177777777777777777L (begin with a **0**)

- **Hex** integer: 0xFF, 0XFFFFffffFFFFFFFFFFL (begin with **0x** or **0X**)

- **Long** integer (unlimited precision): 1234567890123456L (ends with **L** or l) or **long**(1234)

- **Float** (double precision): 3.14e-10, .001, 10., 1E3

- **Complex:** 1J, 2+3J, 4+5j (ends with **J** or **j**, + separates (float) real and imaginary parts)

D.3.6 Sequences

- **Strings** (types `str` and `unicode`) of length 0, 1, 2 (see above) ", '1', "12", 'hello\n'

- **Tuples** (type `tuple`) of length 0, 1, 2, etc: () (1,) (1,2) # parentheses are optional if len > 0

- **Lists** (type `list`) of length 0, 1, 2, etc: [] [1] [1,2]

- Indexing is **0**-based. Negative indices (usually) mean count backwards from end of sequence.

- Sequence **slicing** [starting-at-index : but-less-than-index *[: step]*]. Start defaults to 0, end to len(sequence), step to 1.

```
a = (0,1,2,3,4,5,6,7)
a[3] == 3
a[-1] == 7
a[2:4] == (2, 3)
a[1:] == (1, 2, 3, 4, 5, 6, 7)
a[:3] == (0, 1, 2)
a[:] == (0,1,2,3,4,5,6,7) # makes a \textbf{copy} of the sequence.
a[::2] == (0, 2, 4, 6) # Only even numbers.
a[::-1] = (7, 6, 5, 4, 3 , 2, 1, 0) # Reverse order.
```

TABLE D.4: Operators and Their Evaluation Order

Operator	Comment
, [...] {...} '...'	Tuple, list & dict. creation; string conv.
s[i] s[i:j] s.attr f(...)	indexing & slicing; attributes, fct calls
+x, -x, ~x	Unary operators
x**y	Power
x*y x/y x%y	mult, division, modulo
x+y x-y	addition, substraction
x<<y x>>y	Bit shifting
x&y	Bitwise and
x∧y	Bitwise exclusive or
x|y	Bitwise or
x<y x<=y x>y x>=y x==y x!=y	Comparison
x<>y	identity
x is y x is not y	membership
x in s x not in s	
not x	boolean negation
x and y	boolean and
x or y	boolean or
lambda args: expr	anonymous function

D.3.7 Dictionaries (Mappings)

<u>Dictionaries</u> (type `dict`) of length 0, 1, 2, etc: 1 : 'first' 1 : 'first', 'two': 2, key:value Keys must be of a hashable type; Values can be any type.

D.3.8 Operators and Their Evaluation Order

Operators and Their evaluation order: Table D.4 (page 464)

- Alternate names are defined in module operator (e.g. __add__ and add for +)

- Most operators are overridable

D.4 Basic Types and Their Operations

D.4.1 Comparisons (Defined between Any Types)

Comparisons: Table D.5 (page 465).

TABLE D.5: Comparisons

Comparison	Meaning	Notes
<	strictly less than	(1)
<=	less than or equal to	
>	strictly greater than	
>=	greater than or equal to	
==	equal to	
!= or < >	not equal to	
is	object identity	(2)
is not	negated object identity	(2)

Notes:

- Comparison behavior can be overridden for a given class by defining special method _ _cmp_ _.

- (1) X < Y < Z < W has expected meaning, unlike C

- (2) Compare object identities (i.e. **id**(object)), not object values.

D.4.2 None

- None is used as default return value on functions. Built-in single object with type NoneType. Might become a keyword in the future.

- Input that evaluates to None does not print when running Python interactively.

- None is now a **constant**; trying to bind a value to the name "None" is now a syntax error.

D.4.3 Boolean Operators

Boolean values and operators: Table D.6 (page 466)
Notes:

- Truth testing behavior can be overridden for a given class by defining special method _ _nonzero_ _.

- (1) Evaluate second arg only if necessary to determine outcome.

D.4.4 Numeric Types

Floats, Integers, Long Integers, Decimals.

- Floats (type `float`) are implemented with C doubles.

TABLE D.6:   Boolean Operators

Value or Operator	Evaluates to	Notes
built-in **bool**(*expr*)	**True** if expr is true, **False** otherwise.	see True, False
None, numeric zeros, empty sequences and mappings	considered False	
all other values	considered True	
not x	**True** if x is **False**, else **False**	
x **or** y	if x is **False** then y, else x	(1)
x **and** y	if x is **False** then x, else y	(1)

- Integers (type `int`) are implemented with C longs (signed 32 bits, maximum value is `sys.maxint`)

- Long integers (type `long`) have unlimited size (only limit is system resources).

- Integers and long integers are **unified** starting from release 2.2 (the L suffix is no longer required). `int()` returns a `long` integer instead of raising `OverflowError`. Overflowing operations such as 2<<32 no longer trigger `FutureWarning` and return a long integer.

- Since 2.4, new type `Decimal` introduced (see module: decimal) to compensate for some limitations of the floating point type, in particular with fractions. Unlike floats, decimal numbers can be represented exactly; exactness is preserved in calculations; precision is user settable via the `Context` type [PEP 327].

Operators on all numeric types: *abs, int, long, float, -, +, *, /, //, %, divmod* and ****. See Table D.7 in page 467.

Notes:

- (1) / is still a floor division (1/2 == 0) unless validated by a from `__future__` import division.

- classes may override methods `__truediv__` and `__floordiv__` to redefine these operators.

Bit operators on integers and long integers Bit Operators: See Table D.8 on page 467.
Complex Numbers

- Type `complex`, represented as a pair of machine-level double precision floating point numbers.

TABLE D.7: Operators on All Numeric Types

Operation	Result
abs(x)	the absolute value of x
int(x)	x converted to integer
long(x)	x converted to long integer
float(x)	x converted to floating point
-x	x negated
+x	x unchanged
x + y	the sum of x and y
x - y	difference of x and y
x * y	product of x and y
x / y	true division of x by y: 1/2 -> 0.5 (1)
x // y	floor division operator: 1//2 -> 0 (1)
x % y	x modulo y
divmod(x, y)	the tuple (x//y, x%y)
x ** y	x to the power y (the same as **pow**(x,y))

TABLE D.8: Bit Operators on Integers and Long Integers

Operation	Result
~x	the bits of x inverted
x ∧ y	bitwise exclusive or of x and y
x & y	bitwise and of x and y
x | y	bitwise or of x and y
x << n	x shifted left by n bits
x >> n	x shifted right by n bits

- The real and imaginary value of a complex number z can be retrieved through the attributes z.real and z.imag.

Numeric exceptions

TypeError

raised on application of arithmetic operation to non-number

OverflowError

numeric bounds exceeded

ZeroDivisionError

raised when zero second argument of div or modulo op

D.4.5 Operations on Sequence Types (Lists, Tuples, Strings)

Operations on all sequence types: *in, not in, +, len, min, max, reversed,* and *sorted.* See Table D.9 on page 468.

TABLE D.9:   Operations on All Sequence Types

Operation	Result	Notes
x **in** s	True if an item of s is equal to x, else False	(3)
x **not in** s	False if an item of s is equal to x, else True	(3)
s1 + s2	the concatenation of s1 and s2	
s * n, n*s	n copies of s concatenated	
s[i]	i'th item of s, origin 0	(1)
s[i: j]	s[i: j:step] Slice of s from i (included) to j(excluded). Optional step value, possibly negative (default: 1).	(1), (2)
len(s)	Length of s	
min(s)	Smallest item of s	
max(s)	Largest item of s	
reversed(s)	[2.4] Returns an iterator on s in reverse order. s must be a sequence, not an iterator (use `reversed(list(s))` in this case. [PEP 322]	
sorted(iterable [,cmp] [,cmp=cmpFct] [,key=keyGetter] [,reverse=bool])	[2.4] works like the new in-place list.sort(), but sorts a **new** list created from the iterable.	

Notes:

- (1) if i or j is negative, the index is relative to the end of the string, i.e., len(s)+i or len(s)+j is substituted. But note that -0 is still 0.

- (2) The slice of s from i to j is defined as the sequence of items with index k such that i¡= k ¡ j. If i or j is greater than len(s), use len(s). If j is omitted, use len(s). If i is greater than or equal to j, the slice is empty.

- (3) For strings: before 2.3, x must be a single character string; Since 2.3, x in s is True if x is a substring of s.

D.4.6   Operations on Mutable Sequences (Type `list`)

Operations on mutable sequences: *append, extend, count, index, insert, remove, reverse* and *sort*. See Table D.10 in page 469.

Notes:

- (1) Raises a `ValueError` exception when x is not found in s (i.e. out of range).

- (2) The sort() method takes an optional argument cmp specifying a comparison function taking two list items and returning -1, 0, or 1 de-

TABLE D.10: Operations on Mutable Sequences

Operation	Result	Notes
s[i] =x	item i of s is replaced by x	
s[i:j[:step]] = t	slice of s from i to j is replaced by t	
del s[i:j[:step]]	same as s[i:j] = []	
s.**append**(x)	same as s[len(s) : len(s)] = [x]	
s.**extend**(x)	same as s[len(s):len(s)]= x	(5)
s.**count**(x)	returns number of i's for which s[i] == x	
s.**index**(x[, start[,	returns smallest i such that s[i]==x. start and	(1)
stop]])	stop limit search to only part of the list.	
s.**insert**(i, x)	same as s[i:i] = [x] if i¿= 0. i == -1 inserts	
	before the last element.	
s.**remove**(x)	same as del s[s.index(x)]	(1)
s.**pop**([i])	same as x = s[i]; del s[i]; return x	(4)
s.**reverse**()	reverses the items of s in place	(3)
s.**sort**([**cmp**])	sorts the items of s in place	(2),
s.**sort**([cmp=cmpFct]		(3)
[, key=keyGetter] [,		
reverse=bool])		

pending on whether the 1st argument is considered smaller than, equal to, or larger than the 2nd argument. Note that this slows the sorting process down considerably. Since 2.4, the cmp argument may be specified as a keyword, and 2 optional keywords args are added: key is a fct that takes a list item and returns the key to use in the comparison (**faster** than cmp); `reverse`: If True, reverse the sense of the comparison used. Since Python 2.3, the sort is guaranteed "stable." This means that two entries with equal keys will be returned in the same order as they were input. For example, you can sort a list of people by name, and then sort the list by age, resulting in a list sorted by age where people with the same age are in name-sorted order.

- (3) The `sort()` and `reverse()` methods **modify** the list **in place** for economy of space when sorting or reversing a large list. They don't return the sorted or reversed list to remind you of this side effect.

- (4) The `pop()` method is not supported by mutable sequence types other than lists. The optional argument i defaults to -1, so that by default the last item is removed and returned.

- (5) Raises a `TypeError` when x is not a list object.

D.4.7   Operations on Mappings / Dictionaries (Type `dict`)

Operations on mappings: See Table D.11 on page 470.

TABLE D.11:   Operations on Mappings/Dictionaries

Operation	Result	Notes
len(d)	The number of items in *d*.	
dict()	Creates an empty dictionary.	
dict(**kwargs)	Creates a dictionary init with the keyword	
dict(iterable)	args kwargs.	
dict(d)	Creates a dictionary init with (key, value)	
	pairs provided by iterable.	
	Creates a dictionary which is a copy of dictionary d.	
d.**fromkeys**(*iterable*, *value*=None)	Class method to create a dictionary with keys provided by *iterator*, and all values set to *value*.	
d[k]	The item of *d* with key *k*.	(1)
d[k] = x	Set *d[k]* to x.	
del d[k]	Removes *d[k]* from d.	(1)
d.**clear**()	Removes all items from *d*	
d.**copy**()	A shallow copy of *d*.	
d.**has_key**(*k*) *k* in *d*.	**True** if *d* has key emphk, else **False**	
d.**items**()	A copy of *d's* list of (key, item) pairs	(2)
d.**keys**()	A copy of *d's* list of keys.	(2)
d1.**update**(*d2*)	`for k, v in d2.items(): d1[k] =` v Since 2.4, **update**(**kwargs) and **update**(iterable) may also be used.	
d.**values**()	A copy of *d's* list of values.	(2)
d.**get**(*k, defaultval*)	The item of *d* with key *k*.	(3)
d.**setdefault**(*k[, defaultval]*)	*d[k]* if *k* in *d*, else *defaultval*(also setting it)	(4)
d.**iteritems**()	Returns an iterator over (key, value) **pairs**.	
d.**iterkeys**()	Returns an iterator over the mapping's keys.	
d.**itervalues**()	Returns an iterator over the mapping's **values**.	
d.**pop**(*k[, default]*)	Removes key *k* and returns the corresponding value. If key is not found, *default* is returned if given, otherwise **KeyError** is raised.	
d.**popitem**()	Removes and returns an arbitrary (key, value) pair from *d*	

Notes:

- **TypeError** is raised if key is not acceptable.

- (1) **KeyError** is raised if key k is not in the map.

- (2) Keys and values are listed in random order.

- (3) Never raises an exception if k is not in the map, instead it returns *defaultval*. It is optional, when not provided and k is not in the map, None is returned.

- (4) Never raises an exception if k is not in the map, instead returns defaultVal, and adds k to map with value defaultVal. defaultVal is optional. When not provided and k is not in the map, None is returned and added to map.

D.4.8 Operations on Strings (Types str & unicode)

These string methods largely (but not completely) supersede the functions available in the string module. The str and unicode types share a common base class basestring.

Operations on strings: See Table D.4.8 on page 472.

Notes:

- (1) Padding is done using spaces or the given character.

- (2) If optional argument *start* is supplied, substring s[start:] is processed. If optional arguments *start* and end are supplied, substring s[start:end] is processed.

- (3) Default encoding is sys.getdefaultencoding(), can be changed via sys.setdefaultencoding(). Optional argument errors may be given to set a different error handling scheme. The default for *errors* is **'strict'**, meaning that encoding errors raise a **ValueError**. Other possible values are **'ignore'** and **'replace'**. See also module codecs.

- (4) If optional argument *tabsize* is not given, a tab size of 8 characters is assumed.

- (5) Returns False if string s does not contain at least one character.

- (6) Returns False if string s does not contain at least one cased character.

- (7) A titlecased string is a string in which uppercase characters may only follow uncased characters and lowercase characters only cased ones.

- (8) s is returned if width is less than $\mathbf{len}(s)$

- (9) If the optional argument *maxCount* is given, only the first maxCount occurrences are replaced.

- (10) If *separator* is not specified or None, any whitespace string is a separator. If *maxsplit* is given, at most maxsplit splits are done.

Operation	Result	Notes
s.**capitalize**()	Returns a copy of *s* with its first character capitalized, and the rest of the characters lowercased.	
s.**center**(*width*[, *fillChar*=' '])	Returns a copy of *s* centered in a string of length *width*, surrounded by the appropriate number of *fillChar* characters.	(1)
s.**count**(*sub*[, *start*[, *end*]])	Returns the number of occurrences of substring *sub* in string *s*.	(2)
s.**decode**([*encoding*[, *errors*]])	Returns a unicode string representing the decoded version of str *s*, using the given codec (encoding). Useful when reading from a file or a I/O function that handles only str. Inverse of encode.	(3)
s.**encode**([*encoding*[, *errors*]])	Returns a str representing an encoded version of *s*. Mostly used to encode a unicode string to a str in order to print it or write it to a file (since these I/O functions only accept str). Also used to encode a str to a str. Inverse of decode.	(3)
s.**endswith**(*suffix* [, *start*[, *end*]])	Returns True if s ends with the specified suffix, otherwise return false. Since 2.5 suffix can also be a **tuple** of strings to try.	(2)
s.**expandtabs**([*tabsize*])	Returns a copy of s where all tab characters are expanded using spaces.	(4)
s.**find**(*sub* [,*start*[,*end*]])	Returns the lowest index in s where substring *sub* is found. Returns -1 if *sub* is not found.	(2)
s.**index**(*sub*[, *start*[, *end*]])	like **find**(), but raises ValueError when the substring is not found.	(2)
s.**isalnum**()	Returns True if all characters in *s* are alphanumeric, False otherwise.	(5)
s.**isalpha**()	Returns True if all characters in *s* are alphabetic, False otherwise.	(5)
s.**isdigit**()	Returns True if all characters in *s* are digit characters, False otherwise.	(5)
s.**islower**()	Returns True if all characters in *s* are lowercase, False otherwise.	(6)
s.**isspace**()	Returns True if all characters in *s* are whitespace characters, False otherwise.	(5)
s.**istitle**()	Returns True if string *s* is a titlecased string, False otherwise.	(7)
s.**isupper**()	Returns True if all characters in *s* are uppercase, False otherwise.	(6)
separator.**join**(seq)	Returns a concatenation of the strings in the sequence seq, separated by string separator, e.g.: ",".join(['A', 'B', 'C']) -> "A,B,C"	

s.**ljust** / **rjust** / **center** (*width*[, *fillChar*=' '])	Returns *s* left/right justified/centered in a string of length *width*.	(1), (8)
s.**lower**()	Returns a copy of *s* converted to lowercase.	
s.**lstrip**([*chars*])	Returns a copy of *s* with leading chars (default: blank chars) removed.	
s.**partition**(separ)	Searches for the separator separ in s, and returns a tuple (`head, sep, tail`) containing the part before it, the separator itself, and the part after it.	
s.**replace**(*old, new*[, *maxCount* =-1])	Returns a copy of *s* with the first *maxCount* (-1: unlimited) occurrences of substring *old* replaced by new.	(9)
s.**rfind**(*sub*[, *start*[, *end*]])	Returns the highest index in *s* where substring sub is found. Returns -1 if sub is not found.	(2)
s.**rindex**(sub[, start[, end]])	like **rfind()**, but raises `ValueError` when the substring is not found.	(2)
s.**rpartition**(*separ*)	Searches for the separator *separ* in *s*, starting at the end of *s*, and returns a tuple (`head, sep, tail`) containing the (left) part before it, the separator itself, and the (right) part after it.	
s.**rstrip**([*chars*])	Returns a copy of *s* with trailing *chars*(default: blank chars) removed.	
s.**split**([*separator*[, maxsplit]])	Returns a list of the words in *s*, using *separator* as the delimiter string.	(10)
s.**rsplit**([*separator*[, maxsplit]])	Same as `split`, but splits from the end of the string.	(10)
s.**splitlines**([*keepends*])	Returns a list of the lines in *s*, breaking at line boundaries.	(11)
s.**startswith**(*prefix* [, *start*[, *end*]])	Returns `True` if *s* starts with the specified prefix, otherwise returns `False`. Negative numbers may be used for *start* and *end*. Since 2.5 prefix can also be a **tuple** of strings to try.	(2)
s.**strip**([*chars*])	Returns a copy of *s* with leading and trailing *chars*(default: blank chars) removed.	
s.**swapcase**()	Returns a copy of *s* with uppercase characters converted to lowercase and vice versa.	
s.**title**()	Returns a titlecased copy of *s*, i.e. words start with uppercase characters, all remaining cased characters are lowercase.	
s.**translate**(table [, deletechars])	Returns a copy of *s* mapped through translation table.	(12)
s.**upper**()	Returns a copy of *s* converted to uppercase.	
s.**zfill**(*width*)	Returns the numeric string left filled with zeros in a string of length *width*.	

- (11) Line breaks are not included in the resulting list unless *keepends* is given and true.

- (12) *table* must be a string of length 256. All characters occurring in the optional argument *deletechars* are removed prior to translation.

D.4.9   String Formatting with the % Operator

`formatString % args --> evaluates to a string`

- *formatString* mixes normal text with C printf *format fields*:

  `%[flag][width][.precision] formatCode`

  where *formatCode* is one of c, s, i, d, u, o, x, X, e, E, f, g, G, r, % (see Table D.12).

- The *flag* characters -, +, blank, # and 0 are understood (see table D.12).

- *Width* and *precision* may be a * to specify that an integer argument gives the actual *width* or *precision*. Examples of *width* and *precision*:

  Examples

TABLE D.12:   String Formatting Characters

Format string	Result
'%3d' % 2	' 2'
'%*d' % (3, 2)	' 2'
'%-3d' % 2	'2 '
'%03d' % 2	'002'
'% d' % 2	' 2'
'%+d' % 2	'+2'
'%+3d' % -2	' -2'
'%- 5d' % 2	' 2 '
'%.4f' % 2	'2.0000'
'%.*f' % (4, 2)	'2.0000'
'%0*.*f' % (10, 4, 2)	'00002.0000'
'%10.4f' % 2	' 2.0000'
'%010.4f' % 2	'00002.0000'

- *%s* will convert any type argument to string (uses *str*() function)

- *args* may be a single arg or a tuple of args

```
('%s has %03d quote types.' % ('Python', 2)   ==
'Python has 002 quote types.')
```

- Right-handside can also be a *mapping*:

```
a = '%(lang)s has %(c)03d quote types.' % {'c':2,'lang':'Python'}
```

(vars() function very handy to use on right-handside)

Format codes

TABLE D.13: Format Codes

Code	Meaning
d	Signed integer decimal.
i	Signed integer decimal.
o	Unsigned octal.
u	Unsigned decimal.
x	Unsigned hexadecimal (lowercase).
X	Unsigned hexadecimal (uppercase).
e	Floating point exponential format (lowercase).
E	Floating point exponential format (uppercase).
f	Floating point decimal format.
F	Floating point decimal format.
g	Same as "e" if exponent is greater than -4 or less than precision, "f" otherwise.
G	Same as "E" if exponent is greater than -4 or less than precision, "F" otherwise.
c	Single character (accepts integer or single character string).
r	String (converts any python object using repr()).
s	String (converts any python object using str()).
%	No argument is converted, results in a "%" character in the result. (The complete specification is %%.)

Conversion flag characters: #, *0*, - and +. See Table D.14 on page 476.

D.4.10   String Templating

Since 2.4 [PEP 292], the string module provides a new mechanism to substitute variables into *template* strings. Variables to be substituted begin with a $. Actual values are provided in a dictionary via the substitute or safe_substitute methods (texttttsubstitute throws textttKeyError if a key is missing while safe_substitute ignores it) :

TABLE D.14:   Conversion Flag Characters

Flag	Meaning
#	The value conversion will use the "alternate form".
0	The conversion will be zero padded.
-	The converted value is left adjusted (overrides "-").
	(a space) A blank should be left before a positive number (or empty string) produced by a signed conversion.
+	A sign character ("+" or "-") will precede the conversion (overrides a "space" flag).

```
t = string.Template('Hello $name, you won $$$amount')
# (note $$ to literalize $)
t.substitute({'name': 'Eric', 'amount': 100000})
# -> u'Hello Eric, you won $100000'
```

D.4.11   File Objects

(Type `file`). Created with built-in functions `open()` [preferred] or its alias `file()`. May be created by other modules' functions as well.

Unicode file names are now supported for all functions accepting or returning file names (open, os.listdir, etc...).

D.4.12   Operators on File Objects

File operations: See Table D.15 on page 477.

D.4.13   File Exceptions

EOFError End-of-file hit when reading (may be raised many times, e.g. if f is a tty). *IOError* Other I/O-related I/O operation failure

D.4.14   Sets

Since 2.4, Python has two new *built-in* types with fast C implementations [PEP 218]: `set` and *frozenset* (immutable set). Sets are unordered collections of unique (non duplicate) elements. Elements must be hashable. `frozensets` are hashable (thus can be elements of other *sets* while sets are not). All sets are *iterable*.

Since 2.3, the *classes* `Set` and `ImmutableSet` were available in the module sets. This module remains in the 2.4 std library in addition to the built-in types.

Main Set operations: *len, in, issubset, issuperset, add, remove, discard, pop, clear, intersection, union, difference, symmetric_difference, copy, update.*

TABLE D.15: Operators on File Objects

Operation	Result
f.**close**()	Close file *f*.
f.**fileno**()	Get fileno (fd) for file *f*.
f.**flush**()	Flush file *f*'s internal buffer.
f.**isatty**()	1 if file *f* is connected to a tty-like dev, else 0.
f.**next**()	Returns the next input line of file *f*, or raises `StopIteration` when EOF is hit. Files are their own *iterators*. next is implicitly called by constructs like `for line in f:  print line`.
f.**read**([*size*])	Read at most size bytes from file *f* and return as a string object. If *size* omitted, read to EOF.
f.**readline**()	Read one entire line from file *f*. The returned line has a trailing \n, except possibly at EOF. Return " on EOF.
f.**readlines**()	Read until EOF with **readline**() and return a list of lines read.
f.**xreadlines**()	Return a sequence-like object for reading a file line-by-line without reading the entire file into memory. From 2.2, use rather: **for** line **in** f (see below).
for line **in** f: do something...	Iterate over the lines of a file (using readline)
f.**seek**(offset[, whence=0])	Set file *f*'s position, like "stdio's fseek()". whence == 0 then use absolute indexing. whence == 1 then offset relative to current pos. whence == 2 then offset relative to file end.
f.**tell**()	Return file *f*'s current position (byte offset).
f.**truncate**([*size*])	Truncate *f*'s size. If *size* is present, f is truncated to (at most) that size, otherwise f is truncated at current position (which remains unchanged).
f..**write**(*str*)	Write string to file *f*.
f..**writelines**(*list*)	Write list of strings to file *f*. No EOL are added.

See Table D.16 on page 478.

D.4.15 Date/Time

Python **has no** intrinsic Date and Time types, but provides 2 built-in modules:

- `time`: time access and conversions

- `datetime`: classes `date`, `time`, `datetime`, `timedelta`, `tzinfo`.

See also the third-party module: `mxDateTime`.

TABLE D.16: Main Set Operations

Operation	Result
set/frozenset([iterable=None])	[using built-in types] Builds a set or *frozenset* from the given *iterable* (default: empty), e.g. `set([1,2,3])`, `set("hello")`.
Set/ImmutableSet([*iterable*=None])	[using the `sets` module] Builds a Set or ImmutableSet from the given iterable (default: empty), e.g. `Set([1,2,3])`.
len(*s*)	Cardinality of set *s*.
elt **in** *s* / **not in** *s*	True if element *elt* belongs / does not belong to set *s*.
for *elt* in *s: process elt...*	Iterates on elements of set *s*.
s1.**issubset**(*s2*)	True if every element in *s1* is in *s2*.
s1.**issuperset**(*s2*)	True if every element in *s2* is in *s1*.
s.**add**(*elt*)	Adds element *elt* to set *s* (if it doesn't already exist).
s.**remove**(*elt*)	Removes element *elt* from set *s*. KeyError if element not found.
s.**discard**(*elt*)	Removes element *elt* from set *s* if present.
s.**pop**()	Removes and returns an arbitrary element from set *s*; raises KeyError if empty.
s.**clear**()	Removes all elements from this set (not on immutable sets!).
s1.**intersection**(*s2*) or *s1*&*s2*	Returns a new Set with elements common to *s1* and *s2*.
s1.**union**(*s2*) or *s1*|emphs2	Returns a new Set with elements from both *s1* and *s2*.
s1.**difference**(*s2*) or *s1-s2*	Returns a new Set with elements in *s1* but not in *s2*.
s1.**symmetric_difference**(*s2*) or emphs1∧emphs2	Returns a new Set with elements in either *s1* or *s2* but not both.
s.**copy**()	Returns a shallow copy of set *s*.
s.**update**(*iterable*)	Adds all values from iterable to set *s*.

D.5   Advanced Types

See manuals for more details.

- *Module* objects

- *Class* objects

- *Class instance* objects

- *Type* objects (see module: types)

- *File* objects (see above)

- *Ellipsis* object, used by extended slice notation (unique, named `Ellipsis`)

- *Null* object (unique, named `None`)

- *XRange* objects

- **Callable** types:

  - User-defined (written in Python):
    * User-defined *Function* objects
    * User-defined *Method* objects
  - Built-in (written in C):
    * Built-in *Function* objects
    * Built-in *Method* object

- **Internal** Types:

  - *Code* objects (byte-compile executable Python code: *bytecode*)
  - *Frame* objects (execution frames)
  - *Traceback* objects (stack trace of an exception)

D.5.1 Statements

See Table D.17 for Python statements: *pass, del, print, exec, callable.*

D.5.2 Assignment Operators

Assignment operators
Notes:

- (1) Can unpack tuples, lists, and strings:

```
# equivalent to: first=l[0]; second=l[1]
first, second = l[0:2]
# equivalent to: f=0; s=1
[f, s] = range(2)
# equivalent to: c1='a'; c2='b'; c3='c'
c1,c2,c3 = 'abc'
# equivalent to: a='a'; b='b'; c='c'; d='d'; e='e'; f='f'
(a, b), c, (d, e, f) = ['ab', 'c', 'def']
```

TABLE D.17: Statements

Statement	Result
pass	Null statement
del *name*[, *name*]*	Unbind *name*(s) from object. Object will be indirectly (and automatically) deleted only if no longer referenced.
print[>> *fileobject*,] [*s1* [, *s2*]* [,]	Writes to sys.stdout, or to *fileobject* if supplied. Puts spaces between arguments. Puts newline at end unless statement ends with **comma** [if nothing is printed when using a comma, try calling system.out.flush()]. Print is not required when running interactively, simply typing an expression will print its value, unless the value is None.
exec x [in *globals* [, *locals*]]	Executes *x* in namespaces provided. Defaults to current namespaces. x can be a string, open file-like object or a function object. *locals* can be any mapping type, not only a regular Python dict. See also built-in function execfile.
callable(*value*,... [emphid=value] , [*args*], [**kw*])	Call function *callable* with parameters. Parameters can be passed by name or be omitted if function defines default values. E.g. if *callable* is defined as "def callable(p1=1, p2=2)" "callable()" <=> "callable(1, 2)" "callable(10)" <=> "callable(10, 2)" "callable(p2=99)" <=> "callable(1, 99)" *args* is a tuple of **positional** arguments. ***kw* is a dictionary of **keyword** arguments.

Tip: x,y = y,x swaps *x* and *y*.

- (2) Multiple assignment possible:

```
a = b = c = 0
# list1 and list2 points to the same list (l1 is l2)
list1 = list2 = [1, 2, 3]
```

- (3) Not exactly equivalent - a is evaluated only once. Also, where possible, operation performed in-place - a is modified rather than replaced.

TABLE D.18: Assignment Operators

Operator	Result	Notes
a = b	Basic assignment - assign object b to label a	(1)(2)
a += b	Roughly equivalent to a = a + b	(3)
a -= b	Roughly equivalent to a = a - b	(3)
a *= b	Roughly equivalent to a = a * b	(3)
a /= b	Roughly equivalent to a = a / b	(3)
a //= b	Roughly equivalent to a = a // b	(3)
a %= b	Roughly equivalent to a = a % b	(3)
a **= b	Roughly equivalent to a = a ** b	(3)
a &= b	Roughly equivalent to a = a & b	(3)
a —= b	Roughly equivalent to a = a — b	(3)
a ∧= b	Roughly equivalent to a = a ∧ b	(3)
a >>= b	Roughly equivalent to a = a >> b	(3)
a <<= b	Roughly equivalent to a = a << b	(3)

D.5.3 Conditional Expressions

Conditional *Expressions* (not *statements*) have been added since 2.5 [PEP 308]:

```
result = (whenTrue if condition else whenFalse)
```

is equivalent to

```
if condition:
    result = whenTrue
else:
    result = whenFalse
```

() are not mandatory but recommended.

D.5.4 Control Flow Statements

Control flow Statements: *if, elif, else, while, break, continue, return,* and *yield*. See Table D.19 on page 482

D.5.5 Exception Statements

Exception statements: See Table D.20 on page 483.

- An exception is an *instance* of an *exception class* (before 2.0, it may also be a mere *string*).

- Exception classes must be derived from the predefined class: Exception, e.g.:

TABLE D.19: Control Flow Statements

Statement	Result
if *condition: suite* [**elif** *condition: suite*]* [**else:** *suite*]	Usual if/else statement. See also Conditional Expressions.
while *condition: suite* [**else:** *suite*]	Usual while statement. The `else` *suite* is executed after loop exits, unless the loop is exited with `break`.
for *element* **in** *sequence: suite* [**else:** *suite*]	Iterates over *sequence*, assigning each element to *element*. Use built-in `range` function to iterate a number of times. The `else` *suite* is executed at end unless loop exited with `break`.
break	Immediately exits `for` or `while` loop.
continue	Immediately does next iteration of `for` or `while` loop.
return [*result*]	Exits from function (or method) and returns *result* (use a **tuple** to return more than one value). If no result given, then returns `None`.
yield *expression*	(Only used within the body of a generator function, outside a try of a `try..finally`). "Returns" the evaluated *expression*.

TABLE D.20: Exception Statements

Statement	Result
assert *expr*[, *message*]	*expr* is evaluated. if false, raises exception `AssertionError` with message. Before 2.3, inhibited if `__debug__` is 0.

try:
    block1 [**except** [*exception* [, *value*]]:
        handler]+
[**else**:
    else-block]
	Statements in *block1* are executed. If an exception occurs, look in `except` clause(s) for matching *exception*(s). If matches or bare `except`, execute *handler* of that clause. If no exception happens, *else-block* in `else` clause is executed after *block1*. If *exception* has a value, it is put in variable *value*. exception can also be a **tuple** of exceptions, e.g. `except(KeyError, NameError), e: print e`.

try:
    block1
finally:
    final-block
	Statements in *block1* are executed. If no exception, execute *final-block* (even if *block1* is exited with a **return**, **break** or **continue** statement). If exception did occur, execute *final-block* and then immediately re-raise exception. Typically used to ensure that a resource (file, lock...) allocated before the **try** is freed (in the **final-block**) whatever the outcome of **block1** execution. See also the with statement below.

try:
    block1
[**except** [*exception* [, *value*]]:
        handler1]+
[**else**:
    else-block]
finally:
    final-block
	Unified try/except/finally. Equivalent to a **try**...**except** nested inside a *try..finally* [PEP341]. See also the with statement below.

with *allocate-expression* [**as** *variable*] *with-block*	Alternative to the `try...finally` structure [PEP343]. *allocate-expression* should evaluate to an object that supports the *context management protocol*, representing a resource. This object may return a value that can optionally be bound to *variable* (variable is **not** assigned the result of expression). The object can then run **set-up** code before `with-block` is executed and some **clean-up** code is executed after the block is done, even if the block raised an exception. Standard Python objects such as files and locks support the context management protocol:
	
	 ```# file automatically closed on block exit```
	 ```with open('/etc/passwd', 'r') as f:```
	 ``` for line in f:```
	 ``` print line```
	
	 ```# lock automatically released on block exit```
	 ```with threading.Lock():```
	 ``` do something...```
	
	 - You can write your own context managers. - Helper functions are available in module contextlib. In 2.5, the statement must be enabled by: `from __future__ import with_statement`. The statement will always be enabled in Python 2.6.

raise *exceptionInstance*	Raises an instance of a class derived from `Exception` (**preferred** form of raise).
raise *exceptionClass* [, *value* [, *traceback*]]	Raises *exception* of given class *exceptionClass* with optional value *value*. Arg *traceback* specifies a traceback object to use when printing the exception's backtrace.
raise	A raise statement without arguments re-raises the last exception raised in the current function.

```
class TextException(Exception): pass
try:
    if bad:
        raise TextException()
except Exception:
    # This will be printed because TextException is a
    # subclass of Exception
    print 'Oops'
```

- When an error message is printed for an unhandled exception, the class name is printed, then a colon and a space, and finally the instance converted to a string using the built-in function str().

- All built-in exception classes derive from StandardError, itself derived from Exception.

PEP 352 : Exceptions can now be **new-style classes**, and all built-in ones are. Built-in exception hierarchy slightly reorganized with the introduction of base class BaseException. Raising strings as exceptions is now deprecated (warning).

D.5.6 Name Space Statements

Imported module files must be located in a directory listed in the Python path (sys.path). Since 2.3, they may reside in a **zip** file [e.g. sys.path.insert(0, "aZipFile.zip")]. **Absolute/relative** imports (since 2.5 [PEP328]):

- Feature must be enabled by: from __future__ import absolute_import: will probably be adopted in 2.7.

- Imports are normally *relative*: modules are searched first in the current directory/package, and then in the built-in modules, resulting in possible ambiguities (e.g. masking a built-in symbol).

- When the new feature is enabled:

  – import X will look up for module X in sys.path first (absolute import).

  – import .X (with a dot) will still search for X in the current package first, then in builtins (relative import).

  – import ..X will search for X in the package containing the current one, etc...

Packages (>1.5): a **package** is a name space which maps to a directory including module(s) and the special initialization module __init__.py (possibly empty). Packages/directories can be nested. You address a module's symbol via [package.[package...].module.symbol. [1.51: On Mac and Windows, the case of module file names must now match the case as used in the *import* statement].

Name space statements: See Table D.5.6 on page 487.

D.5.7   Function Definition

```
def funcName ([paramList]):
    suite
```

Creates a function object and binds it to name funcName.

```
paramList ::= [param [, param]*]
param ::= value | id=value | *id | **id
```

- Args are passed by **value**, so only args representing a *mutable* object can be modified (are *inout* parameters).

- Use `return` to return (`None`) from the function, or `return` *value* to return *value*. Use a **tuple** to return more than one value, e.g. `return 1,2,3`

- *Keyword* arguments `arg=value` specify a *default value* (evaluated at function def. time). They can only appear last in the param list, e.g. `foo(x, y=1, s='')`

- Pseudo-arg *emphargs captures a tuple of all remaining non-keyword args passed to the function, e.g. `if def foo(x, *args):   ...` is called `foo(1, 2, 3)`, then args will contain `(2,3)`.

- Pseudo-arg ***kwargs* captures a dictionary of all extra keyword arguments, e.g. `if def foo(x, **kwargs):   ...` is called `foo(1, y=2, z=3)`, then *kwargs* will contain `'y':2, 'z':3`. `if def foo(x, *args, **kwargs):   ...` is called `foo(1, 2, 3, y=4, z=5)`, then *args* will contain `(2, 3)`, and *kwargs* will contain `'y':4, 'z':5`

- *args* and `kwargs` are conventional names, but other names may be used as well.

- **args* and ***kwargs* can be "forwarded" (individually or together) to another function, e.g,

```
def f1(x, *args, **kwargs):
    f2(*args, **kwargs)
```

- See also Anonymous functions (lambdas).

Statement	Result
import *module1* [**as** *name1*] [, *module2*]*	Imports modules. Members of module must be referred to by qualifying with [package.]module name, e.g.:
| | ```
import sys; print sys.argv
import package1.subpackage.module
package1.subpackage.module.foo()
``` |
| | *module1* renamed as *name1*, if supplied. |
| from module import name1 [as othername1][, name2]* | Imports names from module *module* in current namespace. |
| | ```
from sys import argv; print argv
from package1 import module; module.foo()
from package1.module import foo; foo()
``` |
| | *name1* renamed as *othername1*, if supplied. [2.4] You can now put parentheses around the list of names in a `from module import names` statement (PEP 328). |
| **from** module **import** * | Imports **all** names in module, except those starting with "_". **Use sparsely, beware of name clashes!** |
| | ```
from sys import *; print argv
from package.module import *; print x
``` |
| | Only legal at the top level of a module. If *module* defines an __all__ attribute, only names listed in __all__ will be imported. NB: "from **package** import *" only imports the symbols defined in the package's __init__.py file, not those in the package's modules ! |
| **global** *name1* [, *name2*] | Names are from global scope (usually meaning from module) rather than local (usually meaning only in function). E.g. in function without global statements, assuming "x" is name that hasn't been used in function or module so far: - Try to read from "x" -¿ NameError - Try to write to "x" -¿ creates "x" local to function If "x" not defined in fct, but is in module, then: - Try to read from "x", gets value from module - Try to write to "x", creates "x" local to fct But note "x[0]=3" starts with search for "x", will use to global "x" if no local "x". |
```

D.5.8   Class Definition

```
class className [(super_class1[, super_class2]*)]:
    suite
```

Creates a class object and assigns it name *className*. *suite* may contain local "defs" of class methods and assignments to class attributes.
Examples:

```
class MyClass (class1, class2): ...
```

Creates a class object inheriting from both class1 and class2. Assigns new class object to name MyClass.

```
class MyClass: ...
```

Creates a *base* class object (inheriting from nothing). Assigns new class object to name MyClass. Since 2.5, the equivalent syntax **class** MyClass(): ... is allowed.

```
class MyClass (object): ...
```

Creates a *new-style* class (inheriting from object makes a class a *new-style* class -available since Python 2.2-). Assigns new class object to name MyClass.

- First arg to class instance methods (operations) is always the target instance object, called **'self'** by convention.

- Special static method __**new**__(*cls*[,...]) called when instance is created. 1st arg is a class, others are args to __init__(), more details here

- Special method __**init**__() is called when instance is created.

- Special method __**del**__() called when no more reference to object.

- Create instance by *"calling"* class object, possibly with arg (thus instance=apply(aClassObject, args...) creates an instance!)

- Before 2.2, it was not possible to subclass built-in classes like list, dict (you had to "wrap" them, using UserDict & UserList modules); since 2.2, you can subclass them directly (see Types/Classes Unification).

Example:

```
class c (c_parent):
    def __init__(self, name):
        self.name = name
    def print_name(self):
        print "I'm", self.name
    def call_parent(self):
        c_parent.print_name(self)
```

```
instance = c('tom')
print instance.name
'tom'
instance.print_name()
"I'm tom"
```

Call parent's super class by accessing parent's method directly and passing `self` explicitly (see `call_parent` in example above). Many other special methods are available for implementing arithmetic operators, sequence, mapping, indexing, etc.

Types / classes unification **Base types** *int, float, str, list, tuple, dict,* and *file* now (2.2) behave like **classes** derived from base class `object`, and may be **subclassed**:

```
# built-in cast function now a constructor for base type
x = int(2)
# (literals are instances of new base types)
y = 3 # <=> int(3)
# int, int
print type(x), type(y)
# replaces isinstance(x, types.IntType)
assert isinstance(x, int)
# base types derive from base class 'object'.
assert issubclass(int, object)
s = "hello" # <=> str("hello")
assert isinstance(s, str)
f = 2.3  # <=> float(2.3)
class MyInt(int): pass    # may subclass base types
x,y = MyInt(1), MyInt("2")

print x, y, x+y # => 1,2,3

class MyList(list): pass

l = MyList("hello")

print l # ['h', 'e', 'l', 'l', 'o']
```

New-style classes extends `object`. *Old-style* classes don't.

Documentation Strings

Modules, classes and functions may be documented by placing a string literal by itself as the first statement in the suite. The documentation can be retrieved by getting the '**__doc__**' attribute from the module, class or function.

Example:

```
class C:
"A description of C"
    def __init__(self):
        "A description of the constructor"
        # etc.

c.__doc__ == "A description of C".
c.__init__.__doc__ == "A description of the constructor"
```

D.5.9   Iterators

- An *iterator* enumerates elements of a *collection*. It is an object with a single method next() returning the next element or raising StopIteration.

- You get an iterator on *obj* via the new built-in function iter*(obj)*, which calls *obj.*__class__.__iter__().

- A collection may be its **own** iterator by implementing both __iter__() and next().

- Built-in collections (lists, tuples, strings, dict) implement __iter__(); dictionaries (maps) enumerate their keys; files enumerates their lines.

- You can build a list or a tuple from an iterator, e.g. list(anIterator)

- Python implicitly uses iterators wherever it has to **loop** :

    - for elt in collection:
    - if elt in collection: when assigning tuples: x,y,z= collection

D.5.10   Generators

- A *generator* is a function that retains its state between 2 calls and produces a **new** value at *each* invocation. The values are returned (one at a time) using the keyword yield, while return or raise StopIteration() are used to notify the end of values.

- A typical use is the production of IDs, names, or serial numbers. Fancier applications like nanothreads are also possible.

- In 2.2, the feature needs to be **enabled** by the statement: from __future__ import generators (not required since 2.3+)

- To **use** a generator: call the *generator function* to get a generator object, then call generator.next() to get the next value until StopIteration is raised.

- 2.4 introduces *generator expressions* [PEP 289] similar to list compre-hensions, except that they create a generator that will return elements one by one, which is suitable for long sequences :

```
linkGenerator = (link for link in get_all_links() if not<=
link.followed)
    for link in linkGenerator:
        ...process link...
```

Generator expressions must appear between **parentheses**.

PEP342 Generators before 2.5 could only produce **output**. Now values can be **passed** to generators via their method send(value). yield is now an *expression* returning a value, so val = (yield i) will *yield* i to the caller, and will reciprocally evaluate to the value "sent" back by the caller, or None. Two other new generator methods allow for additional control:

- throw(type, value=None, traceback=None) is used to raise an exception inside the generator (appears as raised by the yield expression).

- close() raises a new GeneratorExit exception inside the genera-tor to terminate the iteration.

Example:

```
def genID(initialValue=0):
  v = initialValue
  while v < initialValue + 1000:
    yield "ID_%05d" % v
    v += 1
  return    # or: raise StopIteration()

generator = genID() # Create a generator
for i in range(10): # Generates 10 values
  print generator.next()
```

D.5.11   Descriptors/Attribute Access

- *Descriptors* are objects implementing at least the first of these 3 meth-ods representing the *descriptor protocol*:

  - __get__(self, obj, type=None) --> value

  - __set__(self, obj, value)

  - __delete__(self, obj)

Python now transparently uses *descriptors* to describe and access the attributes and methods of new-style classes (i.e. derived from `object`). [more info])

- Built-in descriptors now allow you to define:

  - **Static methods** : Use `staticmethod(f)` to make method `f(x)` static (unbound).

  - **Class methods**: like a static but takes the Class as 1st argument => Use `f = classmethod(f)` to make method `f(theClass, x)` a class method.

  - **Properties** : A property is an instance of the new built-in type `property`, which implements the descriptor protocol for attributes => Use *propertyName* = property(*fget*=None, *fset*=None, *fdel*=None, *doc*=None) to define a property inside or outside a class. Then access it as *propertyName* or *obj.propertyName*

  - **Slots**. New style classes can define a class attribute `__slots__` to constrain the list of **assignable** attribute names, to avoid typos (which is normally not detected by Python and leads to the creation of new attributes), e.g. `__slots__ = ('x', 'y')` <u>Note:</u> According to recent discussions, the real purpose of slots seems still unclear (optimization?), and their use should probably be discouraged.

D.5.12   Decorators for Functions and Methods

PEP 318 A *decorator* D is noted @D on the line preceding the function/method it decorates :

```
@D
def f(): ...
```

and is equivalent to:

```
def f(): ...
f = D(f)
```

- Several decorators can be applied in cascade :

```
@A
@B
@C
def f(): ...
```

is equivalent to:

```
f = A(B(C(f)))
```

- A decorator is just a function taking the fct to be decorated and returns the same function or some new callable thing.

- Decorator functions can take arguments:

```
@A
  @B
  @C(args)
```

becomes :

```
def f(): ...
  _deco = C(args)
  f = A(B(_deco(f)))
```

- The decorators @staticmethod and @classmethod replace more elegantly the equivalent declarations f = staticmethod(f) and f = classmethod(f).

D.5.13   Miscellaneous

```
lambda [param_list]: returnedExpr
```

Creates an **anonymous** function. *returnedExpr* must be an expression, not a statement (e.g., not "if xx:...", "print xxx", etc.) and thus can't contain newlines. Used mostly for filter(), map(), reduce() functions, and GUI callbacks.
List comprehensions

```
result = [expression for item1 in sequence1  [if condition1]
              [for item2 in sequence2 ... for itemN in sequenceN]
          ]
```

is equivalent to:

```
result = []
for item1 in sequence1:
    for item2 in sequence2:
        ...
        for itemN in sequenceN:
            if (condition1) and further conditions:
                result.append(expression)
```

Nested scopes Since 2.2, *nested scopes* no longer need to be specially enabled by a from __future__ import nested_scopes directive, and are always used.

D.6   Built-in Functions

Built-in functions are defined in a module __builtin__ automatically imported.

Built-in Functions

__import__(*name*[, *globals*[,*locals*[,*from list*]]]): Imports module within the given context (see library reference for more details)

abs(x): Returns the absolute value of the number x.

all(*iterable*): Returns True if bool(x) is True for **all** values x in the iterable.

any(*iterable*): Returns True if bool(x) is True for **any** values x in the iterable.

apply(*f, args*[, *keywords*]): Calls func/method f with arguments args and optional keywords. deprecated since 2.3, replace apply(func, args, keywords) with func(*args, **keywords) [details].

basestring(): Abstract superclass of str and unicode; can't be called or instantiated directly, but useful in: isinstance(obj, basestring).

bool([x]): Converts a value to a Boolean, using the standard truth testing procedure. If x is false or omitted, returns False; otherwise returns True. bool is also a class/type, subclass of int. Class bool cannot be subclassed further. Its only instances are False and True. See also boolean operators.

buffer(*object*[, *offset*[, *size*]]): Returns a Buffer from a slice of *object*, which must support the buffer call interface (string, array, buffer). *Non essential function*, see [details].

callable(x): Returns True if x callable, else False.

chr(i): Returns one-character string whose ASCII code is integer i.

classmethod(*function*): Returns a class method for *function*. A class method receives the class as implicit first argument, just like an instance method receives the instance. To declare a class method, use this idiom:

```
class C:
    def f(cls, arg1, arg2, ...): ...
    f = classmethod(f)
```

Then call it on the class C.f() or on an instance C().f(). The instance is ignored except for its class. If a class method is called for a derived class, the derived class object is passed as the implied first argument. Since 2.4 you can alternatively use the decorator notation:

```
class C:
    @classmethod
    def f(cls, arg1, arg2, ...): ...
```

cmp(*x,y*): Returns negative, 0, positive if $x <, ==, >$ to y respectively.

coerce(*x,y*): Returns a tuple of the two *numeric* arguments converted to a common type. *Non essential function*, see [details].

compile(*string, filename, kind*[*, flags*[*, dont_inherit*]]): Compiles *string* into a code object. *filename* is used in error message, can be any string. It is usually the file from which the code was read, or e.g. ' <string> ' if not read from file. *kind* can be 'eval' if *string* is a single stmt, or 'single' which prints the output of expression statements that evaluate to something else than None, or be 'exec'. New args *flags* and *dont_inherit* concern *future* statements.

complex(*real*[*, image*]): Creates a `complex` object (can also be done using **J** or **j** suffix, e.g. 1+3J).

delattr(*obj, name*): Deletes the attribute named *name* of object *obj* <=> del obj.name

dict([*mapping-or-sequence*]): Returns a new dictionary initialized from the optional argument (or an empty dictionary if no argument). Argument may be a sequence (or anything iterable) of pairs (key,value).

dir([*object*]): Without args, returns the list of names in the current local symbol table. With a module, class or class instance object as *arg*, returns the list of names in its attr. dictionary.

divmod(*a,b*): Returns tuple (*a//b, a%b*)

enumerate(*iterable*): Iterator returning pairs (index, value) of *iterable*, e.g. List(enumerate('Py')) -> [(0, 'P'), (1, 'y')].

eval(*s*[*, globals*[*, locals*]]): Evaluates string *s*, representing a single python *expression*, in (optional) *globals, locals* contexts. *s* must have no NUL's or new lines. s can also be a code object. *locals* can be any mapping type, not only a regular Python dict.

Example:

```
x = 1; assert eval('x + 1') == 2
```

(To execute *statements* rather than a single expression, use Python statement exec or built-in function `execfile`).

execfile(*file*[*, globals*[*,locals*]]): Executes a file without creating a new module, unlike `import`. *locals* can be any mapping type, not only a regular Python dict.

file(*filename*[*,mode*[*,bufsize*]]): Opens a file and returns a new `file` object. Alias for open.

filter(*function,sequence*): Constructs a list from those elements of *sequence* for which *function* returns true. *function* takes one parameter.

float(*x*): Converts a number or a string to floating point.

frozenset([*iterable*]) Returns a frozenset (immutable set) object whose (immutable) elements are taken from iterable, or empty by default. See also Sets.

getattr(*object,name*[*,default*])): Gets attribute called *name* from *object*, e.g. getattr(x, 'f') <=> x.f). If not found, raises `AttributeError` or returns default if specified.

globals(): Returns a dictionary containing the current global variables.

hasattr(*object, name*): Returns true if *object* has an attribute called *name*.

hash(*object*): Returns the hash value of the object (if it has one).

help([*object*]): Invokes the built-in help system. No argument − > interactive help; if *object* is a string (*name* of a module, function, class, method, keyword, or documentation topic), a help page is printed on the console; otherwise a help page on *object* is generated.

hex(*x*): Converts a number *x* to a hexadecimal string.

id(*object*): Returns a unique integer identifier for *object*. Since 2.5 always returns non-negative numbers.

input([*prompt*]): Prints *prompt* if given. Reads input and **evaluates** it. Uses line editing / history if module readline available.

int(*x*[, *base*]): Converts a number or a string to a plain integer. Optional base parameter specifies base from which to convert string values.

intern(*aString*): Enters *aString* in the table of interned strings and returns the string. Since 2.3, interned strings are no longer 'immortal' (never garbage collected), see [details].

isinstance(*obj, classInfo*): Returns true if *obj* is an instance of **class** *classInfo* or an object of **type** *classInfo* (*classInfo* may also be a **tuple** of classes or types). If `issubclass(A,B)` then `isinstance(x,A) => isinstance(x,B)`

issubclass(*class1, class2*): Returns true if *class1* is derived from *class2* (or if *class1* is *class2*).

iter(*obj*[,*sentinel*]): Returns an **iterator** on *obj*. If *sentinel* is absent, *obj* must be a collection implementing either `__iter__()` or `__getitem__()`. If *sentinel* is given, *obj* will be **called** with no arg; if the value returned is equal to *sentinel*, `StopIteration` will be raised, otherwise the value will be returned. See Iterators.

len(*obj*): Returns the length (the number of items) of an object (sequence, dictionary, or instance of class implementing `__len__`).

list([*seq*]): Creates an empty list or a list with same elements as *seq*. *seq* may be a sequence, a container that supports iteration, or an iterator object. If *seq* is already a list, returns a **copy** of it.

locals(): Returns a dictionary containing current local variables.

long(*x*[, *base*]): Converts a number or a string to a long integer. Optional base parameter specifies the base from which to convert string values.

map(*function, sequence*[, *sequence, ...*]): Returns a list of the results of applying *function* to each item from *sequence*(s). If more than one sequence is given, the *function* is called with an argument list consisting of the corresponding item of each sequence, substituting `None` for missing values when not all sequences have the same length. If *function* is `None`, returns a list of the items of the sequence (or a list of tuples if more than one sequence). => You might also **consider using** list comprehensions **instead of map**().

max(*iterable*[, *key*=func]), **max**(*v1, v2, ...*[, *key*=func]): With a single argument *iterable*, returns the **largest** item of a non-empty iterable (such as a string, tuple or list). With more than one argument, returns the largest of the arguments. The optional *key* arg is a function that takes a single argument and is called for every value in the list.

min(*iterable*[, *key*=func]), **min**(*v1, v2, ...*[, *key*=func]): With a single argument *iterable*, returns the **smallest** item of a non-empty iterable (such as a string, tuple or list). With more than one argument, returns the largest of the arguments. The optional *key* arg is a function that takes a single argument and is called for every value in the list.

object(): Returns a new featureless object. object is the base class for all *new style classes*, its methods are common to all instances of new style classes.

oct(*x*): Converts a number to an octal string.

open(*filename* [, *mode*='r', [*bufsize*]]): Returns a new file object. See also alias file(). Use codecs.open() instead to open an **encoded** file and provide transparent encoding / decoding.

- *filename* is the file name to be opened

- *mode* indicates how the file is to be opened:

  - 'r' for reading
  - 'w' for writing (truncating an existing file)
  - 'a' opens it for appending
  - '+' (appended to any of the previous modes) open the file for updating (note that 'w+'truncates the file)
  - 'b' (appended to any of the previous modes) open the file in binary mode
  - 'U' (or 'rU') open the file for reading in *Universal Newline* mode: all variants of EOL (CR, LF, CR+LF) will be translated to a single LF ('\n').

- *bufsize* is 0 for unbuffered, 1 for line buffered, negative or omitted for system default, >1 for a buffer of (about) the given size.

ord(*c*): Returns integer ASCII value of *c* (a string of len 1). Works with Unicode char.

pow(*x, y* [, *z*]): Returns *x* to power *y* [modulo *z*]. See also ** operator.

property([*fget*[, *fset*[, *fdel*[, *doc*]]]]): Returns a property attribute for *new-style* classes (classes deriving from object). *fget, fset,* and *fdel* are functions to get the property value, set the property value, and delete the property, respectively. Typical use:

```
def __init__(self): self.__x = None
def getx(self): return self.__x
def setx(self, value): self.__x = value
def delx(self): del self.__x
x = property(getx, setx, delx, "I'm the 'x' property.")
```

range([*start*,] end [, *step*]): Returns list of ints from >= start and < end. With 1 arg, list from 0..*arg*-1 With 2 args, list from *start*..*end*-1 With 3 args, list from *start* up to *end* by *step*

raw_input([*prompt*]): Prints *prompt* if given, then reads string from std input (no trailing \n). See also input().

reduce(*f, list* [, *init*]): Applies the binary function *f* to the items of *list* so as to reduce the list to a single value. If *init* is given, it is "prepended" to *list*.

reload(*module*): Reparses and reinitializes an already imported module. Useful in interactive mode, if you want to reload a module after fixing it. If module was syntactically correct but had an error in initialization, must import it one more time before calling reload().

repr(*object*): Returns a string containing a printable and if possible **evaluable** representation of an object. <=> 'object' (using backquotes). Class redefinable (__repr__). See also str().

round(*x, n*=0): Returns the floating point value x rounded to *n* digits after the decimal point.

set([*iterable*]): Returns a **set** object whose elements are taken from *iterable*, or empty by default. See also Sets.

setattr(*object, name, value*): This is the counterpart of getattr().setattr(o, 'foobar', 3) <=> o.foobar = 3. **Creates** attribute if it doesn't exist!

slice([*start*,] *stop*[, *step*]): Returns a *slice* object representing a range, with R/O attributes: start, stop, step.

sorted(*iterable*[, *cmp*[, emphkey[, *reverse*]]]): Returns a **new** sorted list from the items in *iterable*. This contrasts with list.sort() that sorts lists **in place** and doesn't apply to immutable sequences like strings or tuples. See sequences.sort method.

staticmethod(*function*): Returns a static method for *function*. A static method does not receive an implicit first argument. To declare a static method, use this idiom:

```
:
class C:
    def f(arg1, arg2, ...): ...
    f = staticmethod(f)
```

Then call it on the class C.f() or on an instance C().f(). The instance is ignored except for its class. Since 2.4 you can alternatively use the decorator notation:

```
:
class C:
    @staticmethod
    def f(arg1, arg2, ...): ...
```

str(*object*): Returns a string containing a nicely printable representation of an object. Class overridable (__str__). See also repr().

sum(*iterable*[, *start=0*]): Returns the sum of a sequence of numbers (**not** strings), plus the value of parameter. Returns *start* when the sequence is empty.

super(*type*[, *object-or-type*]): Returns the superclass of *type*. If the second argument is omitted the super object returned is unbound. If the second argument is an object, isinstance(obj, type) must be true. If the second argument is a type, issubclass(type2, type) must be true. Typical use:

```
:
class C(B):
  def meth(self, arg):
    super(C, self).meth(arg)
```

tuple([*seq*]): Creates an empty tuple or a tuple with same elements as *seq*. *seq* may be a sequence, a container that supports iteration, or an iterator object. If *seq* is already a tuple, returns **itself** (not a copy).

type(*obj*): Returns a *type object* [see module *types*] representing the type of *obj*. Example:

```
import types if type(x) == types.StringType: print 'It is a string'.
```
NB: it is better to use instead:
```
if isinstance(x, types.StringType)....
```

unichr(*code*): Returns a unicode string 1 char long with given *code*.

unicode(*string*[, *encoding*[,*error*]]]): Creates a Unicode string from an 8-bit string, using the given encoding name and error treatment ('strict', 'ignore', or 'replace'). For objects which provide a __unicode__() method, it will call this method without arguments to create a Unicode string.

vars([*object*]): Without arguments, returns a dictionary corresponding to the current local symbol table. With a module, class or class instance object as argument, returns a dictionary corresponding to the object's symbol table. Useful with the "%" string formatting operator.

xrange(*start* [, *end* [, *step*]]): Like range(), but doesn't actually store entire list all at once. Good to use in "for" loops when there is a big range and little memory.

zip(*seq1*[, emphseq2,...]): [No, that's not a compression tool! For that, see module zipfile] Returns a list of tuples where each tuple contains the *n*th element of each of the argument sequences. Since 2.4, returns an empty list if called with no arguments (was raising TypeError before).

D.7 Built-in Exception Classes

BaseException Mother of all exceptions (was Exception before 2.5). New-style class. *exception.args* is a tuple of the arguments passed to the constructor.

KeyboardInterrupt & SystemExit were moved out of Exception because they don't really represent errors, so now a try: ... except Exception: will only catch **errors** while a try: ... except BaseException: (or simply try: ... except:) will still catch **everything**.

- **KeyboardInterrupt** On user entry of the interrupt key (often 'CTRL-C'). Before 2.5 was derived from Exception.

- **SystemExit** On sys.exit(). Before 2.5 was derived from Exception.

- Exception Base of all errors. Before 2.5 was the base of all exceptions.

  - **GeneratorExit** Raised by the close() method of generators to terminate the iteration.

  - **StandardError** Base class for all built-in exceptions; derived from Exception root class.

    * **ArithmeticError** Base class for arithmetic errors.
      · **FloatingPointError** When a floating point operation fails.
      · **OverflowError** On excessively large arithmetic operation.
      · **ZeroDivisionError** On division or modulo operation with 0 as 2nd argument.
    * **AssertionError** When an assert statement fails.
    * textbfAttributeError On attribute reference or assignment failure
    * textbfEnvironmentError [new in 1.5.2] On error outside Python; error arg. tuple is (errno, errMsg...)
      · IOError [changed in 1.5.2] I/O-related operation failure.
      · OSError [new in 1.5.2] Used by the os module's os.error exception.
      · **WindowsError** When a Windows-specific error occurs or when the error number does not correspond to an errno value.
    * **EOFError** Immediate end-of-file hit by input() or raw_input()
    * **ImportError** On failure of import to find module or name.
    * **KeyboardInterrupt** Moved under BaseException.
    * **LookupError** base class for IndexError, KeyError
      · **IndexError** On out-of-range sequence subscript
      · **KeyError** On reference to a nonexistent mapping (dict) key
    * **MemoryError** On recoverable memory exhaustion
    * **NameError** On failure to find a local or global (unqualified) name.

- · **UnboundLocalError** On reference to an unassigned local variable.
- ∗ **ReferenceError** On attempt to access to a garbage-collected object via a weak reference proxy. item **RuntimeError** Obsolete catch-all; define a suitable error instead.
  - · **NotImplementedError** [new in 1.5.2] On method not implemented.
- ∗ **SyntaxError** On parser encountering a syntax error
  - · **IndentationError** On parser encountering an indentation syntax error
  - · **TabError** On improper mixture of spaces and tabs
- ∗ **SystemError** On nonfatal interpreter error - bug - report it !
- ∗ **TypeError** On passing inappropriate type to built-in operator or function.
- ∗ **ValueError** On argument error not covered by TypeError or more precise.
  - · **UnicodeError** On Unicode-related encoding or decoding error.
  - · **UnicodeDecodeError** On Unicode decoding error.
  - · **UnicodeEncodeError** On Unicode encoding error.
  - · **UnicodeTranslateError** On Unicode translation error.
- − **StopIteration** Raised by an iterator's next() method to signal that there are no further values.
- − **SystemExit** Moved under BaseException.
- − **Warning** Base class for warnings (see module warning)
  - ∗ **DeprecationWarning** Warning about deprecated code.
  - ∗ **FutureWarning** Warning about a construct that will change semantically in the future.
  - ∗ **ImportWarning** Warning about probable mistake in module import (e.g. missing __init__.py).
  - ∗ **OverflowWarning** Warning about numeric overflow. Won't exist in Python 2.5.
  - ∗ **PendingDeprecationWarning** Warning about future deprecated code.
  - ∗ **RuntimeWarning** Warning about dubious runtime behavior.
  - ∗ **SyntaxWarning** Warning about dubious syntax.
  - ∗ **UnicodeWarning** When attempting to compare a Unicode string and an 8-bit string that can't be converted to Unicode using default ASCII encoding (raised a UnicodeDecodeError before 2.5).
  - ∗ **UserWarning** Warning generated by user code.

D.8   Standard Methods and Operators Redefinition in Classes

Standard methods and operators map to special methods '__method__' and thus can be **redefined** (mostly in user-defined classes), e.g.,

```
class C:
    def __init__(self, v): self.value = v
    def __add__(self, r): return self.value + r

a = C(3) # sort of like calling C.__init__(a, 3)
a + 4    # is equivalent to a.__add__(4)
```

Special methods for any class

TABLE D.21: Special Methods for Any Class

Method	Description
__new__(*cls*[, ...])	Instance creation (on construction). If __new__ returns an instance of *cls* then __init__ is called with the rest of the arguments (...), otherwise __init__ is not invoked. More details here.
__init__(*self, args*)	Instance initialization (on construction)
__del__(*self*)	Called on object demise (refcount becomes 0)
__repr__(*self*)	repr() and '...' conversions
__str__(*self*)	str() and print statement
__cmp__(*self,other*)	Compares self to *other* and returns <0, 0, or >0. Implements >, <, == etc...
__index__(*self*)	[PEP357] Allows using any object as integer indice (e.g. for slicing). Must return a single integer or long integer value.
__lt__(*self, other*)	Called for *self* < *other* comparisons. Can return anything, or can raise an exception.
__le__(*self, other*)	Called for *self* <= *other* comparisons. Can return anything, or can raise an exception.
__gt__(*self, other*)	Called for *self* > *other* comparisons. Can return anything, or can raise an exception.
__ge__(*self, other*)	Called for *self* >= *other* comparisons. Can return anything, or can raise an exception.
__eq__(*self, other*)	Called for *self* == *other* comparisons. Can return anything, or can raise an exception.
__ne__(*self, other*)	Called for *self* != *other* (and *self* ¡¿ *other*) comparisons. Can return anything, or can raise an exception.
__hash__(*self*)	Compute a 32 bit hash code; hash() and dictionary ops. Since 2.5 can also return a **long** integer, in which case the hash of that value will be taken.
__nonzero__(*self*)	Returns 0 or 1 for truth value testing. when this method is not defined, __len__() is called if defined; otherwise all class instances are considered "true".
__getattr__(*self,name*)	Called when attribute lookup doesn't find *name*. See also __getattribute__.
__getattribute__(*self, name*)	Same as __getattr__ but **always** called whenever the attribute *name* is accessed.
__setattr__(*self, name, value*)	Called when setting an attribute (inside, don't use "*self.name = value*", use instead "*self.__dict__[name] = value*")
__delattr__(*self, name*)	Called to delete attribute <*name*>.
__call__(*self, *args, **kwargs*)	Called when an instance is called as function: obj(arg1, arg2, ...) is a shorthand for obj.__call__(arg1, arg2, ...).

D.8.1   Operators

See list in the `operator` module. Operator function names are provided with **2 variants**, with or without leading and trailing '_' (e.g. __add__ or add).

Numeric Operations Special Methods

TABLE D.22:   Numeric Operations Special Methods

Operator	Special method
self + other	__**add**__(*self, other*)
self - other	__**sub**__(*self, other*)
*self * other*	__**mul**__(*self, other*)
self / other	__**div**__(*self, other*) or __**truediv**__(*self,other*) if __future__.division is active.
self // other	__**floordiv**__(*self, other*)
self % other	__**mod**__(*self, other*)
divmod(self,other)	__**divmod**__(*self, other*)
*self ** other*	__**pow**__(*self, other*)
self & other	__**and**__(*self, other*)
self ∧ other	__**xor**__(*self, other*)
self — other	__**or**__(*self, other*)
self << other	__**lshift**__(*self, other*)
self >> other	__**rshift**__(*self, other*)
bool(self)	__**nonzero**__(*self*) (used in boolean testing)
~self	__**neg**__(*self*)
+self	__**pos**__(*self*)
abs(self)	__**abs**__(*self*)
self	__**invert**__(*self*) (bitwise)
self += other	__**iadd**__(*self, other*)
self -= other	__**isub**__(*self, other*)
*self *= other*	__**imul**__(*self, other*)
self /= other	__**idiv**__(*self, other*) or __**itruediv**__(*self,other*) if __future__.division is in effect.
self //= other	__**ifloordiv**__(*self, other*)
self %= other	__**imod**__(*self, other*)
*self **= other*	__**ipow**__(*self, other*)
self &= other	__**iand**__(*self, other*)
self ∧= other	__**ixor**__(*self, other*)
self —= other	__**ior**__(*self, other*)
self <<= other	__**ilshift**__(*self, other*)
self >>= other	__**irshift**__(*self, other*)

TABLE D.23: Conversions

built-in function	Special method
int(*self*)	**__int__**(*self*)
long(*self*)	**__long__**(*self*)
float(*self*)	**__float__**(*self*)
complex(*self*)	**__complex__**(*self*)
oct(*self*)	**__oct__**(*self*)
hex(*self*)	**__hex__**(*self*)
coerce(*self, other*)	**__coerce__**(*self, other*)

Conversions

Right-hand-side equivalents for all **binary** operators exist (__radd__, __rsub__, __rmul__, __rdiv__, ...). They are called when class instance is on r-h-s of operator:

- a + 3 calls __add__(a, 3)

- 3 + a calls __radd__(a, 3)

Special operations for *containers*

D.9 Special Informative State Attributes for Some Types

Tip: Use Module **inspect** to Inspect Live Objects.

Modules: See page table D.25 on 507.
Classes: See page table D.26 on 507.
Instances: See page table D.27 on 507.
User defined functions: See page table D.28 on 507.
User-defined Methods: See page table D.29 on 508.
Built-in Functions and methods: See page table D.30 on 508.
Codes: See page table D.31 on 508.
Frames: See page table D.32 on 509.

D.10 Important Modules

D.10.1 sys

System-specific parameters and functions. [Full doc]

TABLE D.24: Special Operations for Containers

Operation	Special method	Notes
All sequences and maps : **len**(*self*)	__**len**__(*self*)	length of object, >= 0. Length 0 == false
self[k]	__**getitem**__(*self*, *k*)	Get element at indice /key k (indice starts at 0). Or, if k is a slice object, return a slice.
	__**missing**__(*self*, *key*)	Hook called when key is not found in the dictionary, returns the default value.
self[k] = *value*	__**setitem**__(*self*, *k, value*)	Set element at indice/key/slice *k*.
del *self*[*k*]	__**delitem**__(*self*, *k*)	Delete element at indice/key/slice *k*.
elt **in** *self* *elt* **not in** *self*	__**contains**__(*self*, *elt*) **not** __**contains**__(*self*, *elt*)	More efficient than std iteration thru sequence.
iter(*self*)	__**iter**__(*self*)	Returns an iterator on elements (keys for mappings <=> *self*.**iterkeys**()). See iterators.
self + *other*	__**add**__(*self*, *other*)	(__**concat**__ in the official doc but doesn't work!)
Mappings, general methods, plus: **hash**(*self*)	__**hash**__(*self*)	hashed value of object *self* is used for dictionary keys

TABLE D.25:   Special Informative State Attributes: Modules

Attribute	Meaning
__doc__	(string/None, R/O): doc string (<=> __dict__['__doc__'])
__name__	(string, R/O): module name (also in __dict__['__name__'])
__dict__	(dict, R/O): module's name space
__file__	(string/undefined, R/O): pathname of .pyc, .pyo or .pyd (undef for modules statically linked to the interpreter). Before 2.3 use `sys.argv[0]` instead to find the current script filename.
__path__	(list/undefined, R/W): List of directory paths where to find the package (for packages only).

TABLE D.26:   Special Informative State Attributes: Classes

Attribute	Meaning
__doc__	(string/None, R/W): doc string (<=> __dict__['__doc__'])
__name__	(string, R/W): class name (also in __dict__['__name__'])
__module__	(string, R/W): module name in which the class was defined
__bases__	(tuple, R/W): parent classes
__dict__	(dict, R/W): attributes (class name space)

TABLE D.27:   Special Informative State Attributes: Instances

Attribute	Meaning
__class__	(class, R/W): instance's class
__dict__	(dict, R/W): attributes

TABLE D.28:   Special Informative State Attributes: User-Defined Functions

Attribute	Meaning
__doc__	(string/None, R/W): doc string
__name__	(string, R/O): function name
func_doc	(R/W): same as __doc__
func_name	(R/O, R/W from 2.4): same as __name__
func_defaults	(tuple/None, R/W): default args values if any
func_code	(code, R/W): code object representing the compiled function body
func_globals	(dict, R/O): ref to dictionary of func global variables

D.10.2   os

Miscellaneous operating system interfaces. Full doc: `http://docs.python.org/library/os.html`.

TABLE D.29: Special Informative State Attributes: User Defined Methods

Attribute	Meaning
__doc__	(string/None, R/O): doc string
__name__	(string, R/O): method name (same as im_func.__name__)
im_class	(class, R/O): class defining the method (may be a base class)
im_self	(instance/None, R/O): target instance object (None if unbound)
im_func	(function, R/O): function object

TABLE D.30: Special Informative State Attributes: Built-in Functions and Methods

Attribute	Meaning
__doc__	(string/None, R/O): doc string
__name__	(string, R/O): function name
__self__	[methods only] target object

TABLE D.31: Special Informative State Attributes: Codes

Attribute	Meaning
co_name	(string, R/O): function name
co_argcount	(int, R/0): number of positional args
co_nlocals	(int, R/O): number of local vars (including args)
co_varnames	(tuple, R/O): names of local vars (starting with args)
co_code	(string, R/O): sequence of bytecode instructions
co_consts	(tuple, R/O): literals used by the bytecode, 1st one is function doc (or None)
co_names	(tuple, R/O): names used by the bytecode
co_filename	(string, R/O): filename from which the code was compiled
co_firstlineno	(int, R/O): first line number of the function
co_lnotab	(string, R/O): string encoding bytecode offsets to line numbers.
co_stacksize	(int, R/O): required stack size (including local vars)
co_flags	(int, R/O): flags for the interpreter bit 2 set if fct uses "*arg" syntax, bit 3 set if fct uses '**keywords' syntax

"Synonym" for whatever OS-specific module (nt, mac, posix...) is proper for current environment. This module uses posix whenever possible. See also M.A. Lemburg's utility platform.py (now included in 2.3+).

TABLE D.32: Special Informative State Attributes: Frames

Attribute	Meaning
f_back	(frame/None, R/O): previous stack frame (toward the caller)
f_code	(code, R/O): code object being executed in this frame
f_locals	(dict, R/O): local vars
f_globals	(dict, R/O): global vars
f_builtins	(dict, R/O): built-in (intrinsic) names
f_restricted	(int, R/O): flag indicating whether fct is executed in restricted mode
f_lineno	(int, R/O): current line number
f_lasti	(int, R/O): precise instruction (index into bytecode)
f_trace	(function/None, R/W): debug hook called at start of each source line
f_exc_type	(Type/None, R/W): Most recent exception type
f_exc_value	(any, R/W): Most recent exception value
f_exc_traceback	(traceback/None, R/W): Most recent exception traceback

TABLE D.33: Special Informative State Attributes: Tracebacks

Attribute	Meaning
tb_next	(frame/None, R/O): next level in stack trace (toward the frame where the exception occurred)
tb_frame	(frame, R/O): execution frame of the current level
tb_lineno	(int, R/O): line number where the exception occured
tb_lasti	(int, R/O): precise instruction (index into bytecode)

TABLE D.34: Special Informative State Attributes: Slices

Attribute	Meaning
start	(any/None, R/O): lowerbound, included
stop	(any/None, R/O): upperbound, excluded
step	(any/None, R/O): step value

TABLE D.35: Special Informative State Attributes: Complex Numbers

Attribute	Meaning
real	(float, R/O): real part
imag	(float, R/O): imaginary part

TABLE D.36: Special Informative State Attributes: xranges

Attribute	Meaning
tolist	(Built-in method, R/O): ?

D.10.3   posix

Posix OS interfaces. Full doc: `http://www.python.org/doc/2.4/lib/` `module-posix.html` Do **not** import this module directly, import os instead ! (see also module: shutil for file copy and remove functions). See Table D.41 (page 513) for Variables and Table D.42 (page 514) for Functions.

posix Variables

D.10.4   posixpath

Posix pathname operations. Do **not** import this module directly, import os instead and refer to this module as **os.path**. (e.g. os.path.exists(p)!)

See Table D.43 on page 516.

D.10.5   shutil

High-level file operations (copying, deleting and also, copyfile, copymode, copystat, copy2.). See Table D.44 on page 517.

D.10.6   time

Time access and conversions. (See also module *mxDateTime* if you need a more sophisticated date/time management.) See Table D.45 (page 517) for variables and D.46 (page 518) for functions.

TABLE D.37: Some sys Variables

Variable	Content
argv	The list of command line arguments passed to a Python script. sys.argv[0] is the script name.
builtin_module_names	A list of strings giving the names of all modules written in C that are linked into this interpreter.
byteorder	Native byte order, either 'big'(-endian) or 'little'(-endian).
copyright	A string containing the copyright pertaining to the Python interpreter.
exec_prefix prefix	Root directory where platform-dependent Python files are installed, e.g. 'C:\\Python23', '/usr'.
executable	Name of executable binary of the Python interpreter (e.g. 'C:\\Python23\\python.exe', '/usr/bin/python')
exitfunc	User can set to a parameterless function. It will get called before interpreter exits. Deprecated since 2.4. Code should be using the existing atexit module
last_type, last_value, last_traceback	Set only when an exception not handled and interpreter prints an error. Used by debuggers.
maxint	Maximum positive value for integers. Since 2.2, integers and long integers are unified, thus integers have no limit.
maxunicode	Largest supported code point for a Unicode character.
modules	Dictionary of modules that have already been loaded.
path	Search path for external modules. Can be modified by program. sys.path[0] == directory of script currently executed.
platform	The current platform, e.g. "sunos5", "win32"
ps1, ps2	Prompts to use in interactive mode, normally ">>>" and "..."
stdin, stdout, stderr	File objects used for I/O. One can redirect by assigning a new file object to them (or **any** object: with a method write(string) for stdout/stderr, or with a method readline() for stdin). __stdin__,__stdout__ and __stderr__ are the default values.
subversion	Info about Python build version in the Subversion repository: tuple (interpreter-name, branch-name, revision-range), e.g. ('CPython', 'tags/r25', '51908').
version	String containing version info about Python interpreter.
version_info	Tuple containing Python version info - (major, minor, micro, level, serial).
winver	Version number used to form registry keys on Windows platforms (e.g. '2.2').

TABLE D.38:   Some sys Functions

Function	Result
_current_frames()	Returns the current stack frames for all running threads, as a dictionary mapping thread identifiers to the topmost stack frame currently active in that thread at the time the function is called.
displayhook	The function used to display the output of commands issued in interactive mode - defaults to the builtin repr(). __displayhook__ is the original value.
excepthook	Can be set to a user defined function, to which any uncaught exceptions are passed. __excepthook__ is the original value.
exit(n)	Exits with status n (usually 0 means OK). Raises SystemExit exception (hence can be caught and ignored by program).
getrefcount(object)	Returns the reference count of the object. Generally 1 higher than you might expect, because of object arg temp reference.
getcheckinterval() / setcheckinterval(interval)	Gets / Sets the interpreter's thread switching interval (in number of bytecode instructions, default: 10 until 2.2, 100 from 2.3).
settrace(func)	Sets a trace function: called before each line of code is exited.
setprofile(func)	Sets a profile function for performance profiling.
exc_info()	Info on exception currently being handled; this is a tuple (exc_type, exc_value, exc_traceback). **Warning**: assigning the traceback return value to a local variable in a function handling an exception will cause a circular reference.
setdefaultencoding(encoding)	Change default Unicode encoding - defaults to 7-bit ASCII.
getrecursionlimit()	Retrieve maximum recursion depth.
setrecursionlimit()	Set maximum recursion depth (default 1000).

TABLE D.39: Some os Variables

Variable	Meaning
name	name of O/S-specific module (e.g. "posix", "mac", "nt")
path	O/S-specific module for path manipulations. On Unix, `os.path.split()` <=> `posixpath.split()`
curdir	string used to represent current directory (eg '.')
pardir	string used to represent parent directory (eg '..')
sep	string used to separate directories ('/' or '\'). **Tip**: Use os.path.join() to build portable paths.
altsep	Alternate separator if applicable (None otherwise)
pathsep	character used to separate search path components (as in $PATH), eg. ';' for windows.
linesep	line separator as used in **text** files, ie '\n' on Unix, '\r\n' on Dos/Win, '\r' on Mac.

TABLE D.40: Some os Functions

Function	Result
makedirs(path[, mode=0777])	Recursive directory creation (create required intermediary dirs); os.error if fails.
removedirs(path)	Recursive directory delete (delete intermediary **empty** dirs); fails (os.error) if the directories are not empty.
renames(old, new)	Recursive directory or file renaming; os.error if fails.
urandom(n)	Returns a string containing n bytes of random data.

TABLE D.41: Posix Variables

Variable	Meaning
environ	dictionary of environment variables, e.g. posix.environ['HOME'].
error	exception raised on POSIX-related error. Corresponding value is tuple of errno code and *perror()* string.

TABLE D.42: Posix Functions

Function	Result
chdir(*path*)	Changes current directory to *path*.
chmod(*path*, *mode*)	Changes the mode of *path* to the numeric *mode*
close(*fd*)	Closes file descriptor *fd* opened with posix.open.
_exit(*n*)	Immediate exit, with no cleanups, no SystemExit, etc... Should use this to exit a child process.
execv(*p, args*)	"Become" executable *p* with args *args*
getcwd()	Returns a string representing the current working directory.
getcwdu()	Returns a **Unicode** string representing the current working directory.
getpid()	Returns the current process id.
getsid()	Calls the system call getsid() [Unix].
fork()	Like C's fork(). Returns 0 to child, child pid to parent [Not on Windows].
kill(*pid, signal*)	Like C's kill [Not on Windows].
listdir(*path*)	Lists (base)names of entries in directory *path*, excluding '.' and '..'. If path is a Unicode string, so will be the returned strings.
lseek(*fd, pos, how*)	Sets current position in file *fd* to position pos, expressed as an offset relative to beginning of file (*how*=0), to current position (*how*=1), or to end of file (*how*=2).
mkdir(*path*[, *mode*])	Creates a directory named *path* with numeric *mode* (default 0777).
open(*file, flags, mode*)	Like C's open(). Returns file descriptor. Use file object functions rather than this low level ones.
pipe()	Creates a pipe. Returns pair of file descriptors (r, w) [Not on Windows].
popen(*command, mode*='r', *buf-Size*=0)	Opens a pipe to or from *command*. Result is a file object to read to or write from, as indicated by *mode* being 'r' or 'w'. Use it to catch a command output ('r' mode), or to feed it ('w' mode).
remove(*path*)	See unlink.
rename(*old, new*)	Renames/moves the file or directory *old* to *new*. [error if target name already exists]
renames(*old, new*)	Recursive directory or file renaming function. Works like rename(), except **creation** of any intermediate directories needed to make the new pathname good is attempted first. After the rename, directories corresponding to rightmost path segments of the old name will be pruned away using removedirs().
rmdir(*path*)	Removes the empty directory *path*.

read(*fd, n*)	Reads *n* bytes from file descriptor *fd* and return as string.
stat(*path*)	Returns st_mode, st_ino, st_dev, st_nlink, st_uid, st_gid, st_size, st_atime, st_mtime, st_ctime. [st_ino, st_uid, st_gid are dummy on Windows]
system(*command*)	Executes string *command* in a subshell. Returns exit status of subshell (usually 0 means OK). Since 2.4 use subprocess.call() instead.
times()	Returns accumulated CPU times in sec (user, system, children's user, children's sys, elapsed real time) [3 last not on Windows].
unlink(*path*)	Unlinks ("deletes") the file (not dir!) path. Same as: remove.
utime(*path*, (*aTime*, *mTime*))	Sets the access & modified time of the file to the given tuple of values.
wait()	Waits for child process completion. Returns tuple of pid, exit_status [Not on Windows].
waitpid(*pid*, *options*)	Waits for process *pid* to complete. Returns tuple of *pid*, exit_status [Not on Windows].
walk(*top*[, *top-down*=True [, *onerror*=None]])	Generates a list of file names in a directory tree, by walking the tree either top down or bottom up. For each directory in the tree rooted at directory top (including top itself), it yields a 3-tuple (dirpath, dirnames, filenames) - more info here. See also os.path.walk().
write(*fd, str*)	Writes *str* to file *fd*. Returns nb of bytes written.

TABLE D.43:   Posixpath Functions

Function	Result
abspath(*path*)	Returns absolute *path* for *path*, taking current working dir in account.
commonprefix(*list*)	Returns the longuest path prefix (taken character-by-character) that is a prefix of all paths in list (or " if *list* empty).
dirname/basename (*path*)	directory and name parts of *path*. See also split.
exists(*path*)	True if path is the *path* of an existing file or directory. See also lexists.
expanduser(*path*)	Returns a copy of *path* with "~" expansion done.
expandvars(*path*)	Returns string that is (a copy of) *path* with environment vars expanded. [Windows: case significant; must use Unix: $var notation, not %var%]
getatime(*path*)	Returns last access time of *path* (integer secs. since epoch).
getctime(*path*)	Returns the metadata change time of path (integer secs. since epoch).
getmtime(*path*)	Returns last modification time of *path* (integer secs. since epoch).
getsize(*path*)	Returns the size in bytes of *path*. os.error if file inexistent or inaccessible.
isabs(*path*)	True if *path* is absolute.
isdir(*path*)	True if *path* is a directory.
isfile(*path*)	True if *path* is a regular file.
islink(*path*)	True if *path* is a symbolic link.
ismount(*path*)	True if *path* is a mount point [true for all dirs on Windows].
join(*p*[,*q*[,...]])	Joins one or more *path* components in a way suitable for the current OS.
lexists(*path*)	True if the file specified by *path* exists, whether or not it's a symbolic link (unlike exists).
normcase(*path*)	Normalizes case of *path*. Has no effect under Posix.
normpath(*path*)	Normalizes *path*, eliminating double slashes, etc...
realpath(*path*)	Returns the canonical path for *path*, eliminating any symbolic links encountered in the *path*.
samefile(*f1,f2*)	True if the 2 paths f1 and f2 reference the same file.
sameopenfile(*f1,f2*)	True if the 2 open file objects f1 and f2 reference the same file.
samestat(*s1, s2*)	True if the 2 stat buffers s1 and s2 reference the same file.
split(*p*)	Splits *p* into (head, tail) where tail is last pathname component and head is everything leading up to that. <=> (dirname(p), basename(p))
splitdrive(*p*)	Splits path *p* in a pair ('drive:', tail) [Windows]

| splitext(p) | Splits into (root, ext) where last comp of root contains no periods and ext is empty or starts with a period. |
| walk(p, *visit, arg*) | Calls the function visit with arguments (arg, dirname, names) for each directory recursively in the directory tree rooted at p (including p itself if it's a dir). The argument dirname specifies the visited directory, the argument names lists the files in the directory. The visit function may modify names to influence the set of directories visited below dirname, e.g. to avoid visiting certain parts of the tree. See also os.walk(). |

TABLE D.44: Shutil Functions

Function	Result
copy(*src, dest*)	Copies the contents of file *src* to file *dest*, retaining file permissions.
copytree(*src, dest* [, *symlinks*])	Recursively copies an entire directory tree rooted at *src* into *dest* (which should not already exist). If *symlinks* is true, links in *src* are kept as such in *dest*.
move(*src, dest*)	Recursively moves a file or directory to a new location.
rmtree(*path* [, *ignore_errors*[, *onerror*]])	Deletes an entire directory tree, ignoring errors if *ignore_errors* is true, or calling *onerror*(func, path, sys.exc_info()) if supplied, with arguments *func* (faulty function), and *path* (concerned file). This fact fails when the files are Read Only.

TABLE D.45: Time Access and Conversions: Variables

Variable	Meaning
altzone	Signed offset of local DST time zone in sec west of the 0th meridian.
daylight	Nonzero if a DST time zone is specified.
time zone	The offset of the local (non-DST) time zone, in seconds west of UTC.
tzname	A tuple (name of local non-DST time zone, name of local DST timezone)

TABLE D.46:   Time: Some Functions

Function	Result
clock()	On Unix: current processor time as a floating point number expressed in seconds. On Windows: wall-clock seconds elapsed since the 1st call to this function, as a floating point number (precision < 1μs).
time()	Returns a float representing UTC time in **seconds** since the epoch.
gmtime([*secs*]), **localtime**([*secs*])	Returns a 9-tuple representing time. Current time is used if *secs* is not provided. Since 2.2, returns a struct_time object (still accessible as a tuple) with the following attributes: tm_year, tm_mon, tm_mday, tm_hour, tm_min, tm_sec, tm_wday, tm_yday, tm_isdst
asctime([*timeTuple*]),	24-character string of the following form: 'Mon Apr 03 08:31:14 2006'. *timeTuple* defaults to `localtime()` if omitted.
ctime([*secs*])	equivalent to asctime(localtime(*secs*))
mktime(*timeTuple*)	Inverse of localtime(). Returns a float representing a number of seconds.
strftime(*format*[, timeTuple])	**Formats** a time tuple as a string, according to *format* (see table below). Current time is used if *timeTuple* is omitted.
strptime(*string*[, format])	**Parses** a string representing a time according to *format* (same format as for strftime(), see below), default "%a %b %d %H:%M:%S %Y" = asctime format. Returns a time tuple/struct_time.
sleep(*secs*)	Suspends execution for *secs* seconds. *secs* can be a float.

TABLE D.47: Formatting in strftime() and strptime()

Directive	Meaning
%a	Locale's abbreviated weekday name.
%A	Locale's full weekday name.
%b	Locale's abbreviated month name.
%B	Locale's full month name.
%c	Locale's appropriate date and time representation.
%d	Day of the month as a decimal number [01,31].
%H	Hour (24-hour clock) as a decimal number [00,23].
%I	Hour (12-hour clock) as a decimal number [01,12].
%j	Day of the year as a decimal number [001,366].
%m	Month as a decimal number [01,12].
%M	Minute as a decimal number [00,59].
%p	Locale's equivalent of either AM or PM.
%S	Second as a decimal number [00,61]. Yes, 61 !
%U	Week number of the year (Sunday as the first day of the week) as a decimal number [00,53]. All days in a new year preceding the first Sunday are considered to be in week 0.
%w	Weekday as a decimal number [0(Sunday),6].
%W	Week number of the year (Monday as the first day of the week) as a decimal number [00,53]. All days in a new year preceding the first Sunday are considered to be in week 0.
%x	Locale's appropriate date representation.
%X	Locale's appropriate time representation.
%y	Year without century as a decimal number [00,99].
%Y	Year with century as a decimal number.
%Z	Time zone name (or by no characters if no time zone exists).
%%	A literal "%" character.

TABLE D.48:   Some String Variables

Variable	Meaning
digits	The string '0123456789'.
hexdigits, octdigits	Legal hexadecimal & octal digits.
letters, uppercase, lowercase, whitespace	Strings containing the appropriate characters.
ascii_letters, ascii_lowercase, ascii_uppercase	Same, taking the current locale in account.
index_error	Exception raised by index() if substring (defined between any types) not found.

D.10.7   String

Common string operations. Full doc: `http://www.python.org/doc/lib/module-string.html`. As of Python 2.0, much (though not all) of the functionality provided by the string module has been superseded by built-in string methods. See Operations on strings for details. See Table D.48 on page 520.

D.10.8   re (sre)

Regular expression operations. Full doc: `http://www.python.org/doc/lib/module-re.html`.

Handles Unicode strings. Implemented in new module **sre, re** now a mere front-end for compatibility. Patterns are specified as strings.

Tip: Use **raw** strings (e.g. **r'\w*'**) to literalize backslashes.

See Table D.50 on page 522.

D.10.9   Regular Expression Objects

RE objects are returned by the compile function. See Table D.54 on page 525.

D.10.10   Match Objects

Match objects are returned by the match and search functions. See Table D.56 on page 526.

D.10.11   Math

For complex number functions, see module cmath. For intensive number crunching, see Numerical Python and the Python and Scientific computing page. See math constants in table D.58 (page 526) and math functions in

TABLE D.49: Some String Functions

Function	Result
expandtabs(*s, tabSize*)	Returns a copy of string emphs with tabs expanded.
find/rfind(*s, sub*[, *start*=0[, *end*=0])	Returns the lowest/highest index in *s* where the substring *sub* is found such that *sub* is wholly contained in *s[start:end]*. Return -1 if *sub* not found.
ljust/rjust/center(*s, width*[, *fillChar*=' '])	Returns a copy of string emphs; left/right justified/centered in a field of given width, padded with spaces or the given character. *s* is never truncated.
lower/upper(*s*)	Returns a string that is (a copy of) *s* in lowercase/uppercase.
split(*s*[, *sep*=*whitespace*[, *maxsplit*=0]])	Returns a list containing the words of the string *s*, using the string *sep* as a separator.
rsplit(*s*[, *sep*=*whitespace*[, *maxsplit*=0]])	Same as split above but starts splitting from the end of string, e.g. 'A,B,C'.split(',', 1) == ['A', 'B,C'] but 'A,B,C'.rsplit(',', 1) == ['A,B', 'C']
join(*words*[, *sep*=' '])	Concatenates a list or tuple of words with intervening separators; inverse of split.
replace(*s, old, new*[, *maxsplit*=0]	Returns a copy of string *s* with all occurrences of substring *old* replaced by new. Limits to *maxsplit* first substitutions if specified.
strip(*s*[, *chars*=None])	Returns a string that is (a copy of) *s* without leading and trailing chars (default: whitespace), if any. Also: lstrip, rstrip.

TABLE D.50: Regular Expression Syntax

Form	Description
.	Matches any character (including new line if DOTALL flag specified).
∧	Matches start of the string (of every line in MULTILINE mode
$	Matches end of the string (of every line in MULTILINE mode
*	0 or more of preceding regular expression (as **many** as possible
+	1 or more of preceding regular expression (as **many** as possible
?	0 or 1 occurrence of preceding regular expression.
*?, +?, ??	Same as *, + and ? but matches as **few** characters as possible
{m,n}	Matches from m to n repetitions of preceding RE.
{m,n}?	Idem, attempting to match as **few** repetitions as possible.
[]	Defines character set: e.g. '[a-zA-Z]' to match all letters (see also \w \S).
[∧]	Defines complemented character set: matches if char is NOT in set.
\	Escapes special chars '*?+&$—()' and introduces special sequences (see below). Due to Python string rules, write as '\\' or r'\' in the pattern string.
\\	Matches a literal '\'; due to Python string rules, write as '\\\' in pattern string, or better using raw string: r'\\'.
—	Specifies alternative: 'foo—bar' matches 'foo' or 'bar'.
(...)	Matches any RE inside (), and delimits a group.
(?:...)	Idem but doesn't delimit a group (non capturing parenthesis).
(?P<*name*>...)	Matches any RE inside (), and delimits a **named** group, (e.g r'(?P<id>[a-zA-Z_]\w*)' defines a group named id).
(?P=*name*)	Matches whatever text was matched by the earlier group named name.
(?=...)	Matches if ... matches next, but doesn't consume any of the string e.g. 'Isaac (?=Asimov)' matches 'Isaac' only if followed by 'Asimov'.
(?!...)	Matches if ... **doesn't** match next. Negative of (?=...).
(?<=...)	Matches if the current position in the string is preceded by match for ... that ends at the current position. This is called positive lookbehind assertion.
(?<!...)	Matches if the current position in the string is not preceded by a match for .... This is called a negative lookbehind assertion.
(?(*group*)A—B)	[2.4+] group is either a numeric group ID or a group name defined with (?Pgroup...) earlier in the expression. If the specified group matched, the regular expression pattern A will be tested against the string; if the group didn't match, the pattern B will be used instead.
(?#...)	A comment; ignored.
(?*letters*)	letters is one or more of 'i','L', 'm', 's', 'u', 'x'. Sets the corresponding **flags** (re.I, re.L, re.M, re.S, re.U, re.X) for the entire RE. See the compile() function for equivalent flags.

TABLE D.51: Regular Expression Special Sequences

Sequence	Description
number	Matches content of the *group* of the same number; groups are numbered starting from 1.
\A	Matches only at the start of the string.
\b	Empty str at beginning or end of word: '\bis\b' matches 'is', but not 'his'.
\B	Empty str NOT at beginning or end of word.
\d	Any decimal digit (<=> [0-9]).
\D	Any non-decimal digit char (<=> [∧0-9]).
\s	Any whitespace char (<=> [\t\n\r\f\v]).
\S	Any non-whitespace char (<=> [∧ \t\n\r\f\v]).
\w	Any alphaNumeric char (depends on LOCALE flag).
\W	Any non-alphaNumeric char (depends on LOCALE flag).
\Z	Matches only at the end of the string.

TABLE D.52: Regular Expression Variables

Variable	Meaning
error	Exception when pattern string isn't a valid regexp.

Table D.59 (page 527).

D.10.12   getopt

Parser for command line options. Full doc: http://www.python.org/doc/lib/module-getopt.html.

This was the standard parser until Python 2.3, now superseded by optparse. [See also: Richard Gruet's simple parser getargs.py (shameless self promotion)]

Functions:

```
getopt(list, optstr)
    # -- Similar to C. <optstr> is option letters to look for.
    # Put ':' after letter if option takes arg. E.g.
    # invocation was "python test.py -c hi -a arg1 arg2"
        opts, args = getopt.getopt(sys.argv[1:], 'ab:c:')
    # opts would be
        [('-c', 'hi'), ('-a', '')]
    # args would be
        ['arg1', 'arg2']
```

TABLE D.53: Regular Expression Functions

Function	Result
compile(*pattern*[, *flags*=0])	Compiles a RE pattern string into a *regular expression object*. Flags (combinable by —): *I* or *IGNORECASE* <=> (*?i*) case insensitive matching *L* or *LOCALE* <=> (*?L*) make \w, \W, \b, \B dependent on the current locale *M* or *MULTILINE* <=> (*?m*) matches every new line and not only start/end of the whole string *S* or *DOTALL* <=> (*?s*) '.' matches ALL chars, including new line *U* or *UNICODE* <=> (*?u*) Make \w, \W, \b, and \B dependent on the Unicode character properties database. *X* or *VERBOSE* <=> (*?x*) Ignores whitespace outside character sets
escape(*string*)	Returns (a copy of) *string* with all non-alphanumerics backslashed.
match(*pattern*, *string*[, *flags*])	If 0 or more chars at **beginning** of *string* matches the RE pattern string, returns a corresponding *MatchObject* instance, or None if no match.
search(*pattern*, *string*[, *flags*])	Scans thru *string* for a location matching *pattern*, returns a corresponding *MatchObject* instance, or None if no match.
split(*pattern, string*[, *maxsplit*=0])	Splits *string* by occurrences of *pattern*. If capturing () are used in pattern, then occurrences of patterns or subpatterns are also returned.
findall(*pattern*, *string*)	Returns a list of non-overlapping matches of *pattern* in *string*, either a list of groups or a list of tuples if the pattern has more than 1 group.
finditer(*pattern*, *string*[, *flags*])	Returns an iterator over all non-overlapping matches of *pattern* in *string*. For each match, the iterator returns a match object. Empty matches are included in the result unless they touch the beginning of another match.
sub(*pattern*, *repl*, *string*[, *count*=0])	Returns string obtained by replacing the (*count* first) leftmost non-overlapping occurrences of *pattern* (a string or a RE object) in *string* by *repl*; *repl* can be a string or a function called with a single *MatchObj* arg, which must return the replacement string.
subn(*pattern*, *repl*, *string*[, *count*=0])	Same as sub(), but returns a tuple (newString, numberOfSubsMade).

TABLE D.54: re Object Attributes

Attribute	Description
flags	Flags arg used when RE obj was compiled, or 0 if none provided.
groupindex	Dictionary of {group name: group number} in pattern.
pattern	Pattern string from which RE obj was compiled.

TABLE D.55: re Object Methods

Method	Result
match(*string*[, *pos*][, *ndpos*])	If zero or more characters at the beginning of string match this regular expression, returns a corresponding MatchObject instance. Returns None if the string does not match the pattern; note that this is different from a zero-length match. The optional second parameter pos gives an index in the string where the search is to start; it defaults to 0. This is not completely equivalent to slicing the string; the " pattern character matches at the real beginning of the string and at positions just after a new line, but not necessarily at the index where the search is to start. The optional parameter *endpos* limits how far the string will be searched; it will be as if the string is endpos characters long, so only the characters from pos to endpos will be searched for a match.
search(*string*[, *pos*][, *ndpos*])	Scans through string looking for a location where this regular expression produces a match, and returns a corresponding MatchObject instance. Returns None if no position in the string matches the pattern; note that this is different from finding a zero-length match at some point in the string. The optional pos and *endpos* parameters have the same meaning as for the match() method.
split(*string*[, *maxsplit*=0])	Identical to the split() function, using the compiled pattern.
findall(*string*[, *pos*[, *endpos*]])	Identical to the findall() function, using the compiled pattern.
finditer(*string*[, *pos*[, *endpos*]])	Identical to the finditer() function, using the compiled pattern.
sub(*repl*, *string*[, *count*=0])	Identical to the sub() function, using the compiled pattern.
subn(*repl*, *string*[, *count*=0])	Identical to the subn() function, using the compiled pattern.

TABLE D.56:   Match Object Attributes

Attribute	Description
pos	Value of pos passed to search or match functions; index into string at which RE engine started search.
endpos	Value of endpos passed to search or match functions; index into string beyond which RE engine won't go.
re	RE object whose match or search fct produced this MatchObj instance.
string	String passed to match() or search().

TABLE D.57:   Match Object Functions

Function	Result
group([*g1*, *g2*, ...])	Returns one or more groups of the match. If **one** arg, result is a string; if multiple args, result is a tuple with one item per arg. If *gi* is 0, returns the entire matching string; if $1 <= gi <= 99$, returns string matching group #*gi* (or None if no such group); *gi* may also be a group *name*.
groups()	Returns a tuple of all groups of the match; groups not participating to the match have a value of None. Returns a string instead of tuple if len(tuple)== 1.
start(*group*), end(*group*)	Returns indices of start & end of substring matched by group (or None if group exists but didn't contribute to the match).
span(*group*)	Returns the 2-tuple (start(group), end(group)); can be (None, None) if group didn't contribute to the match.

TABLE D.58:   Math Constants

Name	Value
pi	3.1415926535897931
e	2.7182818284590451

D.11   List of Modules and Packages in Base Distribution

Built-ins and content of python Lib directory. The subdirectory Lib/site-packages contains platform-specific packages and modules. (See Table D.60 on page 528)

[**Main distributions (Windows, Unix)**, some OS specific modules may be missing.]

TABLE D.59: Math Functions

Name	Result
acos(x)	Returns the arc cosine (measured in radians) of x.
asin(x)	Returns the arc sine (measured in radians) of x.
atan(x)	Returns the arc tangent (measured in radians) of x.
atan2(y, x)	Returns the arc tangent (measured in radians) of y/x. The result is between -pi and pi. Unlike atan(y/x), the signs of both x and y are considered.
ceil(x)	Returns the ceiling of x as a float. This is the smallest integral value $>= x$.
cos(x)	Returns the cosine of x (measured in radians).
cosh(x)	Returns the hyperbolic cosine of x.
degrees(x)	Converts angle x from radians to degrees.
exp(x)	Returns e raised to the power of x.
fabs(x)	Returns the absolute value of the float x.
floor(x)	Returns the floor of x as a float. This is the largest integral value $<= x$.
fmod(x, y)	Returns fmod(x, y), according to platform C. x % y may differ.
frexp(x)	Returns the mantissa and exponent of x, as pair (m, e). m is a float and e is an int, such that x = m * 2.**e. If x is 0, m and e are both 0. Else 0.5 $<=$ abs(m) < 1.0.
hypot(x, y)	Returns the Euclidean distance sqrt(x*x + y*y).
ldexp(x, y)	x * (2**i)
log(x[, *base*])	Returns the logarithm of x to the given base. If the base is not specified, returns the natural logarithm (base e) of x.
log10(x)	Returns the base 10 logarithm of x.
modf(x)	Returns the fractional and integer parts of x. Both results carry the sign of x. The integer part is returned as a float.
pow(x, y)	Returns x**y (x to the power of y). Note that for y=2, it is more efficient to use x*x.
radians(x)	Converts angle x from degrees to radians.
sin(x)	Returns the sine (measured in radians) of x.
sinh(x)	Returns the hyperbolic sine of x.
sqrt(x)	Returns the square root of x.
tan(x)	Returns the tangent (measured in radians) of x.
tanh(x)	Returns the hyperbolic tangent of x.

D.12   Workspace Exploration and Idiom Hints

Tips for exploring the Python workspace. See Table D.61 on page 538.

TABLE D.60:   Standard Library Modules

Operation	Result
__builtin__	Provide direct access to all 'built-in' identifiers of Python, e.g. __builtin__.open is the full name for the built-in function open().
__future__	Future statement definitions. Used to progressively introduce new features in the language.

__main__	Represent the (otherwise anonymous) scope in which the interpreter's main program executes – commands read either from standard input, from a script file, or from an interactive prompt. Typical idiom to check if a code was run as a *script* (as opposed to being *imported*):
	
	`if __name__ == '__main__':`
	`  main()  # (this code was run as script)`

aifc	Stuff to parse AIFF-C and AIFF files.
anydbm	Generic interface to all dbm clones. (dbhash, gdbm, dbm, dumbdbm).
array	Efficient arrays of numeric values.
asynchat	A class supporting chat-style (command/response) protocols.
asyncore	Basic infrastructure for asynchronous socket service clients and servers.
atexit	Register functions to be called at exit of Python interpreter.
audiodev	Classes for manipulating audio devices (currently only for Sun and SGI).
audioop	Manipulate raw audio data. 2.5: Supports the a-LAW encoding.
base64	Conversions to/from base64 transport encoding as per RFC-1521.
BaseHTTPServer	HTTP server base class
Bastion	"Bastionification" utility (control access to instance vars).
bdb	A generic Python debugger base class.
binascii	Convert between binary and ASCII.
binhex	Macintosh binhex compression/decompression.
bisect	Bisection algorithms.
bsddb	(Optional) improved BSD database interface [package].
bz2	BZ2 compression.
calendar	Calendar printing functions.
cgi	Wraps the WWW Forms Common Gateway Interface (CGI).

CGIHTTPServer	CGI-savvy HTTP Server.
cgitb	Traceback manager for CGI scripts.
chunk	Read IFF chunked data.
cmath	Mathematical functions for complex numbers. See also math.
cmd	A generic class to build line-oriented command interpreters.
code	Utilities needed to emulate Python's interactive interpreter.
codecs	Lookup existing Unicode encodings and register new ones. 2.5: support for incremental codecs.
codeop	Utilities to compile possibly incomplete Python source code.
collections	high-performance container datatypes. Currently, the only datatype is a double-ended queue. 2.5: Type deque has now a remove method. New type defaultdict.
colorsys	Conversion functions between RGB and other color systems.
commands	Execute shell commands via os.popen [Unix].
compileall	Force "compilation" of all .py files in a directory.
ConfigParser	Configuration file parser (much like windows .ini files).
contextlib	Utilities for with statement contexts.
Cookie	HTTP state (cookies) management.
copy	Generic shallow and deep copying operations.
copy_reg	Helper to provide extensibility for modules pickle/cPickle.
cPickle	Faster, C implementation of pickle.
cProfile	Faster, C implementation of profile.
crypt	Function to check Unix passwords [Unix].
cStringIO	Faster, C implementation of StringIO.
csv	Tools to read comma-separated files (of variations thereof). 2.5: Several enhancements.
ctypes	"Foreign function" library for Python. Provides C compatible data types, and allows to call functions in dlls/shared libraries. Can be used to wrap these libraries in pure Python.
curses	Terminal handling for character-cell displays [Unix/OS2/DOS only].
datetime	Improved date/time types (date, time, datetime, timedelta). 2.5: New method strptime(string, format) for class datetime.
dbhash	(g)dbm-compatible interface to bsdhash.hashopen.
decimal	Decimal floating point arithmetic.

difflib	Tool for comparing sequences, and computing the changes required to convert one into another. 2.5: Improved SequenceMatcher.get_matching_blocks() method .
dircache	Sorted list of files in a dir, using a cache.
dis	Bytecode disassembler.
distutils	Package installation system. 2.5: Function setup enhanced with new keyword parameters requires, provides, obsoletes, and download_url [PEP314].
distutils.command-.register	Registers a module in the Python package index (PyPI). This command plugin adds the register command to distutil scripts.
distutils.debug distutils.emxccompiler distutils.log	
dl	Call C functions in shared objects [Unix].
doctest	Unit testing framework based on running examples embedded in docstrings. 2.5: New SKIP option. New encoding arg to testfile() function.
DocXMLRPCServer	Creation of self-documenting XML-RPC servers, using pydoc to create HTML API doc on the fly. 2.5: New attribute rpc_paths.
dumbdbm	A dumb and slow but simple dbm clone.
dummy_thread dummy_threading	Helpers to make it easier to write code that uses threads where supported, but still runs on Python versions without thread support. The dummy modules simply run the threads sequentially.
email	A package for parsing, handling, and generating email messages. New version 3.0 dropped various deprecated APIs and removes support for Python versions earlier than 2.3. 2.5: Updated to version 4.0.
encodings	New codecs: idna (IDNA strings), koi8_u (Ukranian), palmos (PalmOS 3.5), punycode (Punycode IDNA codec), string_escape (Python string escape codec: replaces non-printable chars w/ Python-style string escapes). New codecs in 2.4: HP Roman8, ISO_8859-11, ISO_8859-16, PCTP-154, TIS-620; Chinese, Japanese and Korean codecs.
errno	Standard errno system symbols. The value of each symbol is the corresponding integer value.
exceptions	Class based built-in exception hierarchy.
fcntl	The fcntl() and ioctl() system calls [Unix].
filecmp	File and directory comparison.

fileinput	Helper class to quickly write a loop over all standard input files. 2.5: Made more flexible (Unicode filenames, mode parameter, etc...)
fnmatch	Filename matching with shell patterns.
formatter	Generic output formatting.
fpectl	Floating point exception control [Unix].
fpformat	General floating point formatting functions.
ftplib	An FTP client class. Based on RFC 959.
functools	tools for functional-style programming. See in particular function partial() [PEP309].
gc	Perform garbage collection, obtain GC debug stats, and tune GC parameters. 2.5: New get_count() function. gc.collect() takes a new generation argument.
gdbm	GNU's reinterpretation of dbm [Unix].
getopt	Standard command line processing. See also optparse.
getpass	Utilities to get a password and/or the current user name.
gettext	Internationalization and localization support.
glob	Filename "globbing" utility.
gopherlib	Gopher protocol client interface.
grp	The group database [Unix].
gzip	Read & write gzipped files.
hashlib	Secure hashes and message digests.
heapq	Heap queue (priority queue) helpers. 2.5: nsmallest() and nlargest() takes a key keyword param.
hmac	HMAC (Keyed-Hashing for Message Authentication).
hotshot.stones	Helper to run the pystone benchmark under the Hotshot profiler.
htmlentitydefs	HTML character entity references.
htmllib	HTML2 parsing utilities
HTMLParser	Simple HTML and XHTML parser.
httplib	HTTP1 client class.
idlelib	(package) Support library for the IDLE development environment.
ihooks	Hooks into the "import" mechanism.
imageop	Manipulate raw image data.
imaplib	IMAP4 client.Based on RFC 2060.
imghdr	Recognizing image files based on their first few bytes.
imp	Access the import internals.
imputil	Provides a way of writing customized import hooks.
inspect	Get information about live Python objects.
itertools	Tools to work with iterators and lazy sequences. 2.5: islice() accepts None for start & step args.
keyword	List of Python keywords.
linecache	Cache lines from files.

locale	Support for number formatting using the current locale settings. 2.5: format() modified; new fcts format_string() and currency()
logging	(package) Tools for structured logging in log4j style.
macpath	Pathname (or related) operations for the Macintosh [Mac].
macurl2path	Mac specific module for conversion between pathnames and URLs [Mac].
mailbox	Classes to handle Unix style, MMDF style, and MH style mailboxes. 2.5: added capability to modify mailboxes in addition to reading them.
mailcap	Mailcap file handling (RFC 1524).
marshal	Internal Python object serialization.
markupbase	Shared support for scanning document type declarations in HTML and XHTML.
math	Mathematical functions. See also cmath
md5	MD5 message digest algorithm. 2.5: Now a mere wrapper around new library hashlib.
mhlib	MH (mailbox) interface.
mimetools	Various tools used by MIME-reading or MIME-writing programs.
mimetypes	Guess the MIME type of a file.
MimeWriter	Generic MIME writer. Deprecated since release 2.3. Use the email package instead.
mimify	Mimification and unmimification of mail messages.
mmap	Interface to memory-mapped files - they behave like mutable strings.
modulefinder	Tools to find what modules a given Python program uses, without actually running the program.
msilib	Read and write Microsoft Installer files [Windows].
msvcrt	File & Console Windows-specific operations [Windows].
multifile	A readline()-style interface to the parts of a multipart message.
mutex	Mutual exclusion – for use with module sched. See also std module threading, and glock.
netrc	Parses and encapsulates the netrc file format.
new	Creation of runtime internal objects (interface to interpreter object creation functions).
nis	Interface to Sun's NIS (Yellow Pages) [Unix]. 2.5: New domain arg to nis.match() and nis.maps().
nntplib	An NNTP client class. Based on RFC 977.
ntpath	Common operations on Windows pathnames [Windows].
nturl2path	Convert a NT pathname to a file URL and vice versa [Windows].

olddifflib	Old version of difflib (helpers for computing deltas between objects)?
operator	Standard operators as functions. 2.5: itemgetter() and attr-getter() now supports multiple fields.
optparse	Improved command-line option parsing library (see also getopt). 2.5: Updated to Optik library 1.51.
os	OS routines for Mac, DOS, NT, or Posix depending on what system we're on. 2.5: os.stat() return time values as floats; new constants to os.lseek(); new functions wait3() and wait4(); on FreeBSD, os.stat() returns times with nanosecond resolution.
os.path	Common pathname manipulations.
os2emxpath	os.path support for OS/2 EMX.
parser	Access Python parse trees.
pdb	A Python debugger.
pickle	Pickling (save/serialize and restore/deserialize) of Python objects (a faster C implementation exists in built-in module: cPickle). 2.5: Value returned by __reduce__() must be different from None.
pickletools	Tools to analyze and disassemble pickles.
pipes	Conversion pipeline templates [Unix].
pkgutil	Tools to extend the module search path for a given package. 2.5: PEP302's import hooks support; works for packages in ZIP format archives.
platform	Get info about the underlying platform.
popen2	Spawn a command with pipes to its stdin, stdout, and optionally stderr. Superseded by module subprocess since 2.4
poplib	A POP3 client class.
posix	Most common POSIX system calls [Unix].
posixfile	(deprecated since 1.5, use fcntl.lockf() instead) File-like objects with locking support [Unix].
posixpath	Common operations on POSIX pathnames.
pprint	Support to pretty-print lists, tuples, & dictionaries recursively.
pre	Support for regular expressions (RE) - see re.
profile	Class for profiling python code. 2.5: See also new fast C implementation cProfile
pstats	Class for printing reports on profiled python code. 2.5: new stream arg to Stats constructor.
pty	Pseudo terminal utilities [Linux, IRIX].
pwd	The password database [Unix].
py_compile	Routine to "compile" a .py file to a .pyc file.
pyclbr	Parse a Python file and retrieve classes and methods.
pydoc	Generate Python documentation in HTML or text for interactive use.
pyexpat	Interface to the Expat XML parser. 2.5: now uses V2.0 of the expat parser.

Queue	A multi-producer, multi-consumer queue.
quopri	Conversions to/from quoted-printable transport encoding as per RFC 1521.
rand	Don't use unless you want compatibility with C's rand().
random	Random variable generators.
re	Regular Expressions.
readline	GNU readline interface [Unix].
repr	Alternate repr() implementation.
resource	Resource usage information [Unix].
rfc822	Parse RFC-8222 mail headers.
rgbimg	Read and write 'SGI RGB' files.
rlcompleter	Word completion for GNU readline 2.0 [Unix]. 2.5: Doesn't depend on readline anymore; now works on non-Unix platforms.
robotparser	Parse robot.txt files, useful for web spiders.
sched	A generally useful event scheduler class.
select	Waiting for I/O completion.
sets	A Set datatype implementation based on dictionaries (see Sets).
sgmllib	A parser for SGML, using the derived class as a static DTD.
sha	SHA-1 message digest algorithm. 2.5: Now a mere wrapper around new library hashlib.
shelve	Manage shelves of pickled objects.
shlex	Lexical analyzer class for simple shell-like syntaxes.
shutil	Utility functions for copying files and directory trees.
signal	Set handlers for asynchronous events.
SimpleHTTPServer	Simple HTTP Server.
SimpleXMLRPCServer	Simple XML-RPC Server. 2.5: New attribute rpc_paths.
site	Append module search paths for third-party packages to sys.path.
smtpd	An RFC 2821 SMTP server.
smtplib	SMTP/ESMTP client class.
sndhdr	Several routines that help recognizing sound.
socket	Socket operations and some related functions. Now supports timeouts thru function settimeout(t). Also supports SSL on Windows. 2.5: Now supports AF_NETLINK sockets on Linux; new socket methods recv_buf(buffer), recvfrom_buf(buffer), getfamily(), gettype() and getproto() .
SocketServer	Generic socket server classes.
spwd	Access to the UNIX shadow password database [Unix].
sqlite3	DB-API 2.0 interface for SQLite databases.

sre	Support for regular expressions (RE). See re.
stat	Constants/functions for interpreting results of os.
statcache	Maintain a cache of stat() information on files.
statvfs	Constants for interpreting statvfs struct as returned by os.statvfs() and os.fstatvfs() (if they exist).
string	A collection of string operations (see Strings).
StringIO	File-like objects that read/write a string buffer (a faster C implementation exists in built-in module cStringIO).
stringprep	Normalization and manipulation of Unicode strings.
struct	Perform conversions between Python values and C structs represented as Python strings. 2.5: faster (new pack() and unpack() methods); pack and unpack to and from buffer objects via methods pack_into and unpack_from.
subprocess	Subprocess management. Replacement for os.system, os.spawn*, os.popen*, popen2.* [PEP324]
sunau	Stuff to parse Sun and NeXT audio files.
sunaudio	Interpret sun audio headers.
symbol	Non-terminal symbols of Python grammar (from "graminit.h").
symtable	Interface to the compiler's internal symbol tables.
sys	System-specific parameters and functions.
syslog	Unix syslog library routines [Unix].
tabnanny	Check Python source for ambiguous indentation.
tarfile	Tools to read and create TAR archives. 2.5: New method TarFile.extractall().
telnetlib	TELNET client class. Based on RFC 854.
tempfile	Temporary files and filenames.
termios	POSIX style tty control [Unix].
test	Regression tests package for Python.
textwrap	Tools to wrap paragraphs of text.
thread	Multiple threads of control (see also threading below).
threading	New threading module, emulating a subset of Java's threading model. 2.5: New function stack_size([size]) allows to get/set the stack size for threads created.
threading_api	(doc of the threading module).
time	Time access and conversions.
timeit	Benchmark tool.
Tix	Extension widgets for Tk.
Tkinter	Python interface to Tcl/Tk.
toaiff	Convert "arbitrary" sound files to AIFF (Apple and SGI's audio format).
token	Token constants (from "token.h").

tokenize	Tokenizer for Python source.
trace	Tools to trace execution of a function or program.
traceback	Extract, format and print information about Python stack traces.
tty	Terminal utilities [Unix].
turtle	LogoMation-like turtle graphics.
types	Define names for all type symbols in the std interpreter.
tzparse	Parse a time zone specification.
unicodedata	Interface to unicode properties. 2.5: Updated to Unicode DB 4.1.0; Version 3.2.0 still available as unicodedata.ucd_3_2_0.
unittest	Python unit testing framework, based on Erich Gamma's and Kent Beck's JUnit.
urllib	Open an arbitrary URL.
urllib2	An extensible library for opening URLs using a variety of protocols.
urlparse	Parse (absolute and relative) URLs.
user	Hook to allow user-specified customization code to run.
uu	Implementation of the UUencode and UUdecode functions.
uuid	UUID objects according to RFC 4122.
warnings	Python part of the warnings subsystem. Issue warnings, and filter unwanted warnings.
wave	Stuff to parse WAVE files.
weakref	Weak reference support for Python. Also allows the creation of proxy objects. 2.5: new methods iterkeyrefs(), keyrefs(), itervaluerefs() and valuerefs().
webbrowser	Platform independent URL launcher. 2.5: several enhancements (more browsers supported, etc...).
whichdb	Guess which db package to use to open a db file.
whrandom	Wichmann-Hill random number generator (obsolete, use random instead).
winsound	Sound-playing interface for Windows [Windows].
wsgiref	WSGI Utilities and Reference Implementation.
xdrlib	Implements (a subset of) Sun XDR (eXternal Data Representation).
xmllib	A parser for XML, using the derived class as static DTD.
xml.dom	Classes for processing XML using the DOM (Document Object Model). 2.3: New modules expatbuilder, minicompat, NodeFilter, xmlbuilder.
xml.etree.ElementTree	Subset of Fredrik Lundh's ElementTree library for processing XML.
xml.parsers.expat	An interface to the Expat non-validating XML parser.

xml.sax	Classes for processing XML using the SAX API.
xmlrpclib	An XML-RPC client interface for Python. 2.5: Supports returning datetime objects for the XML-RPC date type.
xreadlines	Provides a sequence-like object for reading a file line-by-line without reading the entire file into memory. Deprecated since release 2.3. Use for line in file instead. Removed since 2.4
zipfile	Read & write PK zipped files. 2.5: Supports ZIP64 version, a .zip archive can now be larger than 4GB.
zipimport	ZIP archive importer.
zlib	Compression compatible with gzip. 2.5: Compress and Decompress objects now support a copy() method.
xml.etree.ElementTree	Subset of Fredrik Lundh's ElementTree library for processing XML.
xml.parsers.expat	An interface to the Expat non-validating XML parser.
xml.sax	Classes for processing XML using the SAX API.
xmlrpclib	An XML-RPC client interface for Python. 2.5: Supports returning datetime objects for the XML-RPC date type.
xreadlines	Provides a sequence-like object for reading a file line-by-line without reading the entire file into memory. Deprecated since release 2.3. Use for line in file instead. Removed since 2.4
zipfile	Read & write PK zipped files. 2.5: Supports ZIP64 version, a .zip archive can now be larger than 4GB.
zipimport	ZIP archive importer.
zlib	Compression compatible with gzip. 2.5: Compress and Decompress objects now support a copy() method.

TABLE D.61: Workspace Exploration and Idiom Hints

`dir(object)`	List valid attributes of *object* (which can be a module, type or class object).
`dir()`	List names in current local symbol table.
`if __name__ == '__main__':` `main()`	Invoke main() if running as script.
`map(None, lst1, lst2, ...)`	Merge lists; see also zip(lst1, lst2, ...).
`b = a[:]`	create a copy b of sequence a.
`b = list(a)`	If a is a list, create a copy of it.
`a,b,c = 1,2,3`	Multiple assignment, same as a=1; b=2; c=3.
`for key, value in dic.items(): ...`	Works also in this context.
`if 1 < x <= 5: ...`	Works as expected.
`for line in fileinput.input(): ...`	Process each file in command line args, one line at a time.
`_ (underscore)`	In interactive mode, refers to the last value printed.

Appendix E

Answers to Odd-Numbered Questions

E.1   Chapter 2

1. Define: Program, instruction, and variable.

    A **program** is a set of ordered instructions designed to command the computer to do something. An **instruction** is a single order, for example: `print`, `add`, and `append`. A **variable** is a value that may vary during program execution. Note that in Pyhon the term *variable* doesn't fully describe the way data is handled, that is why they are referred as *names*.

3. Name some Python implementations.

    cPython, Jython, IronPython, Stackless, and PyPy.

5. What is the difference between input and raw_input in Python 2.x?

    input expects a valid Python expression as input while **raw_input** reads a line from input and converts it to a string. **input** represents a security risk since a malicious user can enter unchecked code into your program.

7. How do you make a float division in Python 2.x?

    One of the member of the division must be float. To make 1/2, you can do 1.0/2 or float(1)/2. As an alternative, import division from __future__ :

    ```
    >>> from __future__ import division
    >>> 1/2
    0.5
    ```

9. What is a comment in a source code?

    A comment is part of the code that is not executed. It purpose is to annotate what the programmer is attempting to accomplish with the source code.

11. What is a "shebang"?

Is a way to tell the operation system (at leat those linux based) where is the path to Python interpreter. It is used to execute the script without calling the interpreter in a explicit way. When there is more than one interpreter available, a shebang line tells the computer which one to use.

E.2 Chapter 3

1. Which are the principal data types in Python?

String, List, Tuples, Set, and Dictionary.

3. What is a set and when would you use it?

A set is defined as an *unordered collections of unique elements*. And it is used to test membership and to store data that we may apply set operations like *intersection, difference, union, issubset*, and others.

5. What is a dictionary?

It is defined as an unordered set of key:value pairs.

7. What is a "dictionary view"?

Since Python 3, objects returned by `dict.keys()`, `dict.values()`, and `dict.items()` are *dictionary views*. They are dynamic in the sense that when the dictionary changes, this objects also change. Compare it with what is returned by `dict.keys()`, `dict.values()`, and `dict.items()` in Python versions before Python 3.

9. Sort the data types below according to the following criteria:

- Mutable - immutable
- Sorted - unsorted
- Sequence - mapping

Data types to sort: Lists, tuples, dictionaries, sets, strings.

List: Mutable, sorted, and sequence.

Tuple: Immutable, sorted, and sequence.

Dictionary: Mutable, unsorted, and mapping.

Set: Mutable, unsorted, and sequence.

String: Immutable, sorted, sequence.

11. How do you convert any iterable data type into a list?

    With the `list()` function:

    ```
    >>> list('Hello')
    ['H', 'e', 'l', 'l', 'o']
    ```

13. How do you create a list from a dictionary?

    With the `list()` function:

    ```
    >>> list({'a': 1, 'b': 2})
    ['a', 'b']
    ```

E.3   Chapter 4

1. What is a control structure?

   Control structures are instructions to direct the flow of the program.

3. When would you use **for** and when would you use **while**?

   for is used to walk thought an (iterable) object and **while** is used to repeat a block of code until a condition is met.

5. Explain when you would use **pass** and when you would use *break*.

   pass is used as a placeholder where there is a syntactic need to put code but the program requires no action. **break** is used to exit a loop.

7. Make a program that outputs all possible IP addresses, that is, from 0.0.0.0 to 255.255.255.255.

   There are several ways to archive the same result. Here is one:

   ```
   iprange = range(256)
   for i in iprange:
       for j in iprange:
           for k in iprange:
               for l in iprange:
                   print '%s.%s.%s.%s'%(i,j,k,l)
   ```

9. Make a program to check if a given number is a palindrome (that is, it remains the same when its digits are reversed, like 404).

```
n = raw_input("Enter a number: ")
if n == n[::-1]:
    print "Palindrome"
else:
    print "Not palindrome"
```

11. Make a program that converts everything you type into Leetspeak, using the following equivalence: 0 for O, 1 for I (or L), 2 for Z (or R), 3 for E, 4 for A, 5 for S, 6 for G, 7 for T, 8 for B and 9 for P (or Q). So "Hello world!" is rendered as "H3770 w02ld!"

Without the use of a dictionary:

```
x = raw_input('Enter a string: ')
x = x.replace('O','0').replace('o','0').replace('I','1')
x = x.replace('i','1').replace('L','1').replace('l','1')
x = x.replace('Z','2').replace('z','2').replace('R','2')
x = x.replace('r','2').replace('E','3').replace('e','3')
x = x.replace('A','4').replace('a','4').replace('S','5')
x = x.replace('s','5').replace('G','6').replace('g','6')
x = x.replace('T','7').replace('t','7').replace('B','8')
x = x.replace('b','8').replace('P','9').replace('p','9')
x = x.replace('Q','9').replace('q','9')
print x
```

Using a dictionary:

```
e2l = {'O':0,'I':1, 'L':1, 'R':2,'E':3, 'A':4, 'S':5,
       'G':6,'T':7,'B':8,'P':9,'Q':9}
xin = raw_input('Enter a string: ')
xout = ''
for x in xin:
    xout += str(e2l.get(x.upper(),x))
print xout
```

13. Given a protein sequence in the one letter code, calculate the percentage of methionine (M) and cysteine (C). For example from MFKFASAVILC-LVAASSTQA the result must be 10% (1 M and 1 C over 20 amino acids).

```
seq = raw_input('Enter a sequence: ').upper()
m = float(seq.count('M'))
c = seq.count('C')
res = (m+c)/len(seq)*100
print '%2.f%%'%res
```

E.4 Chapter 5

1. What is the difference between "w" and "a" modes if both allow to write files?

   w is used to write a new file while *a* is used to append data to an existing file.

3. Program 5.9 estimates the average in line 6. Instead of dividing over the total number of rows, it does on the total less one. Why?

   Because the first row contains the header and not actual data.

5. Is it possible to parse csv files without `csv` module? If so, how it is done?

   It is possible to parse csv files without `csv` module, using the `split` method.

7. What is the most efficient way to walk through a file line by line?

```
for line in open(FILENAME):
    # do something with line
```

9. Make a program to detect in a text which lines have two consecutive identical words. To detect typos like "the the."

```
filename = # enter a file name here
fin = open(filename,'ru')
for line in fin:
    pword = ' '
    for word in line.replace('\n','').split(' '):
        if len(line.split(' '))>1 and word!='' and word==pword:
            print line
            break
        else:
            pword = word
fin.close()
```

E.5 Chapter 6

1. What is a function?

   A function is a portion of code that is defined once and may be used multiple times. It helps code reusability and maintainance.

3. Can a function be called without any parameters?

Yes. Try for example: **dir()**.

5. Does every function need to know in advance how many parameters will receive?

It depends on how the function was defined. When defined with ***args** or ****kwargs** as argument, it may receive a non-determined number of arguments.

7. Why must all optional arguments in a function be placed at the end in the function call?

Since arguments are put in the same order as originally defined and optional arguments may be omitted, to preserve the order, optional arguments are placed to the end.

9. Why are modules invoked at the beginning of the program?

A module must be invoked before using it. It is invoked at the beginning for convention.

11. How can you test if your code is being executed as a stand alone program or called as a module?

Checking the contents of the __name__ variable:

```
if __name__ == "__main__":
    main()
```

E.6 Chapter 7

1. What is the meaning of LBYL and EAFP? Which one is used in Python?

LBYL stands for "Look Before You Leap" and EAFP stands for "It's Easier to Ask Forgiveness Than Permission." Python support both methods to handling errors.

3. What is an "unhandled exception"?

Is an exception that is not caught by any code.

5. Exceptions are often associated with file handling. Why?

File operations are prone to errors (as disk full, disk write-protected, file or directory not found) to both are usually introduced together.

7. Why is it not advisable to use **except** to catch all kind of exceptions, instead of using, for example, **except IOError**?

   Using **except** to catch all exceptions doesn't allow us to take an apropiate action for a specific error. Another problem is that we may think that we know what error is catched but we may be missing (or overlooking) another type of error without realizing it.

9. What is the purpose of **sys.exc_info()**?

   It gives information about the exception that is currently being handled.

E.7   Chapter 8

1. Why is Python often characterized as a multi-paradigm language?

   Because Python supports both the procedural object paradigm. It doesn't force programmers to use a particular paradigm.

3. Explain the following concepts: Inheritance, Encapsulation, and Polymorphism.

   Inheritance: When methods and attributes are transmitted between related classes (from parent to child classes). Encapsulation: The ability to hide the internal operation of an object and leave access for the programmers only through their public methods. Polymorphism: The ability of different types of objects to respond to the same method with a different behaviour.

5. What is a special method attribute? Name at least four.

   It is a method that is executed under a pre-established condition, for example: __iter__, __len__, __str__, and __setitem__.

7. What is a private method? Are they really private in Python?

   A private method is a method that is intended to be used only inside an object. There is no such a method in Python, but there is a convention that when a method is named with two underscores at the begining (__name), it is considered "private."

9. Define a class that keeps track of how many instances have instantiated.

```
class Foo(object):
    i = 0
    def __init__(self):
        Foo.i += 1
```

E.8   Chapter 9

1. What is a REGEX?

   REGEX is a shorthand for regular expression, that are expressions that sumarize a text pattern.

3. How text patterns search can be applied to biology?

   To look for DNA features that have known patterns like ribosome binding sites, promotors, enhancers, TATAbox and so on. Protein motives are also a good application for REGEX.

5. In Code 9.7 (page 170), the pattern used was "|\d|\n|\t". What other alternative could have been employed?

   "\s \d"

   or

   "[^a-zA-Z]"

7. Make a program to retrieve every e-mail address ending in `.com` present in every file in a given directory.

```
import os
import re

def retrv_email(f):
    for line in open(f):
        if regex.findall(line):
            for email in regex.findall(line):
                emails.append(email)
    return emails

dirname = raw_input('Enter directory name: ')
regex = re.compile('[A-Za-z0-9.-]+@[A-Za-z0-9.-]+.com$')
allfiles = os.listdir(dirname)
emails = []
for f in allfiles:
    if os.path.isfile(os.path.join(dirname,f)):
        emails = retrv_email(os.path.join(dirname,f))

print emails
```

9. Write a REGEX pattern to detect a HindII restriction site. This enzyme recognizes the DNA sequence GTYRAC (where "Y" means "C" or "T" and "R" means "G" or "A").

```
GT[CT]{1}[GA]{1}AC
```

E.9   Chapter 10

1. What is an Alphabet in Biopython? Name at least four.

   An alphabet is a Biopython object used to declare sequence type and letters.

   `DNAAlphabet`, `RNAAlphabet`, `SecondaryStructure`, and `ThreeLetter-Protein`.

3. What advantage provides a Seq object over a string?

   On Seq objects you can apply operations that are permitted to sequences. Seq objects have specific methods for dealing with common sequence manipulation (like **transcribe** and **translate**).

5. What is a MutableSeq object?

   Is a Seq object that is not "read-only".

7. Name the methods of the SeqIO module.

   `parse`, `read`, `to_alignment`, `to_dict` and `write`.

9. Name five functions found in SeqUtils.

   `GC`, `molecular_weight`, `nt_search`, `reverse`, and `six_frame_translations`.

11. What module would you use to retrieve data from the NCBI Web server?

    The `Bio.Entrez` module.

E.10   Chapter 11

1. What is CGI?

   CGI, Common Gateway Interface, is a protocol to connect an application, written in any language with a Web server.

3. How do you use **cgi.FieldStorage** to retrieve values sent over an HTML form?

   First instanciate the **FieldStorage** class and then call the **getvalue** method (in this case, the field to retrieve is called **username**):

```
form = cgi.FieldStorage()
name = form.getvalue("username","NN")
```

5. What is WSGI? Why is it the recommended choice for Web programming?

WSGI, **W**eb **S**erver **G**ateway **I**nterface) is a specification for Web servers and application servers to communicate with Web applications. It is the recommended choice because it can be easily deployed in any WSGI compatible server.

7. Python includes a limited Web server. Why would you use such a web server if there are free full featured Web servers like Apache?

This server is meant for testing purposes. It requires no installation and it is available in any platform that Python runs.

9. Why is client-side data validation not useful as server-side data validation?

Because you should never trust on data sent by the client. Client-side data validation can be by-passed.

E.11   Chapter 12

1. What does the OpenOffice format have in common with RSS feeds and GoogleEarth's geographic coordinates?

All those files use XML to store data.

3. When you will not use XML?

For simple configuration files, XML is an overkill. XML is very verbose, so is no the best container for large amounts of data.

5. Distinguish between the terms: tag, element, attribute, value, DTD, and Schema.

Tag: Keywords written between angled brackets that defines elements.

Element: An element is the information from the beginning of the start tag to the end of the end tag, including all that lies in between. For example, in: `<data>23</data>`, `<data>23</data>` is the element, 23 is the element content and `<data></data>` are start and end tags.

Attribute: Is an optional information that is related to an element. For example, in: `<seq len="8" checksum="F188A">acggtcga</seq>`, both `len` and `checksum` are atributes.

Value: Is referred to the value of the attribute, in `<seq len="8" checksum ="F188A">acggtcga</seq>` there are two values: 8 and F188A.

DTD: Document Type Definition. It contains information about the particular structure of the XML file: permited tags and attributes, as well as where they can be found.

Schema: XML Schemas contain the same type of information as a DTD file, but is based in XML.

7. What is the difference between the SAX and DOM models of XML file processing?

   SAX is based on events and doesn't build a tree in memory while DOM works by reading the whole document and creating a tree based on the XML document.

9. In **cElementTree.iterparse** there are both **start** and **end** event types. By default it returns only **end** event. When would you use the information in a **start** event?

   I would you use the information in a **start** event when I need only the event name and attributes.

E.12   Chapter 13

1. What is a database?

   A database is an ordered collection of related data.

3. What is a relational database?

   A relational database is a database that groups data using common attributes.

5. What is SQL?

   SQL is a database computer language designed for managing data in relational database management systems (RDBMS).

7. Translate this query into English:
   `SELECT LastName,Score FROM Student,Scores WHERE Scores.Score>3;`

   Shows the information on the fields `LastName` and `Score` from the tables `Student` and `Scores`, but only when the field `Score` from the table `Scores` is greater than three.

9. When is it appropriate to use SQLite?

Where the data you want to store doesn't require a full feature database, when your application has few concurrent users and when you don't want distribute an external database server with your application.

E.13   Chapter 14

1. What is version control software?

   It is a program that allows a developer or a group of developers to handle multiple revisions of the same set of documents. With this kind of software a group of developers can coordinate their work.

3. Why would a single programmer may use such a program?

   If a programmer works on different locations (like home and work) and want to keep track of different versions. To go back to a previous point. To have multiple versions of the same program.

5. Define (in the context of version control): repository, branch, commit, merge, and check-out.

   Repository: The place where all the shared files and complete revision history are stored.

   Branch: A set of files under version control which may be branched or forked at a point in time so that, from that time forward, two copies of those files may be developed at different speeds or in different ways independently of the other.

   Commit: When a change made by a programmer is written into the repository (either personal or shared).

   Merge: A merge brings together two sets of changes from a set of files into a unified version of these files.

   Check-out: Creates a copy of the code from the repository. Usually the latest version is requested, but also a specific version can be retrieved if needed.

7. What kind of server is needed to publish a branch using Bazaar?

   There is no special need on the server side to publish a bransh using Bazaar. As long as users has access to those files, any kind of service can be used (such as FTP, sFTP, Remote Directory).

9. What is a patch file and how do you submit one?

   A patch is a file that contains information on what is changed from una version of the code to another version. A patch can be submited by e-mail, uploaded to a web interface, or applied to the target source code.

Appendix F

Python Style Guide

F.1 Introduction

The notion of coding style was introduced on page 5 and it is expected that all code listing in the book has served the reader to get the feeling of how code should look like. This reference chapter has a more formal approach to this subject.

Remember that keeping a consistent coding style helps to keep your code clean and makes bugs easier to spot. There are programs like pep8.py[1] to help you format your code properly.

This document is based on coding conventions posted for the Python code in the One Laptop Per Child project, which in turn was adapted from several sources: Guido's original Python Style Guide essay,[2] with some additions from Barry's style guide.[3]

Copyright notice for this chapter (Python Style Guide):

Portions of this text are from PEP 8 that are in the public domain. Other portions are under a CC-Attribution license, that means you are free to distribute the contents of this chapter as long as you attribute authorship to OLPC and link it to this URL: `http://wiki.laptop.org/go/Python_Style_Guide`. For a complete description of this license, please see: `http://creativecommons.org/licenses/by/2.5`.

A Foolish Consistency Is the Hobgoblin of Little Minds

One of Guido's key insights is that code is read much more often than it is written. The guidelines provided here are intended to improve the readability

---

[1]Source code available at `http://svn.browsershots.org/trunk/devtools/pep8/pep8.py`.
[2]`http://www.python.org/doc/essays/styleguide.html` and `http://www.python.org/dev/peps/pep-0008/`.
[3]`http://barry.warsaw.us/software/STYLEGUIDE.txt`

of code and make it consistent across the wide spectrum of Python code. As PEP 20 says, "Readability counts."[4]

A style guide is about consistency. Consistency with this style guide is important. Consistency within a project is more important. Consistency within one module or function is most important.

But most importantly: know when to be inconsistent – sometimes the style guide just doesn't apply. When in doubt, use your best judgment. Look at other examples and decide what looks best. And don't hesitate to ask!

Two good reasons to break a particular rule:

1. When applying the rule would make the code less readable, even for someone who is used to reading code that follows the rules.

2. To be consistent with surrounding code that also breaks it (maybe for historic reasons)—although this is also an opportunity to clean up someone else's mess (in true XP style).

A Note on Consistency

When you are interfacing with another library and providing a Python wrapping for its functions, you should always adopt the naming style of that library.

If you are changing the style of a piece of code, this should be done all at once and no other changes should be made at the same time. Whitespace changes in particular should be done separate from even naming changes.

F.2   Code Lay-Out

Indentation

Use 4 spaces per indentation level. Do not use tabs.

The number of spaces used can be easily changed with a script. I think we should give serious consideration to reducing this to 2 spaces per indentation level to minimize the number of line breaks needed and also minimize the whitespace on a screenful of code. Admittedly, lots of people, using 19 and 21 inch monitors, currently use a 4-space standard, but that can be easily fixed with a simple script. Python has a built-in parser module that can be used to do this. If all the code lives in a repository such as SVN, then this can be done as part of the code check-in process without anyone needing to

---

[4]http://www.python.org/dev/peps/pep-0020, also available by typing "import this" at the Python shell.

think about it. However, the end-users of the laptop, working on their small screens, will thank you for it.

Maximum Line Length

Limit all lines to a maximum of 79 characters.

There are still many devices around that are limited to 80 character lines. The default wrapping on such devices looks ugly. Plus, limiting windows to 80 characters on a large display makes it possible to have several windows side-by-side.

Therefore, please limit all lines to a maximum of 79 characters. For flowing long blocks of text (docstrings or comments), limiting the length to 72 characters is recommended.

The preferred way of wrapping long lines is by using Python's implied line continuation inside parentheses, brackets and braces. If necessary, you can add an extra pair of parentheses around an expression, but sometimes using a backslash looks better. Make sure to indent the continued line appropriately. Some examples,

```python
class Rectangle(Blob):

    def __init__(self, width, height,
                 color='black', emphasis=None, highlight=0):
        if width == 0 and height == 0 and \
           color == 'red' and emphasis == 'strong' or \
           highlight > 100:
            raise ValueError("sorry, you lose")
        if width == 0 and height == 0 and (color == 'red' or
                                           emphasis is None):
            raise ValueError("I don't think so")
        Blob.__init__(self, width, height,
                      color, emphasis, highlight)
```

Assert statements in particular tend to go over the line boundaries; so generally asserts should look like this:

```python
assert value is not None, (
    "value should not be None")
```

Blank Lines

Vertical whitespace (blank lines) are not that important to readability. For the most part this can be left to the developer's discretion. As a general guideline,

- Separate top-level function and class definitions with two blank lines.

- Method definitions inside a class are separated by a single blank line.

- Extra blank lines may be used (sparingly) to separate groups of related functions. Blank lines may be omitted between a bunch of related one-liners (e.g. a set of dummy implementations).

- Use blank lines in functions, sparingly, to indicate logical sections.

Encodings (PEP 263)

Python source must contain a Unicode UTF-8 encoding declaration, which looks like:

```
# coding: UTF8
```

Only UTF8 should be used even if you are not using non-ASCII characters in your code. The reason is to make it easy for others to take up any Python file, make modifications and add comments in their own language.

As a special case a file with the UTF8 signature '\xef\xbb\xbf' at the beginning of the file will be detected by Python as a UTF8 file. Do not use or rely on this signature since some editors will remove it. Always include the UTF-8 encoding declaration.

Note that you cannot use unicode in any identifiers in Python; the encoding only applies to Unicode strings like u"a string" and comments. Long strings of text (that are not English) should be in localization files, not in the code itself.

Imports

Imports should usually be on separate lines, e.g.,

```
Yes: import os
     import sys
```

```
No:  import sys, os
```

it's okay to say this though:

```
from subprocess import Popen, PIPE
```

[note: this is a soft requirement]

Imports are always put at the top of the file, just after any module comments and docstrings, and before module globals and constants.

Imports should be grouped in the following order: standard library imports, related third party imports, and application/library specific imports.

You should put a blank line between each group of imports. [note: I don't care about the blank line, and consider the ordering to be only a suggestion]

Put any relevant __all__ specification after the imports.

Relative imports for intra-package imports are highly discouraged.

Always use the absolute package path for all imports. If or until we settle on Python 2.5 we cannot use PEP 328, and so cannot do explicit relative imports.

"from x import *" is generally discouraged.

You should only import this way from packages that are intended to be used like this (the packages generally define __all__).

You should never use "import *" more than once in a file. If you use it more than once then there is no way to know (without leaving the file) exactly where a name comes from. So long as "import *" is used just once, one can assume when no other source can be found for a name that it must come from this import.

When importing a class from a class-containing module

It's usually okay to spell this:

```
from myclass import MyClass
from foo.bar.yourclass import YourClass
```

If this spelling causes local name clashes, then spell them

```
import myclass
import foo.bar.yourclass
```

and use `myclass.MyClass` and `foo.bar.yourclass.YourClass`

In summary, a file should generally look like this:

```
# -*- coding: UTF8 -*-   (MUST always be used)
"""
docstring: may also be a unicode or 'raw' string
If you are using doctest then a raw string is recommended
(prefix the string with an r)
[are unicode strings generally preferred for docstrings?
that would give a prefix or u or ur]
"""
from __future__ ...
import stdlib modules
import external modules
import internal modules
__all__ = [...]   # If you use __all__
constants...
functions and classes...
```

F.3   __init__.py Files

__init__.py files should generally contain no substantive code. Instead they should import from other modules. Importing from other modules is done so that a package can provide a front-facing set of objects and functions it exports, without exposing each of the internal modules in the package. Note however that this causes the submodules to be eagerly imported; if this is likely to cause unnecessary overhead then the import in __init__.py should be reconsidered.

F.4   Whitespace in Expressions and Statements

Pet Peeves

Avoid extraneous whitespace in the following situations:
Immediately inside parentheses, brackets, or braces.

```
Yes: spam(ham[1], {eggs: 2})
No:  spam( ham[ 1 ], { eggs: 2 } )
```

Immediately before a comma, semicolon, or colon:

```
Yes: if x == 4: print x, y; x, y = y, x
No:  if x == 4 : print x , y ; x , y = y , x
```

[Note: if you do not put a space after a comma, it is harder to visually distinguish from: e.g., foo(a,b) and foo(a.b). Please use spaces after commas!]
Immediately before the open parenthesis that starts the argument list of a function call:

```
Yes: spam(1)
No:  spam (1)
```

Immediately before the open parenthesis that starts an indexing or slicing:

```
Yes: dict['key'] = list[index]
No:  dict ['key'] = list [index]
```

More than one space around an assignment (or other) operator to align it with another.

Yes:

```
x = 1
y = 2
long_variable = 3
```

No:

```
x             = 1
y             = 2
long_variable = 3
```

[note: I'm soft on this one, though less soft on the others]

F.5 Other Recommendations

Always surround these binary operators with a single space on either side: assignment (=), augmented assignment (+=, -= etc.), comparisons (==, <, >, !=, <>, <=, >=, in, not in, is, is not), Booleans (and, or, not).

Use spaces around arithmetic operators:

Yes:

```
i = i + 1
submitted += 1
x = x * 2 - 1
hypot2 = x * x + y * y
c = (a + b) * (a - b)
```

No:

```
i=i+1
submitted +=1
x = x*2 - 1
hypot2 = x*x + y*y
c = (a+b) * (a-b)
```

Don't use spaces around the '=' sign when used to indicate a keyword argument or a default parameter value.

Yes:

```
def complex(real, imag=0.0):
```

```
        return magic(r=real, i=imag)
```

No:

```
    def complex(real, imag = 0.0):
        return magic(r = real, i = imag)
```

[note: this is really helpful to make the code more readable; please use this convention. Keyword arguments aren't assignments, and this makes that visually clear.]

Compound statements (multiple statements on the same line) are strongly discouraged.

Yes:

```
    if foo == 'blah':
        do_blah_thing()
    do_one()
    do_two()
    do_three()
```

Rather not:

```
    if foo == 'blah': do_blah_thing()
    do_one(); do_two(); do_three()
```

Don't be lazy, just hit enter!

if/else expressions and list comprehensions should not be deeply nested.

Yes:

```
    if x>5 and t>10 and m<20:
        print x,t,m
```

No:

```
    if x>5:
        if t>10:
            if m<20:
                print x,t,m
```

F.6 Comments

Comments that contradict the code are worse than no comments. Always make a priority of keeping the comments up-to-date when the code changes!

Comments should go before the thing they are commenting on, like:

```
# match will be the regex match object:
match = None
```

Or sometimes inside an if statement or other control structure:

```
if match is None:
    # None of our attempts to match worked
    raise ValueError("Nothing matched!")
```

Comments should be complete grammatically correct sentences.

If a comment is short, the period at the end can be omitted. Block comments generally consist of one or more paragraphs built out of complete sentences, and each sentence should end in a period.

Regardless of the language you use, you should write clear and easily understandable sentences. If you use English, many readers will only understand basic English. If you use your native language, many readers will be children who are still learning their language.

When choosing the language for comments, think of who will have to read these comments. If you are writing code that will be used by people in many countries, then English is probably the best choice.

Block Comments

Block comments generally apply to some (or all) code that follows them, and are indented to the same level as that code. Each line of a block comment starts with a # and a single space (unless it is indented text inside the comment).

Paragraphs inside a block comment are separated by a line containing a single #.

Inline Comments

Use inline comments sparingly.

An inline comment is a comment on the same line as a statement. Inline comments should be separated by at least two spaces from the statement. They should start with a # and a single space.

Inline comments are unnecessary and in fact distracting if they state the obvious. Don't do this:

```
x = x + 1                        # Increment x
```

But sometimes, this is useful:

```
x = x + 1                        # Compensate for border
```

Generally comments on separate lines are easier to edit:

```
# Compensate for border:
x = x + 1
```

F.7   Documentation Strings

Conventions for writing good documentation strings ("docstrings") are immortalized in PEP 257.[5]

Write docstrings for all public modules, functions, classes, and methods. Docstrings are not necessary for nonpublic methods, but you should have a comment that describes what the method does. This comment should appear after the "def" line.

PEP 257 describes good docstring conventions. Note that most importantly, the """ that ends a multiline docstring should be on a line by itself, and preferably preceded by a blank line, e.g.:

```
"""Return a foobang

Optional plotz says to frobnicate the bizbaz first.

"""
```

For one liner docstrings, it's okay to keep the closing """ on the same line. Avoid using ' ' ' for docstrings.

F.8   Naming Conventions

Descriptive: Naming Styles

There are a lot of different naming styles. It helps to be able to recognize what naming style is being used, independently from what they are used for.

The following naming styles are commonly distinguished:

---

[5]`http://www.python.org/dev/peps/pep-0257`

- b (single lowercase letter)

- B (single uppercase letter)

- lowercase

- lower_case_with_underscores

- UPPERCASE

- UPPER_CASE_WITH_UNDERSCORES

- CapitalizedWords (or CapWords, or CamelCase – so named because of the bumpy look of its letters). This is also sometimes known as StudlyCaps. (Note: When using abbreviations in CapWords, capitalize all the letters of the abbreviation. Thus HTTPServerError is better than HttpServerError.)

- mixedCase (differs from CapitalizedWords by initial lowercase character!)

- Capitalized_Words_With_Underscores (ugly!)

There's also the style of using a short unique prefix to group related names together. This is not used much in Python, but it is mentioned for completeness. For example, the os.stat() function returns a tuple whose items traditionally have names like st_mode, st_size, st_mtime and so on. (This is done to emphasize the correspondence with the fields of the POSIX system call struct, which helps programmers familiar with that.)

The X11 library uses a leading X for all its public functions. In Python, this style is generally deemed unnecessary because attribute and method names are prefixed with an object, and function names are prefixed with a module name.

In addition, the following special forms using leading or trailing underscores are recognized (these can generally be combined with any case convention):

_single_leading_underscore:

Weak "internal use" indicator. E.g. "from M import *" does not import objects whose name starts with an underscore.

single_trailing_underscore_:

used by convention to avoid conflicts with Python keyword, e.g.

```
Tkinter.Toplevel(master, class_='ClassName')
```

__double_leading_underscore:

When naming a class attribute, invokes name mangling (inside class FooBar, __boo becomes _FooBar__boo; see below).

__double_leading_and_trailing_underscore__:

"magic" objects or attributes that live in user-controlled namespaces. E.g. __init__, __import__ or __file__. Never invent such names; only use them as documented.

Prescriptive: Naming Conventions

Names to Avoid

Never use the characters 'l' (lowercase letter el), 'O' (uppercase letter oh), or 'I' (uppercase letter eye) as single character variable names.

In some fonts, these characters are indistinguishable from the numerals one and zero. When tempted to use 'l', use 'L' instead.

Do not abbreviate names by removing vowels. Instead truncate the name.

Yes:

```
func
decl
```

No:

```
fnctn
dcln [note: these aren't very good examples, because they are just
    *too* ugly to be plausible...]
```

F.8.1 Module Names

Modules should have short, lowercase names, without underscores.

This naming convention distinguishes modules from both functions and classes. This is important; consider this example from Zope 2:

```
from DateTime.DateTime import DateTime
```

In Zope 2 the DateTime package contained a DateTime module with a DateTime class. As a result when you see "DateTime" in the source you can't be sure if it's referring to the package, module, or class. If the module had been named datetime it would be obvious when you were referring to the module and when you were referring to the class. Similar confusion can exist with functions, which is the motivation for leaving underscores out of module names (but using them in function names).

When an extension module written in C or C++ has an accompanying Python module that provides a higher level (e.g. more object oriented) interface, the C/C++ module has a leading underscore (e.g. _socket).

Like modules, Python packages should have short, all-lowercase names, without underscores.

Class Names

Almost without exception, class names use the CapWords convention. Classes for internal use have a leading underscore in addition.

Exception Names

Because exceptions should be classes, the class naming convention applies here. However, you should use the suffix "Error" on your exception names (if the exception actually is an error).

Global Variable Names

(Let's hope that these variables are meant for use inside one module only.) The conventions are about the same as those for functions.

Modules that are designed for use via "from M import *" should use the __all__ mechanism to prevent exporting globals, or use the older convention of prefixing such globals with an underscore (which you might want to do to indicate these globals are "module nonpublic").

Many modules are not really intended to be used with "from M import *" and will export many unintended objects (like other modules). Generally you should not use "import *" unless a module is intended to be used like that, and the presence of __all__ is a good indication if a module is intended to be used that way.

Function Names

Function names should be lowercase, with words separated by underscores as necessary to improve readability.

mixedCase is allowed only in contexts where that's already the prevailing style (e.g. threading.py).

Function and Method Arguments

Always use 'self' for the first argument to instance methods.

Always use 'cls' for the first argument to class methods.

Always use 'metacls' for the first argument to metaclass method. These are technically class methods of the metaclass, but if you don't distinguish metaclasses from classes you will confuse readers terribly.

If a function argument's name clashes with a reserved keyword, it is generally better to append a single trailing underscore rather than use an abbreviation or spelling corruption. Thus print_ is better than prnt. (Perhaps better is to avoid such clashes by using a synonym.)

Method Names and Instance Variables

Use the function naming rules: lowercase with words separated by underscores as necessary to improve readability.

Use one leading underscore only for nonpublic methods and instance variables.

Do *not* use two leading underscores. Python mangles these names with the class name: if class Foo has an attribute named __a, it cannot be accessed by Foo.__a. (An insistent user could still gain access by calling Foo._Foo__a.) If you have some reason to want to avoid name clashes in subclasses, you should use *explicit* name mangling by using an explicit prefix in front of your attributes or functions, like Foo._Foo_a.

Designing for Inheritance

[note: this is rather complex; generally I think designing for inheritance should be avoided except in specific cases where it provides real benefits. In many cases first class functions and other techniques are easier to understand and manage than subclassing.]

Always decide whether a class's methods and instance variables (collectively: "attributes") should be public or nonpublic. If in doubt, choose nonpublic; it's easier to make it public later than to make a public attribute nonpublic.

Public attributes are those that you expect unrelated clients of your class to use, with your commitment to avoid backward incompatible changes. Nonpublic attributes are those that are not intended to be used by third parties; you make no guarantees that nonpublic attributes won't change or even be removed.

We don't use the term "private" here, since no attribute is really private in Python (without a generally unnecessary amount of work).

Another category of attributes are those that are part of the "subclass API" (often called "protected" in other languages). Some classes are designed to be inherited from, either to extend or modify aspects of the class's behavior. When designing such a class, take care to make explicit decisions about which attributes are public, which are part of the subclass API, and which are truly only to be used by your base class.

With this in mind, here are the Pythonic guidelines:

- Public attributes should have no leading underscores.

- If your public attribute name collides with a reserved keyword, append a single trailing underscore to your attribute name. This is preferable to an abbreviation or corrupted spelling. (However, notwithstanding this rule, 'cls' is the preferred spelling for any variable or argument which is known to be a class, especially the first argument to a class method.) Note 1: See the argument name recommendation above for class methods.

- For simple public data attributes, it is best to expose just the attribute name, without complicated accessor/mutator methods. Keep in mind that Python provides an easy path to future enhancement, should you find that a simple data attribute needs to grow functional behavior.

In that case, use properties to hide functional implementation behind simple data attribute access syntax. Note 1: Properties only work on new-style classes. Note 2: Try to keep the functional behavior side-effect free, although side-effects such as caching are generally fine. Note 3: Avoid using properties for computationally expensive operations; the attribute notation makes the caller believe that access is (relatively) cheap.

F.9   Programming Recommendations

Code should be written in a way that does not disadvantage other implementations of Python (PyPy, Jython, IronPython, Pyrex, Psyco, and such).

For example, do not rely on CPython's efficient implementation of in-place string concatenation for statements in the form a+=b or a=a+b. Those statements run more slowly in Jython. In performance sensitive parts of the library, the .join() form should be used instead. This will assure that concatenation occurs in linear time across various implementations.

[note: I think we can be softer about this, as we need to target more than just CPython but the performance characteristics of our particular software and hardware stack.]

Comparisons to singletons like None should always be done with 'is' or 'is not', never the equality operators.

Note is and is not compare the identity of an object. == can be overridden and does more complex comparisons, and so there is a small performance penalty. There is and only will ever be one None.

Also, beware of writing "if x" when you really mean "if x is not None" – e.g. when testing whether a variable or argument that defaults to None was set to some other value. The other value might have a type (such as a container) that could be false in a boolean context!

Use class-based exceptions.

String exceptions in new code are strongly discouraged, as are deprecated (since Python 2.5) and then (in Python 3000 or perhaps sooner) removed.

Modules or packages should define their own domain-specific base exception class, which should be subclassed from the built-in Exception class. Always include a class docstring.

E.g.:

```
class MessageError(Exception):
    """Base class for errors in the email package."""
```

Class naming conventions apply here, although you should add the suffix "Error" to your exception classes, if the exception is an error. Non-error exceptions need no special suffix.

When raising an exception, use `raise ValueError('message')` **instead of the older form** `raise ValueError, 'message'`.

The paren-using form is preferred because when the exception arguments are long or include string formatting, you don't need to use line continuation characters thanks to the containing parentheses. The older form will be removed in Python 3000.

Use string methods instead of the string module.

String methods are always much faster and share the same API with unicode strings. Override this rule if backward compatibility with Pythons older than 2.0 is required.

[note: we can be strict here. string.Template is an exception, which is the only reason the string module should be used at all.]

Use .startswith() and .endswith() instead of string slicing to check for prefixes or suffixes.

startswith() and endswith() are cleaner and less error-prone. For example:

```
Yes: if foo.startswith('bar'):
```

```
No:  if foo[:3] == 'bar':
```

Object type comparisons should always use isinstance() instead of comparing types directly.

```
Yes: if isinstance(obj, int):
```

```
No:  if type(obj) is type(1):
```

When checking if an object is a string, keep in mind that it might be a unicode string too! In Python 2.3, str and unicode have a common base class, basestring, so you can do:

```
if isinstance(obj, basestring):
```

In Python 2.2, the types module has the StringTypes type defined for that purpose, e.g.:

```
from types import StringTypes
if isinstance(obj, StringTypes):
```

In Python 2.0 and 2.1, you should do:

```
\begin{verbatim}
from types import StringType, UnicodeType
if isinstance(obj, StringType) or \
   isinstance(obj, UnicodeType) :
```

[note: obviously we can just use basestring, though we need to be careful about distinguishing str and unicode. It is valid and perhaps preferred for us to be careful in distinguishing these two values. `assert isinstance(value, unicode)` is probably an assert we should use liberally]

The exception is if your code must work with Python 1.5.2 (but let's hope not!). [note: clearly we don't]

For sequences, (strings, lists, tuples), use the fact that empty sequences are false.

```
Yes: if not seq:
     if seq:
```

```
No: if len(seq)
    if not len(seq)
```

Don't write string literals that rely on significant trailing whitespace.

Such trailing whitespace is visually indistinguishable and some editors (or more recently, reindent.py) will trim them.

[note: this only applies to multiline/triple-quoted strings]

Don't compare boolean values to True or False

Using:

```
Yes:    if greeting:
```

```
No:     if greeting == True:
```

```
Worse: if greeting is True:
```

F.10   Strings and Unicode

Generally there are three types of strings:

1. 8-bit strings ("str") that contain binary data

2. Unicode strings that contain textual data

3. Encoded strings, represented as 8-bit strs, that contain textual data

The third form can cause problems. Python is encoding agnostic; the only encoding it does automatically is ASCII. When using ASCII text, an encoded and unicode string look very similar; they compare as equal, they hash to the same value, and str() and unicode() will convert cleanly between the two. Once non-ASCII text is introduced this all breaks.

We should avoid encoded strings when possible. When we expect to receive unicode strings, it is acceptable and even encouraged to do `assert isinstance(value, unicode)`.

F.11   Internationalization and Localization

If you are writing code for use in many countries then all user-visible strings should be in English and should be translatable. You do this like so:

```
from gettext import gettext as _
import getpass

print _("Hello %(name)s!") % {'name': getpass.getuser()}
```

Note that string substitutions should be done after the translation via _(). Also, named values should be used. You may find string.Template preferable to %-based substitution; you can use it like:

```
import string
print string.Template(_("Hello $name!")).substitute(name=getpass<=
.getuser())
```

There's a long document on internationalizing Pylons, most of which applies to any Python i18n code.

F.12   Testing

The Testing Tool Taxonomy[6] provides a long and comprehensive list of test systems available for Python.

There are three core packages that can be used for testing:

doctest

doctest[7] is a standard library module, and a testing system. It's probably the simplest test system to use and read.

This is a common pattern for testing a package:

---

[6]http://pycheesecake.org/wiki/PythonTestingToolsTaxonomy
[7]http://python.org/doc/current/lib/module-doctest.html

```
if __name__ == '__main__':
    import doctest
    doctest.testmod()
    doctest.testfile('test_this_module.txt')
```

While this works, it's very easy to forget to run tests after making changes. It's also easy to forget to test for regressions. Because of this, you should provide a way to run all of your tests.

Note that you can put tests in your file's docstrings, or in an external text file. For tests that don't have documentation value an external text file is best (it won't clutter your source or the helpful information in your docstrings). For extended examples external files are also best; inline docstring doctests are mostly best to simply confirm those examples are correct, not to do extensive testing of your routines.

unittest

unittest[8] is the "standard" standard library testing module. It is modeled after SUnit, JUnit, etc. Tests using this tend to be somewhat long-winded, and not very readable (this is Ian's personal opinion, but he holds it very strongly).

When a project is already using unittest, you should use it for new tests to maintain consistency. Note that doctest can produce unittest-compatible tests.[9] When creating new tests, seriously consider using doctest, as the resulting tests are usually much more readable. This is less true for tests that contain considerable logic (especially things like stress testing, or using fuzzed input).

If you are using unittest-based tests you should provide a test runner as part of your code; this is a script that will run all the tests in your code. While some people use the same kind of __name__ == '__main__' trick for unittest that they do for doctest, this is not desirable (for all the same reasons).

nose

nose[10] is a (nonstandard) library/script for finding and running tests. It is based on unittest, and provides the tests collection that the other two modules are missing. It also can run doctests directly (without having to explicitly wrap them as unittests) and has some improved features over typical unittest test runners (like showing detail about failed assertions, and dropping into a

---

[8]http://python.org/doc/current/lib/module-unittest.html

[9]http://python.org/doc/current/lib/doctest-unittest-api.html.

[10]http://somethingaboutorange.com/mrl/projects/nose

debugger on failure). It has features very similar to py.test,[11] but is easier to install and is more compatible with unittest-based tests than py.test.

Nose also lets you use simpler tests than unittest's class-based tests. Functions with names starting with `test_` will be run.

If you use this test runner, it is recommended that you include a shell script or Python script to run nose with your project; this will make it easy for other developers to see how you run your tests.

File Names

Except for embedded doctests, tests should generally go in files separate from the module they are testing. This way importing the module will not load the tests and won't add any overhead unless you are actually running the tests.

Tests should be named test_modulename.... You can add more to the name if you have multiple files associated with one module. Use .py for Python-based files (of course), and .txt for external doctests. Tests are sometimes put in a subpackage called tests (note that test is unfortunately used by a very boring standard library module,[12] and it can lead to confusing situations if you use that name). It's also fine to simply put tests right beside the modules they test.

External doctest files that have documentation value should be named the same as the module (with .txt), and should not have a test_ prefix; their primary value is not the testing they do, but the information they convey. Ideally all programmer documentation will use doctest, so that the accuracy of the documentation can be easily confirmed.

F.13   Documentation

Deprecations and Warnings

When other people use code of yours, you will have to support them as you update your code. Even if you mark your package as being "version 0.1," it doesn't matter—if your code is useful, and someone uses it, then you'll need to start thinking about backward compatibility, or else make life difficult for your users.

Deprecations and warnings are specifically meant to deal with this. Warnings should seldom go in new code. For instance, you could do:

---

[11] http://codespeak.net/py/current/doc/test.html

[12] http://python.org/doc/current/lib/module-test.html

```
def send_content(dest, data):
    if not isinstance(data, str):
        warnings.warn('You should only send str data')
        data = str(data)
```

But because there are no current users (if this is new code), this should simply be an error:

```
def send_content(dest, data):
    assert isinstance(data, str), (
        "data should be a str, not %r" % data)
```

Then callers will see this error and call str(data) on their end, removing any potential ambiguity.

When you want to use warnings is when in the past you've allowed non-str data, and you want to change that. There is no firm rule about when you should simply turn something into an error, and when you should provide warnings.

If you provide a warning, it should be in this form:

```
import warnings
def send_content(dest, data):
    if not isinstance(data, str):
        # Deprecated since 2005-05-01
        warnings.warn('send_data(dest, data=%r) should only be<=
passed a str value for data',
                        DeprecationWarning, stacklevel=2)
        data = str(data)
```

DeprecationWarning is a category of warnings. You can disable warnings by category, or turn them into errors. stacklevel=2 means that the bad behavior happened at stack level 2 (the immediate caller of this function). This will show the caller's filename and line number in the warning. You might have to increase this number if you are using more indirection in your code.

Including the date of the deprecation in a comment makes it easier to determine when the deprecated usage should be turned into an error (after some time one can assume all callers have fixed their code).

When a function has been moved or removed, you should start with a warning and then turn it into an error like:

```
def send_content(dest, data):
    # Moved on 2005-07-10
    raise NotImplementedError(
        'The send_content function has been moved to mypkg.<=
content_sending.send_content')
```

You should not simply remove a public function; by putting in an error you tell callers exactly how they should update their code. Like warnings, these should eventually be removed. There should always be a stage where you make it an explicit error, as some users may ignore warnings entirely until it is turned into an error.

str, unicode, and repr

There are three ways to coerce an object to text: *str(obj)*, *unicode(obj)* and *repr(obj)*.

str coerces an object to its non-unicode textual representation. Though this is very commonly used, *unicode(obj)* should be preferred as it creates unicode text. As an example, here's code that can cause a problem:

```
class User(object):
    def __init__(self, name):
        self.name = name
    def __str__(self):
        return 'User %s' % self.name
u = User(u'Itrntinliztin')
# This works fine:
print repr(unicode(u))
# This won't work:
print str(u)
# This won't work either:
print u
# This won't work either:
'Hi ' + str(u)
```

So what happened there? Well, self.name was a unicode string. When you do 'User %s' % self.name it returns a unicode string as well (str strings are turned into unicode when used with % – this itself is a little scary). Then str() calls u.__str__(). It sees unicode, and it tries u'Internationalizatión'.encode('ascii') ('ascii' is sys.getdefaultencoding()).

What is scary is here is that if you do your testing with this:

```
u = User(u'Bob')
```

then everything will work, because u'Bob'.encode('ascii') succeeds.

The moral of this story? If you had implemented __unicode__ there would have been no problem. (Well, calling code could still be broken, but your code would not be broken.)

So:

- If you are dealing with textual data, use __unicode__ and unicode(obj).

- If you are really just dealing with binary data, __str__ is okay to use, but it is usually preferable to just use another method that returns the string/binary form of the object. It is not necessary to overload every magic method just because you can.

repr()

The repr(obj) form (and its __repr__ magic method) are intended for Programmer representations of objects. These are handy representations that you can use to see what kind of object it is. Sometimes it is true that eval(repr(obj)) == obj, but this should never be relied upon (in fact eval() should generally never be used). More to the point, the repr() of an object should show useful or interesting information about an object. It should never be confused for a textual description. It should never be shown to a user (unless that user is acting as a programmer). If you want to override the default repr() for an object (and that is encouraged!) The general form for this method is:

```
class User(object):
    def __init__(self, name):
        self.name = name
    def __repr__(self):
        return '<%s %s name=%r>' % (
            self.__class__.__name__, hex(id(self)), self.name)
```

Note: we use self.__class__.__name__ so that subclasses won't lie about their class. We use hex(id(self)) so that two similar objects will look distinct (it is often important when there are two different objects with the same data). The id is not necessary for Value Objects.[13] Lastly, any instance variables you want to expose are done through %r, which puts the repr() of those objects into the string. This is very helpful when, for instance, a new line is embedded in the value. Because repr() is most useful when there are bugs, you shouldn't assume all instance variables contain well formed data.

It is also acceptable (especially for Value Objects) to use:

```
def __repr__(self):
    return '%s(name=%r)' % (self.__class__.__name__, self.name)
```

You should be careful that repr() return values are not too long. If you sometimes have long data in a value, you might do this:

```
def __repr__(self):
    bio_repr = repr(self.bio)
```

---

[13]http://c2.com/cgi/wiki?ValueObject

```
if len(bio_repr) > 20:
    bio_repr = bio_repr[:15]+'...'+bio_repr[-5:]
return '<%s bio=%s>' % (self.__class__.__name__, bio_repr)
```

Index